Distributed Machine Learning Patterns

A Patterns-First Manual for Architects, Engineers, and Technical Leads

Jazper Carter

Cybersoft Publishing LLC

Fort Washington, MD 20744

Drclaude.net

First Edition, April 2026

Contents

1 Introduction

At 3:14 on a Sunday morning, a training job that has been running for nineteen hours on a 1,024-GPU cluster goes silent. No error message appears. No process exists with a nonzero code. The job tracker reports all workers as healthy. The accelerator utilization dashboards show every device at 98%. From the outside, everything looks normal. Inside the cluster, something has gone quietly and completely wrong.

The on-call engineer who picks up the alert six minutes later begins with the obvious checks: network saturation, storage I/O, and memory pressure. All are within bounds. She pulls the gradient norms from the metrics store. They are zero. Every worker on every node computes a gradient of exactly zero and passes it to the all-reduce collective without complaint. The optimizer is updating the weights to 0 on every step. The loss has not changed over 4,000 iterations. The job is burning compute at full rate and producing nothing. It has been doing so for eleven hours.

The root cause, when she finds it three hours later, is a single misconfigured gradient clipping threshold set to 1e-12 in a configuration file that was merged during a late-Friday refactor. The value was legal. The training framework accepted it without warning. The job launched, distributed the configuration across all 1,024 workers, and clipped every gradient to a value indistinguishable from zero before the collective communication even began. No single node had enough context to recognize the failure. The system's distributed nature was precisely what made the failure invisible.

This scenario, assembled from patterns that recur across organizations and workload classes, captures the central difficulty of distributed machine learning: failures in distributed systems are not just larger versions of single-node failures. They have a different shape. They require different detection strategies, debugging disciplines, and design patterns to prevent them in the first place. A team that learned ML on a single workstation, or even on a small multi-GPU server, carries a mental model that does not transfer cleanly to a 1,024-GPU job, a multi-region inference cluster, or a sharded parameter server serving ten thousand requests per second. The patterns in this book exist to close that gap.

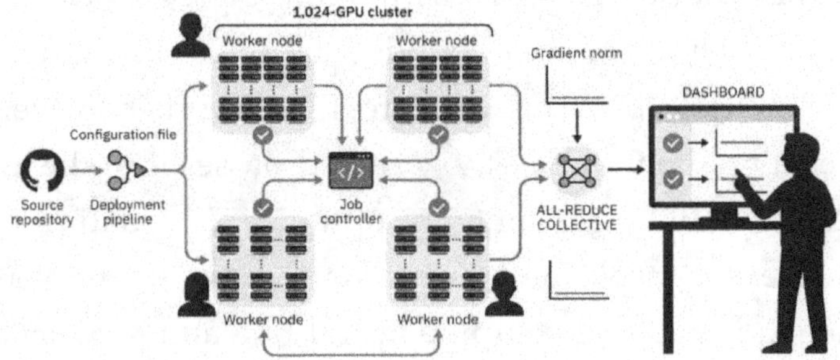

Why This Book, Why Now

Distributed machine learning has undergone a structural shift in the past five years. Training large language models and foundation models at scale is no longer confined to a handful of research laboratories. The economics of pre-trained model weights, combined with the practical advantages of fine-tuning on domain-specific data, have pulled organizations of every size into distributed training workflows that would have been considered frontier infrastructure just a few years ago. At the same time, serving those models under production latency and throughput requirements has created an entirely new class of inference infrastructure problems: multi-replica deployments, model parallelism for inference, KV cache management, continuous batching, and multi-model serving across heterogeneous accelerator pools.

The engineering challenge is not merely that the systems are larger: it is that the design space has expanded faster than the field's shared vocabulary for navigating it. Data parallelism, model parallelism, pipeline parallelism, and tensor parallelism each solve distinct problems and introduce different classes of failure. Sharded optimizers reduce memory pressure but add communication cost. Spot instances cut training costs but require fault-tolerant patterns that most teams have never needed. Navigating these choices correctly requires a framework, not just familiarity with a specific framework's API.

This is why patterns matter more than tools at this moment. The ML infrastructure landscape rotates every two to three years: what is standard today was experimental yesterday, and frameworks that do not exist yet will be standard two years from now. A team organized around a specific tool's documentation is perpetually one platform cycle away from obsolescence. A team organized around patterns can evaluate each new tool against a stable set of engineering criteria, adopt it deliberately rather than reactively, and carry their hard-won operational knowledge forward regardless of which names appear in the configuration files. The patterns are a durable investment. The tools are the current implementation.

The scale economics driving this shift are worth naming directly. Training a large language model requires months of wall-clock time across thousands of accelerators, with compute costs that can reach millions of dollars per run. Cost overruns in distributed training are not line-item nuisances; they are program-ending events.

Reliability failures in distributed inference are revenue events. The stakes of getting distributed ML infrastructure right now belong in the same conversation as the modeling decisions themselves.

The foundation model era has also surfaced patterns previously irrelevant to most practitioners: reinforcement learning from human feedback infrastructure, retrieval-augmented generation pipelines, mixture-of-experts routing at serving time, and speculative decoding to reduce latency. Chapter 10 addresses them directly, positioned after the foundational patterns that make them comprehensible. Hence, the reader arrives with the vocabulary to understand both what the patterns do and why they are necessary at that scale.

1.1.1 The Serious AI Manager Series — A Roadmap

This book is the third volume of the Serious AI Manager Series, a five-book program published by Cybersoft Publishing for the architects, engineers, and technical leads who carry production machine learning from prototype to scale. The series is deliberately sequenced. Each volume assumes the discipline established by those before it, and the present book sits at the structural center, where the operational foundations of the earlier volumes meet the system-design pressure of the later ones.

The series opens with *The Enterprise MLOps Playbook: How to Deploy, Govern, and Scale Machine Learning Systems in the Real World* by Jasper Carter, which establishes the operational substrate every distributed system in this book runs on — deployment pipelines, governance, lineage, FinOps, and the organizational accountability that keeps a fleet of models maintainable rather than merely launchable. That book answers how a machine learning practice is run; this one assumes that practice exists and asks how its training and serving workloads are scaled.

Production Data Engineering for Machine Learning by Claude J. Louis-Charles, PhD, supplies the second prerequisite: the data pipelines, feature systems, and ingestion architectures that feed every distributed training run described here. Distributed training is only as sound as the data movement underneath it, and that book treats the movement as a first-class engineering problem. Readers who arrive at the data-parallelism patterns of Chapter 3 without that grounding will find the present treatment compressed, because it assumes the pipeline already exists.

The fourth volume, *Managing Production Large Language Models: Designing, Deploying, and Operating LLM Systems at Scale* by Jordan O'Neal, picks up where this book's LLM chapters leave off, taking an operations view of the same serving and training patterns that this book frames at the architectural level. Where Chapter 10 here names the distributed patterns for foundation models, that book operates them.

The series closes with *Designing Data-Intensive AI Applications: An Architect's Blueprint for AI Systems Under Real-World Scale, Latency, and Compliance Pressure*, which generalizes the patterns of this book into the broader class of data-intensive systems, treating latency, scale, and compliance as the three constant pressures every production AI architecture must answer to. It is the synthesizing volume; this book is one of its load-bearing inputs.

The recommendation is explicit. Readers should absorb Book 1 for its MLOps foundations and Book 2 for its data-engineering depth before working through this volume. Book 3 assumes both. It does not re-derive the deployment, governance, or data-pipeline material that those books establish, because doing so would dilute the distributed-systems argument that is this book's reason for existing.

Who This Book Is For

This book is written for four roles that converge on a single distributed ML system: ML architects, ML and data engineers, technical team leads, and platform and SRE engineers. Each role brings a different vantage point to the same infrastructure decisions, and the book is structured to serve all four without forcing any of them to translate the others' vocabulary. Every chapter pairs the architect's design view with the engineer's implementation view, and the team discussion sections at the end of each chapter are designed for conversations that cross role boundaries.

ML architects will find a framework for making design decisions that age well across platform cycles. When does a workload cross the threshold where model parallelism becomes necessary? When does sharding the optimizer save enough memory to justify the added communication cost? When does a spot-instance strategy reduce costs without introducing a reliability risk that the business cannot accept? The book provides the criteria for those decisions, not just conclusions drawn from a specific configuration. A mission-aligned distributed ML architecture is one designed from the start to absorb the changes that inevitably accompany scale, team growth, and platform evolution.

ML and data engineers will find implementation patterns grounded in the realities of production systems: how to build training loops that checkpoint correctly across heterogeneous failure modes, how to design data pipelines that saturate accelerator bandwidth without becoming the bottleneck, how to instrument a distributed job so that an on-call engineer can localize a failure in minutes rather than hours. The technical checklists at the end of each chapter are designed to be used directly as release gates: specific, actionable, and grounded in the failure modes each chapter names.

Technical team leads will find a shared language to align engineers and architects on reliability standards, a structure for onboarding new team members into existing distributed systems, and a set of discussion guides to surface architectural misalignments before they become production incidents. The pilot-to-production pathway this book describes is one that a team can document, audit, and repeat, rather than one that lives only in the engineers' tribal knowledge from the first time.

Platform and SRE engineers will find the operational layer of each pattern explicitly named: what the cluster scheduler needs to know about ML workloads, how multi-tenancy and resource isolation interact with GPU topology, which observability signals matter for distributed training and serving, and where security boundaries belong in a shared ML platform.

How This Book Is Organized

The twelve chapters are organized into four logical parts, each building on the previous and each covering a distinct phase of the distributed ML lifecycle. Reading the parts in order gives a complete picture of the discipline from first principles to frontier practice. Reading individual chapters gives actionable guidance on a specific problem without requiring the others as prerequisites. The four appendices provide reference material that is useful long after the first read.

Part I, Foundations (Chapters 1 through 4), establishes the vocabulary and mental models on which the rest of the book depends. Chapter 1 introduces distributed ML as a system design problem, frames the two foundational axes of parallelism, and defines what "production" means in the context of distributed ML. Chapter 2 covers compute parallelism patterns in depth: data parallelism, model parallelism, pipeline parallelism, tensor parallelism, and the hybrid strategies that compose them. It is the chapter in which the vocabulary of parallelism is established and the trade-offs among communication cost, memory efficiency, and compute utilization are precisely articulated. Chapter 3 covers data sharding and I/O optimization, addressing the reality that distributed training is frequently bottlenecked not by compute but by the data pipeline that feeds it. Chapter 4 covers orchestration and scheduling: how containers become a distributed system, how cluster schedulers allocate accelerators, and how workflow managers coordinate multi-stage ML pipelines across heterogeneous infrastructure.

Part II, Training at Scale (Chapter 5), applies the Part I foundations to the production concerns of large-scale training runs: checkpointing strategies, fault-tolerance patterns, gradient correctness under partial failures, elastic training for preemptible clusters, and the cost-reliability trade-offs that govern spot-instance decisions.

Part III, Serving and Operations (Chapters 6 through 9), covers the production infrastructure surrounding a deployed model. Chapter 6 addresses serving at scale and inference model parallelism: KV cache management, continuous batching, and multi-model serving across accelerator pools. Chapter 7 covers resource management and cost optimization: GPU provisioning, spot and preemptible strategies, autoscaling, and cost attribution. Chapter 8 covers observability and debugging, naming the distributed failure modes that standard application monitoring misses. Chapter 9 covers security and multi-tenancy for shared ML clusters.

Part IV, Frontier Patterns and Synthesis (Chapters 10 through 12), extends the book's patterns to the current shape of the field and then synthesizes them into a complete reference architecture. Chapter 10 covers the distributed patterns specific to large language models and foundation models: tensor parallelism at inference time, mixture-of-experts routing, speculative decoding, retrieval-augmented generation infrastructure, and RLHF training pipelines. Chapter 11 presents composite case studies and reference architectures that show how the individual patterns compose into production systems for representative workload classes. Chapter 12 synthesizes the entire arc into a map of the modern distributed ML platform and a maturity model that the reader can use to assess their organization and plan the next stage of investment.

Four appendices complete the reference layer. Appendix A provides a tool comparison matrix and selection guide for orchestrators, schedulers, training frameworks, serving stacks, and observability platforms, framed as decision guides rather than buyer's guides. Appendix B provides framework cheat sheets covering configuration shape, common patterns, common pitfalls, and diagnostic signals. Appendix C provides reusable configuration templates and reference scripts annotated with the decisions they embody. Appendix D provides a glossary of distributed systems and ML terminology, as well as a curated list of recommended readings organized by chapter, favoring durable references over tool-specific tutorials.

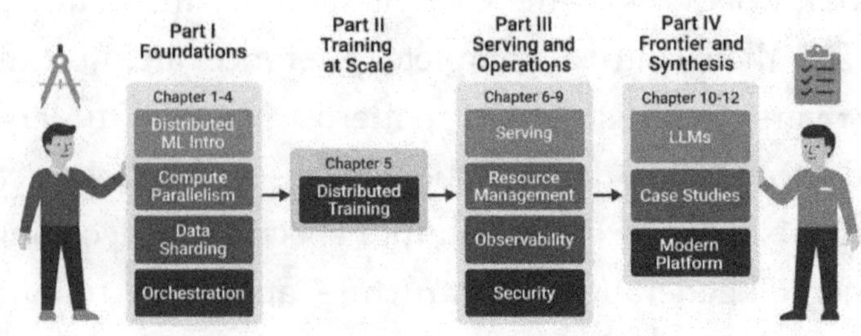

How to Read This Book

Architects and team leads new to distributed ML infrastructure will benefit most from reading cover to cover, in chapter order. The arc is cumulative: parallelism vocabulary introduced in Chapter 2 is applied in Chapter 5, extended to inference in Chapter 6, and revisited for foundation models in Chapter 10. Orchestration patterns from Chapter 4 underpin the fault-tolerance patterns in Chapter 5 and the cost-optimization patterns in Chapter 7. Reading sequentially builds the mental model that lets later chapters land with full force rather than as isolated techniques.

ML and data engineers with an existing production system and a specific problem will find each chapter a usable standalone reference. Every chapter opens with a concrete composite scenario that situates the chapter's subject in a recognizable operational context, followed by a "Why It Matters" section that connects the technical patterns to workflow-level impact and organizational risk. The core technical content is organized into named patterns with clear criteria for when to apply them. Each chapter closes with a Technical Checklist the team can run immediately, a Team Conversation guide designed for the next architecture review or team meeting, and a Key Takeaway that distills the chapter's essential argument into a single actionable position.

For architects building a new distributed ML platform from a clean slate, the recommended path is Chapters 1, 2, 4, 5, and 6 in order, followed by Chapter 12 for the synthesis. This path covers the foundations of parallelism, the orchestration layer, training at scale, and serving infrastructure, and then synthesizes them into a complete platform view. Chapters 7 and 8 can follow immediately to add cost management and observability to the architecture.

For engineers inheriting an existing distributed training system tasked with improving reliability or reducing costs, the recommended path starts with Chapter 5 on fault-tolerance patterns, Chapter 7 on cost optimization, and Chapter 8 on observability and debugging. Those three chapters address the most common operational pain points in mature distributed training environments. They can be read independently of the earlier foundational chapters if the system's parallelism strategy is already established.

For platform and SRE engineers supporting ML workloads they did not design, the recommended entry point is Chapter 4 for orchestration and scheduling, Chapter 9 for security and multi-tenancy, and Chapter 8 for observability. Chapter 1 supplies the distributed ML vocabulary needed to reason about ML workloads as a distinct workload class. For technical leads oriented toward the frontier, Chapter 10 on LLMs and foundation models can be read early as a motivation for the foundational patterns, then revisited after Chapters 5 and 6 have established the training and serving foundations it builds on.

1.1.2 Reading Paths by Background

Distributed machine learning draws practitioners from at least four distinct backgrounds, and each arrives missing a different prerequisite. A linear reading serves the architect for whom the series was designed, but it wastes the time of a reader who already commands half the material and lacks the other half. The diagnostic below routes four common backgrounds through the chapters in the order that builds competence fastest, given what each reader already knows and what they do not.

1.1.2.1 Path: Researcher arriving from PyTorch / JAX

This reader understands single-node deep learning intimately — autograd, optimizers, mixed precision — but has never operated a system in which the failure of one machine corrupts the work of forty others. The danger is overconfidence: the mathematics is familiar, so the system's problems read as incidental. Start at Chapter 1 and do not skip the failure-modes section, because it reframes correctness as a distributed-systems property rather than a numerical one. Then jump directly to Chapter 5 on fault-tolerant training before attempting the full parallelism taxonomy in Chapter 2; running a real distributed job with real checkpoints and real preemptions makes the taxonomy concrete rather than abstract. Return to Chapter 2 only after that first job has run. Readers who lack any distributed-systems background should read Appendix E first as a refresher; it supplies the network, consistency, and failure vocabulary the rest of the book assumes.

1.1.2.2 Path: Platform / SRE engineer arriving from Kubernetes

This reader commands orchestration, scheduling, and observability, but has not internalized what makes a machine learning workload different from a stateless web service — namely, that a training job is a single long-lived computation whose state cannot be horizontally replicated. Start at Chapter 4, where orchestration is familiar terrain, and the ML-specific scheduling constraints stand out against a known background. Move next to Chapter 8 on observability, where the instinct to instrument is already present, and only the metrics need translating. Then go back to Chapters 1 through 3 to acquire the machine learning vocabulary — parallelism, gradients, accelerators, feature pipelines — that the earlier chapters used in passing. Defer Chapter 10's LLM-specific patterns until a foundation-model workload actually lands on the cluster; the material is dense and best read against a concrete need rather than in the abstract.

1.1.2.3 Path: Data engineer arriving from batch/streaming pipelines

This reader understands data movement, partitioning, and backpressure, but has never reasoned about accelerator topology or the bandwidth of an intra-node interconnect. Start at Chapter 3, the home turf, where distributed data loading and sharding map cleanly onto familiar pipeline concepts. Then read Chapter 1 for the system framing that situates those pipelines inside a training cluster, followed by Chapter 5 on fault tolerance, which connects checkpoint design to the durability guarantees a data engineer already values. Defer Chapter 2's parallelism taxonomy until the work demands designing an actual training run; the taxonomy is the analytical core of the book, but it lands hardest when there is a concrete topology to reason about rather than a hypothetical one.

1.1.2.4 Path: Architect (the series' core reader)

This reader has absorbed Books 1 and 2 and approaches this volume as it was written to be read. Read in order, front to back, and treat Chapter 2 as the analytical core to which every later chapter refers back. Spend extra time on Chapter 11's case studies, which exercise the framework end-to-end against six composite organizations, and on Chapter 12's maturity model, which converts the patterns into an assessment instrument. The architect's payoff is cumulative; nothing here is safely skipped, because the later trade-off arguments assume the vocabulary the earlier chapters establish.

Every path converges at Chapter 12. The maturity model is where the patterns stop being a catalog and become a way to assess and advance a real organization. It reads the same whether the reader arrived through the mathematics, the orchestration layer, the data pipeline, or the architect's front door.

A Note on Tool Neutrality

The patterns in this book are described independently of any specific platform, framework, or vendor. Distributed training frameworks, cluster orchestrators, workflow managers, model serving systems, observability platforms, and vector databases are referenced as illustrative examples in callouts, not as the spine of the content. The spine is the underlying patterns: the decisions that must be made, the failure modes that must be anticipated, and the operational disciplines that must be institutionalized, regardless of which tools a team currently uses or will adopt in the next platform cycle.

This choice is deliberate and consequential. The ML infrastructure landscape has shifted significantly over the past five years. Frameworks that were experimental two years ago are now the de facto standard for certain workload classes. Frameworks that were the de facto standard two years ago have since been superseded. A book organized around specific tools risks becoming a historical document within a single platform generation. A book organized around patterns remains useful across platform generations because the patterns describe problems that do not change when the tool names do.

All-reduce collective communication is a fundamental pattern, whether the framework is written in Python, C++, or Rust. Checkpoint design for fault-tolerant training is a pattern regardless of the checkpoint format. KV cache memory management is a pattern whether the serving runtime is a research prototype or a production system. The goal is a reader who can evaluate any new tool against a clear set of engineering criteria.

When specific tools are named, they appear to make abstract patterns concrete rather than to endorse a specific vendor or to constrain an implementation choice. The same pattern will be implemented differently across frameworks and organizations, and the book's value lies precisely in naming the pattern above its implementation so that the reader can recognize it in any context. Teams should feel free to substitute the tools they actually use for any callout that names a specific example.

Conventions Used in This Book

Each chapter follows a consistent structure. The chapter opens with a composite scenario drawn from the shapes of real production incidents, followed by a "Why It Matters" section that ties the chapter's patterns to mission-level risk and workflow-level impact. The core technical content is organized into named patterns with clear criteria for when each applies and what failure modes it addresses.

Each chapter closes with three structured artifacts. The Technical Checklist is a specific, actionable list of conditions the team should verify before declaring a system production-ready against the chapter's topic; it is designed to be used as a release gate or architecture review checklist rather than as a summary of the chapter's content. The Team Conversation section provides a discussion guide for the next team meeting or architecture review, with questions designed to surface misalignments among architects, engineers, and platform teams before they surface as production incidents. The Key Takeaway distills the chapter's essential argument into a single position the team can adopt and defend.

Diagram placeholders appear throughout each chapter as inline Normal-style paragraphs, numbered by chapter and sequence (for example, Diagram 2.3 is the third diagram in Chapter 2; Diagram I.1 is the first diagram in the Introduction). Each placeholder includes a title and a description specific enough for a designer to implement without ambiguity. Diagrams follow a consistent visual style: a blue color palette, a white background, a flat vector look, clean sans-serif typography, and human figures interacting with the systems being described. Red is used only to indicate errors or failure states.

Code samples, configuration fragments, and command-line examples are formatted in a monospace style and presented as illustrative examples rather than prescriptive templates. Every code sample includes inline commentary explaining the decision it embodies and the modification expected when adapting it to a different cluster topology or workload class. Appendix C compiles the most reusable configuration templates into a single reference section, annotated for adaptation.

Callouts appear occasionally to name a specific tool or framework as an illustrative example of the pattern being described. They are visually distinct from the main body text and should be read as one concrete instance of the pattern, not as a prescription. The pattern's applicability extends well beyond any single example named in a callout.

Glossary terms are defined in Appendix D and introduced in context in the chapter where they matter most. Chapter 1 introduces the broadest vocabulary, and later chapters add to it incrementally. Readers who arrive at a specific chapter without the earlier vocabulary can use Appendix D to look up definitions without having to read back through the foundational material.

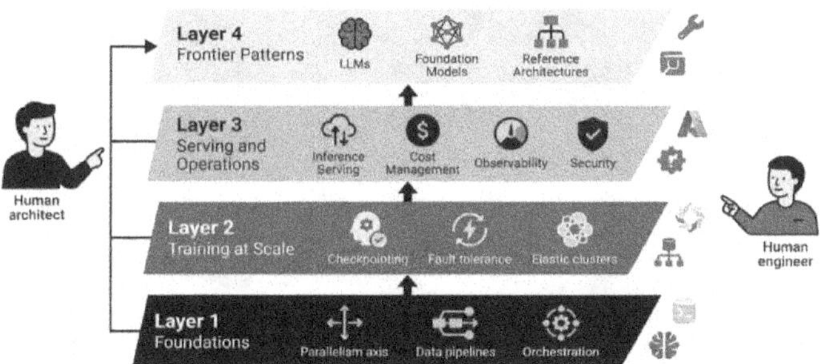

Distributed machine learning is the discipline that bridges the complexity of distributed systems and the stochastic nature of ML workloads. It rewards operational clarity: teams that have named their patterns and documented their failure modes are consistently more reliable and cost-efficient than teams that have accumulated their knowledge one painful production incident at a time.

The goal of this book is to compress hard-won lessons from distributed ML engineering into a pattern catalog that any team can adopt in the order that fits their current situation and return to as a reference long after the first read. Each chapter ends where the next begins. The patterns compose. The vocabulary grows precise enough to make the next incident shorter.

The first Chapter is where the foundations begin.

2 Introduction to Distributed Machine Learning

Opening Scenario

On a Tuesday morning in the third quarter, a senior ML engineer at a fast-growing fintech opens a ticket from the platform team. The recommendation model that took six hours to train on a single high-memory GPU last quarter now requires thirty-one hours. The training dataset has grown sixfold since the previous release cycle, and the model has been expanded with two additional representation layers to capture richer signals from that larger corpus. The ticket is politely worded, but the message is clear: something has to change before the next release gate.

The team's first instinct is to request a bigger GPU. The platform team responds that the next instance size up is unavailable in their cloud region, and that even if it were available, the model's parameter count would exceed the memory ceiling of any single device they operate. The following conversation spans 48 hours and involves the ML architect, the data engineer who owns the pipeline, the infrastructure lead, and the cost center owner. The question on the table is no longer which GPU but which parallelism strategy, on which cluster topology, with which fault-tolerance guarantees, at which cost ceiling, within which latency budget for the downstream serving layer.

That pivot from a hardware question to a systems question is the moment every team eventually faces. The engineers who navigate it well are the ones who have internalized a small library of patterns: how to decompose a workload across compute and data axes, how to reason about gradient synchronization and its cost, how to size a cluster against a network topology, how to checkpoint across nodes so a partial failure does not erase days of compute, and how to connect the training cluster to a serving cluster without rebuilding the pipeline from scratch. Those patterns are not framework-specific; they are portable abstractions that survive the next tool rotation.

The fintech team in this scenario eventually settled on a hybrid approach: data parallelism across eight GPU nodes for the bulk of training, model-parallel sharding of the two expanded representation layers that exceeded per-device memory limits, and an all-reduce ring for gradient synchronization over the intra-cluster fabric. The time-to-train dropped from 31 hours to 4 hours and 20 minutes. The cost per run increased by a factor of 2.3, but this was offset by a 12-hour reduction in the team's wait time for results and a material improvement in the model's offline metrics. The decision was deliberate rather than accidental, and it was documented in a single-page architecture decision record that the team could revisit six months later.

This chapter uses that incident shape as its anchor. Every concept introduced here maps onto a question the team had to answer, and every later chapter in the book provides the canonical pattern for answering that class of question at production scale. By the time you finish this chapter, you will have the vocabulary and the mental models to read the rest of the book as a coherent whole rather than as a collection of independent topics.

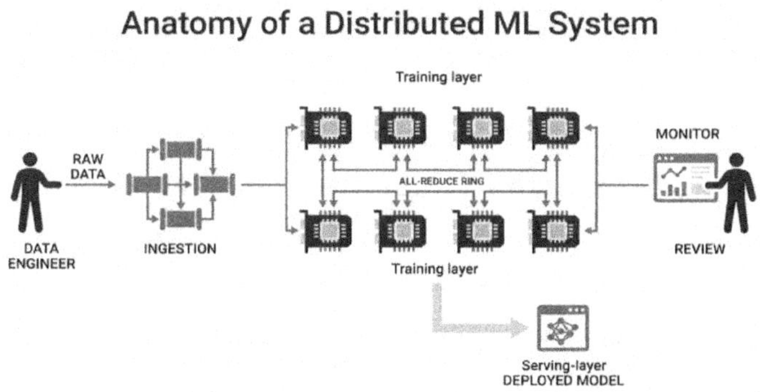

Anatomy of a Distributed ML System

Why It Matters

Distributed machine learning is no longer an advanced specialty practiced only by a handful of research labs. The growth in model parameter counts, dataset volumes, and latency expectations for downstream applications has pushed even mid-sized engineering teams toward multi-node training and multi-replica serving. The gap between teams that understand distributed patterns and those that do not is now one of the largest single determinants of cost, reliability, and time-to-production in applied ML. A team that cannot reason about parallelism strategies will overbuild, underbuild, or fail to ship within the constraints the business actually cares about.

The workflow-level impact is concrete. When a training job takes three days on a single node, and the team has a weekly release cadence, distribution is not an optimization; it is a prerequisite. When a serving endpoint must handle 100,000 requests per second with a 99th-percentile latency under 50 milliseconds, a single replica is not an option. The same logic applies to cost: a team that runs eight-GPU nodes at 50% utilization because they do not understand how to load-balance across replicas is paying twice per experiment. These are not edge cases; they are the daily operating conditions for any organization shipping modern ML at commercial scale.

The mission-aligned reason to learn these patterns is durability. Frameworks rotate every two to three years, but the underlying patterns do not. The parameter server pattern that governed large-scale distributed training in the mid-2010s gave way to ring all-reduce architectures, which gave way to hybrid patterns that combine sharded state with asynchronous communication. In each transition, teams that understood the pattern level adapted quickly; teams that had optimized only for a specific tool had to rebuild from scratch. The investment this book asks for is pattern literacy, not tool mastery, and that literacy compounds rather than depreciates across every platform cycle.

The Growth Curve: Why Distribution Becomes Mandatory

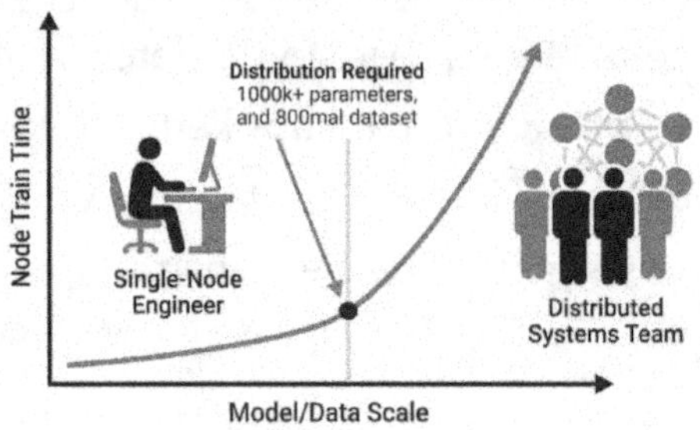

Why Distribution Matters for Modern ML

The forces that push workloads from single-node to distributed are three: memory, compute time, and serving throughput. Each force has a rough quantitative boundary that the engineer can use to locate their own workload on the curve.

Memory is the first boundary. A single accelerator device today ships with between 16 and 141 gigabytes of high-bandwidth memory, depending on the hardware class. A model with 7 billion parameters, stored at 16-bit precision, occupies approximately 14 gigabytes of memory for the weights alone. Adding optimizer state, gradients, and activation checkpoints increases the working memory footprint to three to four times the raw weight size during training. A seven-billion-parameter model trained with a standard optimizer therefore requires roughly 56 to 80 gigabytes of device memory, which already exceeds the memory available in many single-device configurations. Models with more than roughly 3 billion parameters are effectively mandatory candidates for model-parallel distribution under standard training regimes, and that boundary drops further when the optimizer or activation footprint is large.

Compute time is the second boundary. A rough rule of thumb is that training a modern transformer-class model to convergence requires approximately six times the parameter count in floating-point operations per token, multiplied by the total token count in the training corpus. A ten-billion-parameter model trained on one trillion tokens requires on the order of sixty zettaflops of compute. A single high-performance accelerator operating at a peak throughput of, say, 312 teraflops would require approximately 192,000 hours to complete that run, which is not compatible with any plausible release schedule. Distribution across hundreds or thousands of accelerators brings that time down to days to weeks, and the parallelism strategy determines how efficiently those accelerators are used.

Serving throughput is the third boundary. A single replica, at 200 milliseconds per request, can serve at most 5 requests per second. A product endpoint with 10,000 simultaneous users, each making 1 request every 10 seconds, requires 2,000 requests per second, which demands a large fleet of replicas unless batching and model parallelism are applied. Serving cost, latency, and reliability are all determined at the distribution layer, not at the model layer.

Accelerator economics reinforce all three boundaries. The cost per teraflop continues to decline, but the highest-memory single-device options have not declined at the same rate. Distributing across more commodity-class devices is often more cost-effective than focusing on a single premium device, and teams that understand that trade-off capture real budget advantages.

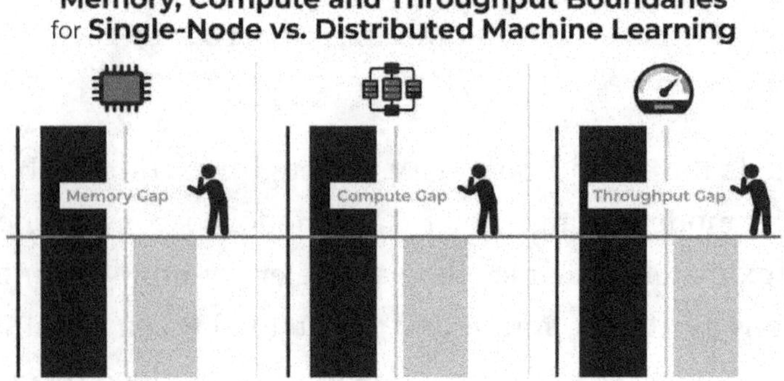

Compute Parallelism and Data Parallelism Fundamentals

Every distributed ML workload is a composition of two foundational axes of parallelism: data parallelism and compute (model) parallelism. Understanding these two axes and the language used to describe them is a prerequisite for every subsequent chapter in this book. All of the advanced patterns covered in later chapters, including pipeline parallelism, tensor parallelism, sequence parallelism, and expert parallelism, are refinements or extensions of these two axes.

Data parallelism is the simpler of the two axes. In data parallelism, the full model is replicated across multiple devices or nodes, and each replica processes a different subset of the training data (called shards or mini-batches) simultaneously. Each replica computes a gradient on its local mini-batch, and the gradients from all replicas are aggregated into a single global gradient, which is then applied to update all replicas in lockstep. The aggregation step is the critical communication operation, and the dominant pattern for executing it is all-reduce: a collective communication primitive in which every node both contributes to and receives the final reduced result, with no single coordinator node becoming a bottleneck. The bandwidth cost of all-reduce scales with model size rather than data size, making data parallelism most efficient when models are small enough to fit on a single device and the training data is large.

Model parallelism is the second axis. When a model's parameter count exceeds the memory capacity of a single device, the model itself must be partitioned. Tensor parallelism splits individual weight matrices across devices, with each device holding a column or row shard of the matrix, and the outputs are concatenated or reduced after each forward pass. Pipeline parallelism partitions the model's layers into sequential stages, assigns each stage to one or more devices, and overlaps computation across stages using micro-batches to reduce the idle time (called the pipeline bubble) that would otherwise be spent at stage boundaries. Fully Sharded Data Parallelism, commonly known as FSDP, combines data parallelism with model sharding by sharding not only the model parameters but also the optimizer state and gradients across the data-parallel workers, dramatically reducing the per-device memory footprint without abandoning the data-parallel programming model.

The parameter server is an older yet still relevant architectural pattern that separates the roles of parameter storage and updates from those of gradient computation. In a parameter server architecture, a set of server nodes holds the canonical model parameters and applies updates as they arrive from worker nodes that compute gradients on mini-batches. The pattern supports asynchronous updates, which can improve throughput by eliminating the synchronization barrier of all-reduce, but at the cost of gradient staleness: workers may be computing gradients on parameters that other workers have already updated. The tradeoff between synchronous consistency and asynchronous throughput is one of the defining design decisions in distributed training.

Sharding is the general term for partitioning any large data structure (model weights, optimizer state, datasets, embedding tables, KV caches) across multiple storage or compute units. Every parallelism pattern in the book involves some form of sharding. The choice of what to shard, along which dimension, and with what replication factor determines the system's bandwidth and memory requirements, as well as its failure surface. The vocabulary of sharding, rank (the index of a given worker in the global process group), world size (the total number of workers), and process group (a named subset of workers that participate in a collective operation) will appear throughout the book and should be treated as first-class vocabulary from this point forward.

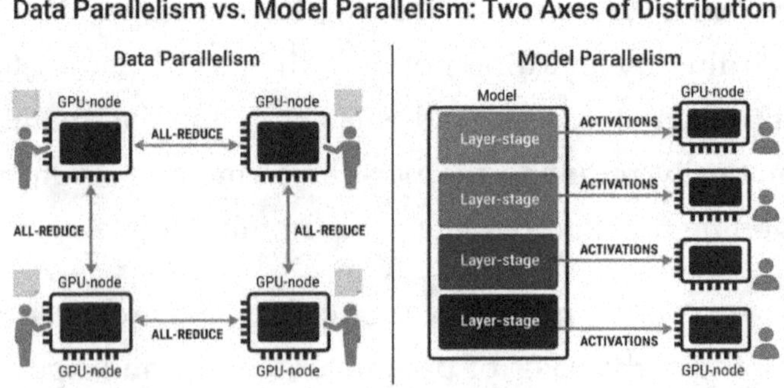

Common Failure Modes in Distributed Systems

Distributed systems fail in ways that single-node systems do not, and ML workloads introduce failure modes that general distributed systems lack. Understanding the catalog of failure modes before encountering them in production is one of the highest-return investments a distributed ML team can make.

Partial failure is the defining failure mode of distributed systems. In a single-node system, the system either works or it does not. In a distributed system, any subset of nodes can fail independently while the others continue operating. A training job running across 100 nodes can lose one node to a hardware fault, another to a network partition, and a third to an out-of-memory condition, all within the same training run. The naive response is to treat any node failure as a total job failure, but that approach makes long runs on large clusters impractical. The correct response is fault tolerance: a combination of periodic checkpointing, failure detection, and restart protocols that allow the job to recover from partial failures without losing the full accumulated compute budget.

Stragglers are nodes that execute significantly more slowly than the rest of the group, typically because of hardware heterogeneity, thermal throttling, or interference from other workloads. In synchronous training, the full cluster must wait for the slowest node to complete each step before the gradient aggregation can proceed. A single straggler at the ninety-ninth percentile of the step-time distribution can reduce effective cluster utilization to a fraction of its theoretical peak. Straggler mitigation patterns include backup workers (launching redundant copies of the slowest-executing work), asynchronous gradient aggregation (accepting gradients as they arrive rather than waiting for all), and straggler detection with automatic node eviction and replacement.

Network partitions occur when a network fault isolates one group of nodes from another while both groups continue operating independently. In a synchronous training job, a partition typically causes the job to hang at the collective communication barrier, waiting for nodes that will never respond. Detecting partitions quickly and distinguishing them from slowness is a hard problem. Timeout-based detection is common but requires careful tuning: timeouts that are too short cause false positives under load; timeouts that are too long mean minutes or hours of cluster idle time before recovery begins.

Head-of-line blocking is a latency pattern in which a slow or stuck request at the front of a queue blocks all subsequent requests from being processed, even if they would complete quickly. In a distributed serving system, head-of-line blocking occurs when a long inference request occupies a compute slot, preventing shorter requests from being batched or executed. Continuous batching and preemption-aware schedulers address this pattern in serving contexts.

Tail latency is the phenomenon by which the worst-case latency in a distributed system grows faster than the median as the number of components increases. A system with one hundred components, each with a one-percent probability of a slow response, will see a slow response on approximately sixty-three percent of requests that touch all one hundred components. Tail latency is a multiplier on the number of parallel operations in the critical path, which means that distributed systems must be designed to explicitly tolerate and hedge against it rather than relying on the average case.

ML-specific failure modes compound the failures of distributed systems. Gradient correctness failures occur when numerical issues, wrong shard assignments, or inconsistent floating-point rounding cause the aggregated gradient to diverge from the correct value. Checkpoint corruption occurs when a node fails mid-write, leaving a partial file that restores to an inconsistent model state. Stochasticity failures occur when workers share the same random seed and produce identical rather than independently varying augmentation, biasing the gradient estimate. All three are silent: the job keeps running and produces output; the output is wrong.

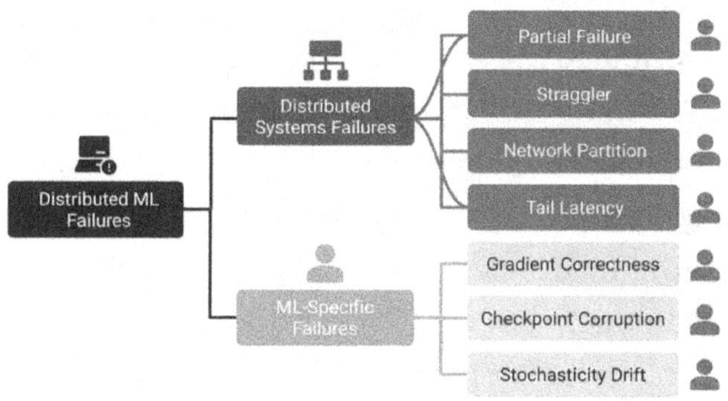

Key Trade-offs: Consistency, Performance, and Cost

Every distributed ML decision is a negotiation across three axes: consistency, performance, and cost. No single configuration optimizes all three simultaneously, and the ML architect's role is to choose deliberately rather than by default. This section provides working definitions of each axis as they apply to distributed ML and identifies the levers that move each.

Consistency in distributed ML refers to the degree to which all workers agree on the model's state at any given point in training. Fully synchronous training, in which all workers compute gradients on the same step's mini-batches and aggregate before updating, is the strongest consistency mode: every worker's model is identical after every step. Asynchronous training, in which workers apply updates as they arrive without waiting for a global barrier, sacrifices consistency for throughput: at any given moment, workers may be operating on parameters that are one, two, or many steps out of date relative to the most recently updated version. The tradeoff is empirically workload-dependent. For convex or nearly-convex loss surfaces, asynchrony introduces noise that may be tolerable or even beneficial as a regularizer. For complex non-convex surfaces with sharp minima, gradient staleness can prevent convergence or produce qualitatively worse models.

Performance in distributed ML has two distinct meanings that must not be conflated. Training throughput is measured in samples or tokens per second and determines how quickly a model can converge. Serving throughput is measured in requests per second and determines how many users the deployed model can serve within a latency budget. Both are subject to the fundamental constraint of the hardware topology: the peak compute of the accelerators limits the throughput ceiling, and the network bandwidth between nodes limits the efficiency ceiling for any workload that requires inter-node communication. A training job that spends forty percent of its wall-clock time waiting for the all-reduce of gradients is operating at sixty percent of its theoretical peak, and the correct intervention is to reduce the communication volume (by increasing the local batch size or applying gradient compression) or to overlap communication with computation (by pipelining).

Cost is a function of accelerator hours, storage I/O, and network egress, in roughly that order of magnitude. Device count, time-to-train, and utilization efficiency determine the number of accelerator hours. A 50% model flops utilization produces half as much useful work per dollar as a fully-utilized cluster. The highest-leverage cost lever is therefore utilization, not instance count, and utilization is primarily a function of batch size, data pipeline throughput, and communication overhead. Storage costs matter at scale because frequent writes to a remote object store incur significant I/O charges. Network egress between cloud regions adds a secondary cost that surprises teams whose training and serving pipelines span different regions.

The three axes interact. Improving consistency by moving to synchronous training reduces throughput (because of the synchronization barrier) and increases cost (because stragglers waste accelerator idle time). Increasing batch size improves throughput and hardware utilization, reduces the cost per sample, but may degrade model quality through reduced generalization. Reducing cost by cutting the cluster size lengthens the training run, which may increase the window of exposure to hardware faults and increase the need for more checkpoint overhead. The architect who understands these interactions can choose deliberately; the engineer who does not understand them will optimize one axis by accident while degrading another without realizing it.

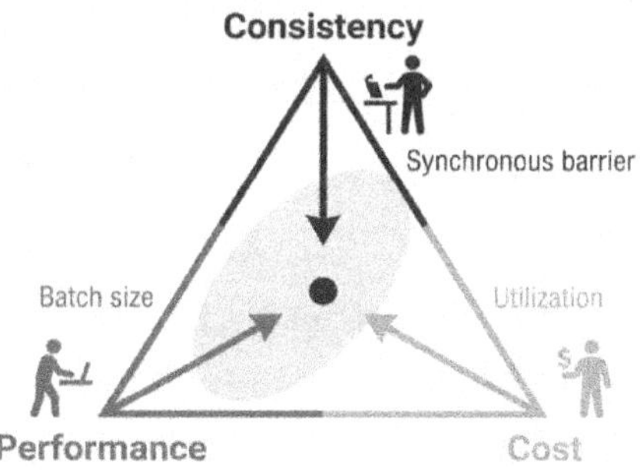

Hardware Topology and the Network as a First-Class Concern

The network is almost always the bottleneck in distributed training throughput, not the accelerator. This is a counterintuitive claim for engineers coming from a single-node background, and understanding it requires understanding the multi-level memory and communication hierarchy that a distributed ML cluster actually presents.

Within a single node, accelerator devices communicate over a high-bandwidth intra-node interconnect. The most common example is NVLink, a proprietary interconnect that connects GPU devices within a server chassis, with bandwidth ranging from 300 gigabytes per second in older generations to over 900 gigabytes per second in current high-end configurations. All-reduce within a single node over NVLink is fast enough that it rarely appears in the critical path of a well-tuned job. The intra-node interconnect is also used for direct peer-to-peer transfers between GPU memories, bypassing the host CPU and reducing latency for small tensor transfers between co-located devices.

Between nodes, communication must traverse the inter-node fabric. The two dominant technologies in high-performance ML clusters are InfiniBand and RDMA over Converged Ethernet (RoCE). InfiniBand provides low-latency, high-bandwidth links (commonly 200 to 400 gigabits per second per port in current deployments) with hardware-assisted remote direct memory access, allowing one node to read or write another node's memory without involving the remote CPU. RoCE achieves similar semantics over standard Ethernet infrastructure with appropriate hardware support, but with somewhat higher latency and lower peak bandwidth than InfiniBand, at a significantly lower cost and with greater compatibility with existing data center networking fabrics.

The bandwidth asymmetry between intra-node and inter-node communication is the source of the network bottleneck in distributed training. A typical NVLink-connected eight-GPU node has approximately 600 gigabytes per second of total GPU-to-GPU bandwidth. The inter-node fabric connecting that node to the rest of the cluster might provide 200 gigabits (25 gigabytes) per second, a factor of twenty-four lower. Any operation that requires communication across nodes, including gradient all-reduce in data-parallel training and activation transfers in pipeline-parallel training, incurs latency and throughput losses due to this asymmetry. The first-order implication for architecture is that inter-node communication volume must be minimized and overlapped with computation wherever possible.

The topology of the network fabric also matters. A fat-tree or dragonfly topology provides all-to-all bandwidth that scales with the number of nodes, making it suitable for all-reduce patterns. A ring topology, in which nodes are arranged in a logical circle, provides predictable bandwidth per node but scales total all-reduce time linearly with the number of nodes. Understanding the cluster topology determines which collective communication algorithm is most efficient and which parallelism strategies will hit the bandwidth ceiling first.

Bisection bandwidth captures the worst-case inter-cluster communication capacity: the bandwidth available when the cluster is split into two halves, with all traffic crossing the bisection. Low bisection bandwidth is a hidden bottleneck that appears only when the communication pattern is adversarial to the physical topology, for example, a ring all-reduce mapped onto a network with a bottleneck in the middle of the logical ring. Architects who ignore topology discover the bottleneck only after the cluster is provisioned and the job is underperforming.

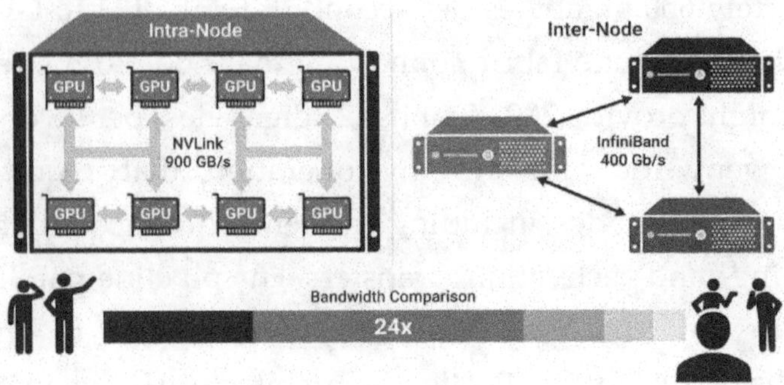

The Distributed ML Lifecycle

A distributed ML workload does not begin at model training and end at model deployment. It spans a lifecycle that starts with data ingestion, proceeds through feature engineering and dataset preparation, moves on to distributed training, covers evaluation and hyperparameter optimization, then moves to packaging and serving, and extends through monitoring, retraining, and eventual model retirement. Each stage of the lifecycle has its own distributed patterns, failure modes, and cost profile. Understanding the lifecycle as a whole is essential to designing systems that do not create bottlenecks upstream that become crises downstream.

Data ingestion is where distribution begins. Large-scale ML datasets are not stored in a single file; they are partitioned across distributed storage systems such as object stores or distributed file systems, and the ingestion pipeline must read from multiple partitions in parallel without creating a single point of contention at the storage layer. The data pipeline's throughput must match the training loop's; a pipeline that cannot saturate the GPU with data is a waste of GPU cycles and money. Data I/O patterns and sharding strategies are covered in depth in Chapter 3.

Distributed training is the stage most engineers associate with distributed ML, and it receives the most attention in the book across Chapters 2, 5, and 10. But training is only one stage, and it is often not the most expensive one when the total system cost is accounted for. Hyperparameter optimization, which requires running many training experiments in parallel, can consume more accelerator hours than the final training run itself if it is not managed carefully. Chapter 5 covers fault-tolerant training at scale; Chapter 7 covers cost management for training and HPO jobs.

Model serving is where the trained model meets the users who motivated its creation. A serving system must handle the full inference workload at the latency and throughput targets specified by the product or service, with sufficient headroom to absorb traffic spikes without degrading. At scale, serving a single model requires a fleet of replicas behind a load balancer. At a larger scale, serving a single very large model may require tensor parallelism or pipeline parallelism across multiple devices per replica, in addition to data parallelism across replicas. Chapter 6 covers serving at scale, and Chapter 10 covers the specific challenges of serving large language models.

Observability spans the entire lifecycle. A distributed ML system that cannot be observed cannot be debugged, and a system that cannot be debugged produces silent failures. Production observability requires instrumentation at four levels: the training loop (step time, gradient norm, loss), the serving layer (request latency, token throughput), the infrastructure (GPU utilization, memory, network bandwidth), and the data pipeline (read throughput, queue depth). Chapter 8 covers observability and debugging in detail.

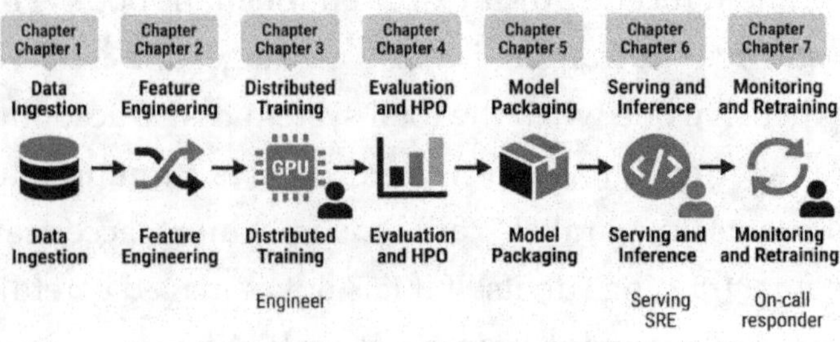

Roles, Responsibilities, and the Architect-Engineer-Lead Triangle

Distributed ML projects fail at the seams between roles as often as they fail at the technical level. The three primary roles in a distributed ML project are the ML architect, the ML or data engineer, and the technical team lead, each with distinct responsibilities and decision rights. Platform and SRE engineers partner with all three, and the seams between them are where the most consequential decisions are made, either deliberately or by accident.

The ML architect is responsible for the system-level design: choosing the parallelism strategy, sizing the cluster, specifying the network topology requirements, defining the fault-tolerance model, and setting the interface contracts between the training system, the data pipeline, and the serving system. The architect makes decisions that are expensive to reverse: the choice between synchronous and asynchronous training, the choice of a collective communication library, and the design of the checkpoint format all propagate through every subsequent engineering choice. The architect must also own the cost model: estimating accelerator, storage, and network costs before the cluster is provisioned, rather than discovering them at the end of the billing cycle.

The ML engineer or data engineer is responsible for implementation: writing the training loop, configuring the data pipeline, instrumenting the job, managing checkpoints, and debugging failures. The engineer works within the architecture specified by the architect, but implementation choices within the spec have real consequences for throughput, cost, and reliability. A training loop that does not overlap data loading with compute, or a checkpoint routine that blocks the training thread, can halve effective throughput without violating any architectural constraint. Operational clarity is also part of the engineer's role: runbooks for expected failure modes, monitoring alerts, and enough documentation for an on-call engineer who is not an ML specialist to diagnose and recover from common failures.

The technical team lead is responsible for alignment: ensuring that the architect and engineer agree on the design, that the platform and SRE teams understand what the job requires from the infrastructure, that the cost model is visible to the cost-center owner, and that the timeline accounts for the iteration cycles that distributed ML projects require. The lead does not need to understand every implementation detail, but must understand the major trade-off decisions well enough to communicate them upward and to protect the team from pressure to skip steps that would create technical debt, particularly around fault tolerance, observability, and reproducibility.

Platform and SRE engineers partner with all three roles. They provision the cluster, manage the scheduler, own the storage and networking layers, and respond to infrastructure incidents. The seam between the ML team and the platform team is where many projects encounter their first friction: the ML team knows what the job needs but cannot express it in terms the platform team can meet, or the platform team meets the stated requirement but not the underlying need. Clear interface contracts, written down before the cluster is provisioned, are the primary mitigation.

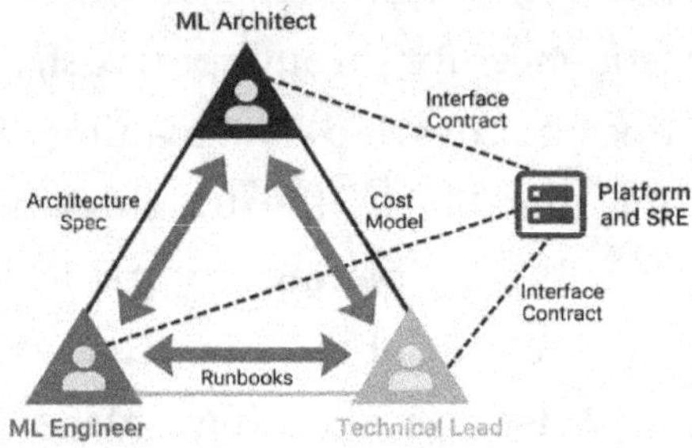

Why a Pattern Mindset Outlasts Tools

The distributed ML tooling landscape has cycled completely at least twice in the past decade. The parameter server architecture that dominated large-scale training in the early 2010s was displaced by ring all-reduce frameworks by 2017. Those frameworks have since been supplemented by hybrid parallelism frameworks optimized for very large models. The serving infrastructure built around 2021 has been redesigned to meet the memory bandwidth requirements of large language models. In each transition, teams that had built knowledge on tool-specific tutorials found that their expertise did not transfer. Teams built on the underlying patterns found that the transition required learning new syntax rather than new concepts.

A pattern is an abstract description of a recurring solution to a recurring problem in a specific context. The all-reduce pattern is a solution to the problem of aggregating gradients across distributed workers without a single-coordinator bottleneck; it is implemented differently across frameworks and network substrates, but the core structure remains the same. The checkpoint-and-restart pattern is a solution to the problem of recovering from partial failures in a long training run; the checkpoint format and the restart protocol differ across frameworks, but the structure and the tradeoffs are invariant. Learning the pattern means learning something that applies to every past and future implementation of that pattern.

The pattern mindset also improves communication across teams. When an ML architect and a platform engineer can both refer to 'the ring all-reduce pattern' and know what the other means, the design conversation is faster and more precise than when each is describing a tool configuration. Shared vocabulary makes architectural decision records more durable: a record that says 'we chose ring all-reduce over parameter server because our topology provides uniform bandwidth' remains readable five years later, regardless of which specific implementation was in use.

This book teaches patterns first and names tools only as illustrative callouts. When a framework or platform is named, it is to anchor the pattern in something the reader may have encountered, not to recommend it for any particular workload. The architect should be able to take the patterns to any framework. The engineer should be able to recognize them in any codebase. The team lead should be able to use them as a vocabulary for technical direction-setting that holds across the next platform cycle.

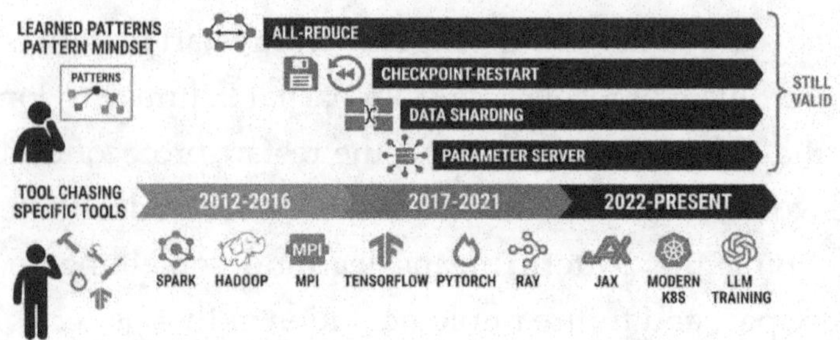

How to Read This Book

The book is structured to follow the lifecycle of a distributed ML workload from foundations to production, so chapters build on one another when read in order. However, the book is also designed to serve as a reference, so any chapter can be opened and read independently. This section describes the chapter template and offers a short guide to non-linear reading for practitioners who need specific answers quickly.

Every chapter follows the same template. It opens with an Opening Scenario: a composite incident drawn from real production failure shapes, designed to make the chapter's topic concrete before the technical content begins. The scenario is followed by the core technical sections, each of which names a canonical pattern, explains the problem it solves, describes the solution's structure, and identifies the trade-offs involved. The chapter closes with a Technical Checklist, a Team Conversation section with open-ended discussion questions, and a Key Takeaway section that synthesizes the chapter's main points. The Technical Checklist is designed to serve as a release gate: the team should be able to answer every item on the checklist before declaring a system production-ready for that chapter's topic.

For practitioners who need to address a specific class of problem immediately, the following chapter guide is provided. If your primary concern is the parallelism strategy for a workload that does not fit on a single device, start with Chapter 2 (Compute Parallelism Patterns) and Chapter 5 (Distributed Training at Scale). If your primary concern is data pipeline throughput and storage costs, start with Chapter 3 (Data Sharding and I/O Optimization). If your primary concern is cluster scheduling and resource utilization, start with Chapter 4 (Orchestration and Scheduling). If your primary concern is serving latency and throughput, start with Chapter 6 (Serving at Scale). If your primary concern is GPU cost and budget, start with Chapter 7 (Resource Management and Cost Optimization). If your primary concern is debugging a distributed job that is failing silently, start with Chapter 8 (Observability and Debugging). If your primary concern is large language model infrastructure, start with Chapter 10 (Distributed Patterns for LLMs and Foundation Models).

Regardless of the entry point, the vocabulary established in this chapter is assumed throughout the book. The terms defined here, including data parallelism, model parallelism, pipeline parallelism, all-reduce, sharding, parameter server, rank, world size, and process group, appear in every subsequent chapter without re-definition. Appendix D provides quick-reference definitions for all technical terms used throughout the book.

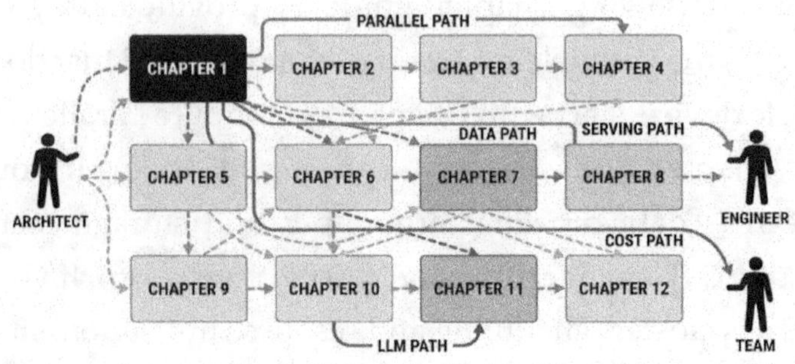

A Working Definition of Production Distributed ML

The word 'production' is used loosely in ML contexts to mean anything from 'it ran once without crashing' to 'it meets the full operational, reliability, and cost requirements of a commercial service.' This book uses the word precisely, and that precision matters because every chapter implicitly builds toward this definition. A distributed ML system is production-ready when it satisfies five properties: reproducibility, observability, fault tolerance, cost discipline, and a defensible security posture. These five properties are the acceptance criteria for every workload, and every chapter in this book adds a piece of the machinery required to satisfy them.

Reproducibility means that given the same inputs (data, hyperparameters, random seeds, and hardware configuration), the system produces the same outputs within an acceptable tolerance. In distributed ML, reproducibility is harder than in single-node training because the order of gradient updates varies with network timing, floating-point arithmetic is not generally commutative, and data sharding interacts with the random seed in ways that differ across frameworks. A system that cannot be reproduced cannot be debugged. Reproducibility is not always achievable at the bit-for-bit level, but it must be achievable at the level of statistically equivalent model quality metrics across independent runs.

Observability means the team can infer the system's internal state from its external signals, without modifying the system to extract them. A distributed ML system that is not observable is one in which the team discovers that training has diverged only after the evaluation metrics are computed, or in which a node failure goes undetected for hours because there is no alert on job-level heartbeat or step-time degradation. Production observability for distributed ML requires instrumentation at four levels: the model training loop, the data pipeline, the infrastructure layer, and the serving layer. Each level produces different signals at different timescales, and the on-call engineer must be able to correlate signals across levels to diagnose a failure quickly enough to contain its impact.

Fault tolerance means that the system can survive the failure of any single component without losing the full accumulated work product. In a training context, checkpoints at regular intervals are sufficient to restart the job after a node failure. In a serving context, the loss of one or more replicas does not take the endpoint dark; traffic is redistributed to surviving replicas while new ones are provisioned. Fault tolerance does not mean zero data loss or zero downtime; it means graceful degradation and automatic or low-touch recovery.

Cost discipline means that the team actively manages and optimizes the cost of operating the distributed system, that cost is visible to the people accountable for it, and that cost-reducing decisions are made deliberately rather than discovered on a billing statement. It requires understanding the per-experiment cost of training runs, the per-request cost of serving, and the storage and I/O consumed by the pipeline. It also requires a process for approving large training runs before they start, so a misconfigured job that would consume ten times the intended budget is caught before it runs, not after.

A defensible security posture means that the system handles training data, model weights, and inference outputs in a way consistent with the organization's data governance requirements, that multi-tenant access to shared cluster resources is appropriately isolated, and that the team can demonstrate that posture to an auditor. Security is the property most often deferred in early-stage distributed ML projects, and it is the most expensive to retrofit. Responsible by design means building security into the architecture from the beginning, not bolting it on after a compliance review. Chapter 9 covers security and multi-tenancy.

Five Properties of a Production Distributed ML System

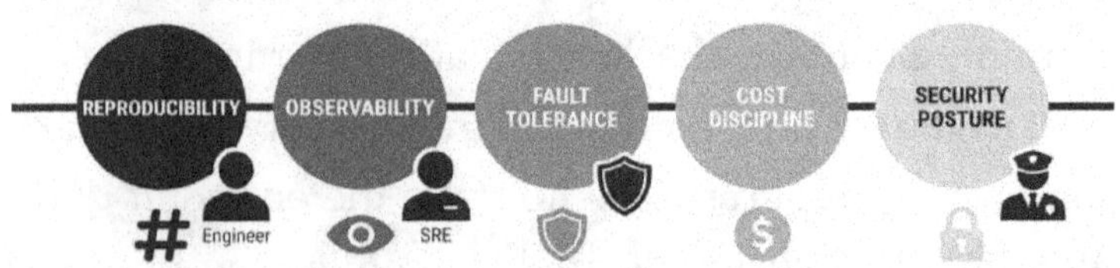

Canonical Vocabulary Reference

The terms introduced in this chapter are assumed throughout the rest of the book. The following definitions are working definitions; each term is explored in more depth in the chapters that cover the pattern it names.

Data parallelism replicates the full model across multiple devices and processes different data shards in parallel, aggregating gradients after each step. It scales throughput with the replica count but does not reduce per-device model memory.

Model parallelism partitions the model across devices when the model exceeds the per-device memory limit. The two principal forms are tensor parallelism (splitting individual weight tensors across devices) and pipeline parallelism (assigning layer groups to sequential pipeline stages).

Pipeline parallelism divides model layers into stages, each assigned to a different device. It passes micro-batches through the pipeline to overlap computation across stages, minimizing the idle gap at stage boundaries (the pipeline bubble).

All-reduce is the collective communication operation used to aggregate gradients across data-parallel workers. Every participant contributes a tensor and receives the elementwise reduction. The ring all-reduce variant achieves this in O(N) steps with each step transferring a constant fraction of the total data volume.

The parameter server pattern separates parameter storage (server nodes) from gradient computation (worker nodes). It supports asynchronous updates at the cost of gradient staleness, or synchronous updates at the cost of throughput limited by the slowest worker.

Sharding partitions a large data structure (dataset, model weights, optimizer state, embedding table, KV cache) across multiple storage or compute units. The shard key determines how partitions are created; the replication factor determines how many copies exist for redundancy.

Fully Sharded Data Parallelism (FSDP) extends data parallelism by sharding model parameters, optimizer state, and gradients across workers. Each worker gathers the full tensor only when needed and immediately discards it, substantially reducing peak per-device memory. Zero Redundancy Optimizer (ZeRO) implements the same approach in three progressive stages: sharding optimizer state, gradients, and parameters. ZeRO Stage 3 and FSDP are functionally equivalent in memory profile.

Rank is the integer index of a process in its process group (starting from zero). World size is the total number of processes. A process group is a named subset of all processes that participate in collective operations together. All-reduce and other collectives operate within a process group, not across the full cluster unless the full cluster is a single group.

The pilot-to-production pathway is the sequence of stages a distributed ML workload moves through, from proof of concept to a fully production-hardened deployment. Each stage has acceptance criteria tied to the five production properties: reproducibility, observability, fault tolerance, cost discipline, and security. It is a continuous incremental process, not a single project milestone.

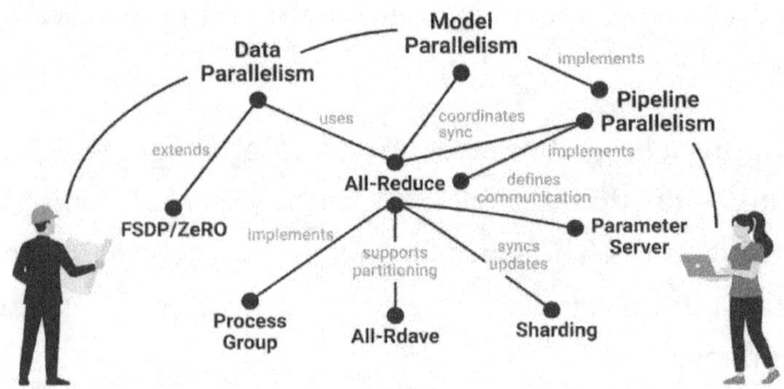

Technical Checklist

Use this checklist as a release gate before declaring any distributed ML workload production-ready against the standards established in this chapter.

- The team can articulate, in writing, why the workload requires distribution rather than a larger single node, including a quantitative argument based on memory, compute time, or serving throughput boundaries.
- The ML architect has classified the workload along the compute parallelism and data parallelism axes, named the binding constraint (memory, compute, or network bandwidth), and documented the classification in an architecture decision record.
- The platform team has documented the cluster's network topology, including the intra-node interconnect type and bandwidth, the inter-node fabric type and bandwidth, and the bisection bandwidth of the cluster.
- The team has named the consistency, performance, and cost trade-offs the current design makes explicitly, and has listed the trade-offs being deliberately deferred to a later iteration.
- Roles for the ML architect, ML engineer, and technical team lead are written down for the project, with one named owner per role and the decision rights of each role specified.
- A baseline single-node implementation of the workload exists, is reproducible end to end, and has been used to establish reference metrics for model quality and training throughput.
- A short list of the failure modes the job is expected to encounter has been written and shared with the on-call rotation, along with runbook links for each expected failure mode.
- Acceptance criteria for what production means for this specific workload have been agreed in writing across the ML team, the

platform team, and the team lead, covering reproducibility, observability, fault tolerance, cost ceiling, and security posture.

- A checkpoint frequency has been chosen and tested: the team can demonstrate that a job interrupted at the most recent checkpoint can be restarted and resume training correctly.
- The data pipeline has been benchmarked to confirm that it can saturate the GPU cluster at the target batch size, and any I/O bottleneck has been identified and addressed.
- The cost model for the training run (accelerator hours, storage I/O, network egress) has been estimated, reviewed by the cost-center owner, and approved before the full training run is initiated.
- All vocabulary terms introduced in this chapter (data parallelism, model parallelism, pipeline parallelism, all-reduce, sharding, FSDP, ZeRO, rank, world size, process group) have a shared definition within the team that is consistent with the definitions in this book.
- Observability instrumentation is in place at the training loop level (step time, loss, gradient norm), the infrastructure level (GPU utilization, memory, network bandwidth), and the data pipeline level (read throughput, queue depth) before the first full training run is initiated.
- The team has reviewed the pilot-to-production pathway and agreed on which properties (reproducibility, observability, fault tolerance, cost discipline, security) are fully satisfied and which are planned for a future iteration.

Team Conversation

Use these questions in a team meeting to pressure-test alignment on the concepts introduced in this chapter. There are no correct answers; the goal is to surface disagreements and gaps before they become production incidents.

1. Where on the single-node-to-distributed curve does our most important workload sit today, and where will it sit in twelve months, given current data growth rates and model size trends?
2. Which of the three core trade-offs (consistency, performance, cost) are we currently optimizing for, and which are we silently sacrificing? Does the team agree on which is which?
3. Which of the failure modes cataloged in this chapter have already cost us a production incident, and which ones are we still flying blind on because we lack observability?
4. Do our architects and engineers have a shared, written definition of who owns the parallelism strategy, who owns the cluster sizing decision, who owns the cost ceiling, and who owns fault-tolerance design? If not, what is the process for resolving a disagreement in each of those areas?
5. If our current training or serving framework were deprecated or superseded tomorrow, how much of our team's working knowledge would survive the transition intact? Which parts are tied to the tool and which parts are tied to the patterns underneath?
6. What is the smallest change we could make to our current pipeline this sprint to move it meaningfully closer to the production definition in this chapter? What is blocking that change?
7. Who on our team should read which chapters of this book first, and what is the mechanism by which we will share what we learn across the team, rather than siloing it in one person?
8. What does production mean, specifically, for our most critical workload? Can we write down the acceptance criteria for reproducibility, observability, fault tolerance, cost, and security today, before we need them in a crisis?

9. How do we currently communicate with our platform and SRE teams about what a distributed training or serving job needs from the infrastructure? Is that communication happening before provisioning or after a failure?
10. What is our current checkpoint frequency, and has anyone tested that a checkpoint restart actually works correctly? When was the last time that test was run?

Key Takeaway

Distribution is now the default operating mode for serious machine learning, not an advanced specialty or an optional optimization. The three forces that push workloads from single-node to distributed (memory constraints, compute time requirements, and serving throughput demands) apply to any team whose models and datasets are growing, which is most teams working on ML in a commercial context. The teams that adapt well are the ones that understand distribution as a library of patterns rather than as a collection of specific tools, and that have invested in the vocabulary that makes those patterns legible across roles and across the infrastructure boundary.

Every distributed ML decision is a negotiation across three axes: consistency, performance, and cost. No configuration optimizes all three, and the engineer or architect who does not understand this will inadvertently optimize one axis while degrading another without realizing it. The network is almost always the binding constraint on training throughput, not the accelerator, and understanding the cluster's hardware topology is a prerequisite for understanding why a job is underperforming. The five properties that define a production-ready distributed ML system (reproducibility, observability, fault tolerance, cost discipline, and security posture) are the acceptance criteria that every subsequent chapter in this book is building toward. A pilot-to-production pathway that does not address all five properties will encounter an expensive surprise at some point in the production lifecycle.

The pattern mindset is the single most durable investment a distributed ML team can make. Frameworks rotate; the underlying patterns do not. The vocabulary established in this chapter (data parallelism, model parallelism, pipeline parallelism, all-reduce, sharding, parameter server, FSDP, ZeRO, rank, world size, process group) will be assumed throughout the rest of this book. Fluency in that vocabulary allows an engineer to read a new framework's documentation and immediately understand which patterns it implements and which trade-offs it has made. The fintech team from the opening scenario did not solve their training-time problem by learning a new tool; they solved it by understanding the patterns well enough to choose the right combination of those patterns for their specific cluster topology, memory constraints, and cost ceiling. That is the capability this book is designed to build.

3 Compute Parallelism Patterns

Opening Scenario

The message that arrived at 11:47 p.m. on a Tuesday read: training crashed again. The applied research team had been attempting to scale a sequence model from 7 billion to 14 billion parameters. The architecture was sound, the dataset was ready, and the cluster had 64 high-end GPUs sitting idle. What the team had not done was model the memory requirements before they wrote the launch script. The first run died with an out-of-memory error before a single gradient was computed.

The team's first instinct was to spread the work across more GPUs using data parallelism. They configured eight replicas, each holding the full model, and restarted training. The job launched this time, but the same crash recurred within two minutes: each replica still held the complete 14-billion-parameter model, and no amount of horizontal replication could fix a per-device memory limit. Data parallelism replicates the computation, not the model. The engineers had reached for the familiar tool without checking whether it addressed the actual constraint.

Their second attempt was naive model parallelism. They split the model's layers evenly across four GPUs, assigned forward passes in sequence, and restarted. Training ran, but hardware utilization hovered near 25 percent. The problem was pipeline stalls: each GPU waited for the previous one to finish before computing, serializing the work rather than parallelizing it. End-to-end throughput was lower than that of a single well-optimized machine.

Two engineers spent a week reading the framework documentation before discovering, through a combination of forum posts and profiler traces, that the correct approach was a two-dimensional combination of pipeline parallelism and tensor parallelism, sized to match the specific cluster topology and the model's attention headcount. They iterated on micro-batch size, pipeline schedule, and tensor-parallel degree for another week before finding a configuration that saturated the interconnect without overflowing device memory. The model eventually shipped, but the schedule was blown, and the budget was gone.

The lesson that the team learned is the premise of this chapter: parallelism is a design problem, not a configuration problem. The right strategy follows directly from the model's memory profile, its computation graph, and the cluster's network topology. When those inputs are known, the answer is usually unambiguous. When they are unknown, any choice is a guess, and most guesses are expensive. This chapter identifies the canonical patterns, explains when each applies, and provides a decision framework, so no team has to discover the answer by accident.

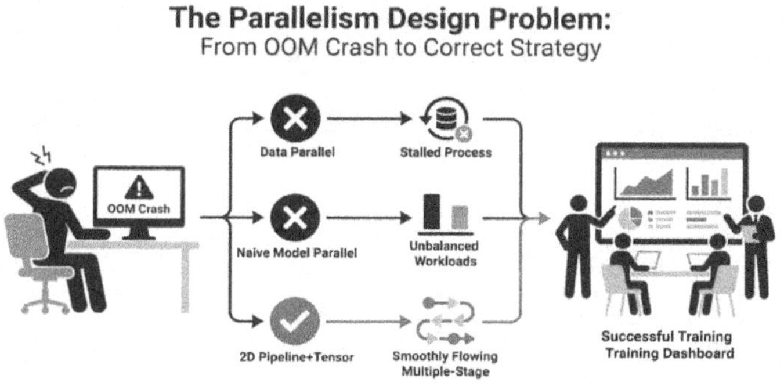

The Parallelism Design Problem:
From OOM Crash to Correct Strategy

Why It Matters

Parallelism decisions are among the few technical choices in a distributed ML project that are genuinely irreversible during training. Changing a parallelism strategy mid-run means stopping training, re-sharding or re-replicating the model, and restarting from a compatible checkpoint, assuming one exists. For a run that takes days or weeks, that cost is not academic. An incorrect parallelism choice made at launch is a budget and schedule event, not merely a configuration inconvenience.

The workflow-level impact extends beyond training. Parallelism choices in training constrain the serving architecture. A model trained with tensor parallelism of degree 8 requires at least 8 devices to load during inference because the weights were never stored in a form that fits on fewer devices. A model trained with data parallelism and a large global batch size may exhibit convergence properties that differ from those of a smaller-batch baseline, affecting how the team evaluates quality. Parallelism is not an infrastructure concern isolated from the rest of the ML workflow; it is a first-class design input that shapes everything downstream.

The mission-aligned goal of this chapter is operational clarity: a team that can name the parallelism primitives they are using, explain why they chose them, and describe the trade-offs they accepted is a team that can reason about failures, estimate costs, and make confident decisions when the system needs to change. That level of clarity does not come from reading framework documentation. It comes from understanding the patterns themselves, independent of any specific implementation.

Data Parallelism: Strategies and Trade-offs

Data parallelism is the default starting point for distributed training, and for good reason. The mental model is simple: place a complete copy of the model on each worker, split each training batch across workers, compute gradients independently on each worker, and then synchronize gradients before updating weights. Every worker always holds the same model state, so, from the optimizer's perspective, the distributed system behaves like a single large-batch training run. The implementation contract is straightforward, the failure modes are well-understood, and the pattern scales gracefully as long as the model fits in the memory of a single device.

The synchronous variant of data parallelism requires every worker to complete its local gradient computation before any weight update is applied. The coordination mechanism is an all-reduce collective: each worker contributes its gradient tensor, the tensors are summed, and the result is broadcast back so every worker applies an identical update. The training loss curve is mathematically equivalent to a single-worker run with a proportionally larger batch size, making synchronous data parallelism the baseline against which all other patterns are compared.

The asynchronous variant relaxes the synchronization barrier. Workers push gradients to a central parameter server as soon as local computation finishes and pull updated parameters at the start of each step. The throughput gain is significant when workers have heterogeneous speeds or when network latency is high. The cost is gradient staleness: slow workers compute gradients that refer to a model state that faster workers have already overwritten. Staleness degrades convergence at large worker counts, particularly where the loss landscape is sharp. Asynchronous data parallelism remains common in recommendation systems, where gradients are sparse, and some degree of staleness is acceptable. For dense transformer training, synchronous variants dominate.

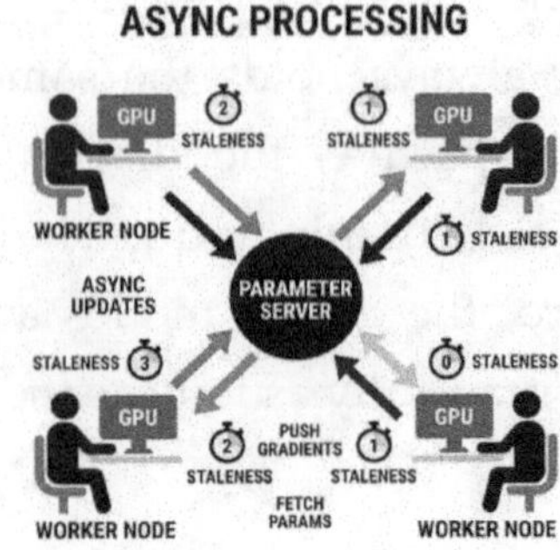

The ring all-reduce algorithm is the dominant communication pattern for synchronous data parallelism at scale. Rather than routing all gradient traffic through a central coordinator, ring all-reduce arranges workers in a logical ring. Each worker sends a fraction of its gradient to the next worker and receives a fraction from the previous worker, accumulating partial sums as data travels around the ring. The per-worker communication volume is roughly 2(N-1)/N times the gradient size, which approaches 2 times the gradient size as N grows, but does not scale with N. This bandwidth efficiency is the primary reason data parallelism scales well on high-bandwidth interconnects.

The tree-all-reduce algorithm is an alternative for hierarchical network topologies, where the bandwidth between racks is lower than within a rack. Workers within a rack first perform a local all-reduce over the fast intra-rack network, producing a single reduced gradient per rack. The rack-level results are then all-reduced over the slower inter-rack network. Tree all-reduce reduces traffic on the expensive inter-rack links by a factor equal to the rack size, at the cost of two sequential reduction phases instead of a single one. Clusters built with a fat-tree or two-tier spine-leaf topology benefit from hierarchical all-reduce; flat high-bandwidth clusters like NVLink meshes typically do not.

In the parameter server architecture, as an alternative to ring all-reduce, workers (which compute gradients) are separated from servers (which store and update parameters). Workers push gradients to the servers, the servers apply the update, and workers pull the new parameters. The architecture decouples communication from computation more explicitly than the ring all-reduce, making it easier to implement asynchronous training and sparse-update patterns. The failure surface differs: the parameter server is a coordination point whose failure halts the entire job unless it is replicated. At very large worker counts, the parameter server's network bandwidth becomes the bottleneck, and partitioning the server tier across multiple machines becomes necessary. Modern large-scale training systems have largely moved away from classic parameter server designs in favor of sharding patterns discussed later in this chapter. Still, parameter servers remain relevant for specific workload shapes.

Model Parallelism: Tensor and Layer Splitting

Model parallelism is the collective name for strategies that distribute the model's parameters across multiple devices, so that no single device needs to hold the entire model. The distinction from data parallelism is fundamental: data parallelism replicates the model and distributes the data; model parallelism distributes the model and routes the data through it. The two categories address different constraints. Data parallelism is constrained by communication bandwidth and batch size. Model parallelism is constrained by how parameters, activations, and inter-device activation traffic are arranged to minimize idle time and avoid memory overflow.

Tensor parallelism splits individual operations across devices. The canonical case is a matrix multiplication in a transformer's attention or feed-forward layer. A weight matrix of shape [M, N] can be partitioned column-wise into K shards of shape [M, N/K], one shard per device. Each device computes the partial matrix product for its shard using its local activations, and the results are gathered and combined across devices before the next layer. The communication pattern for column-wise tensor parallelism is an all-gather on the input activations before the operation and a reduce-scatter on the output activations after it. Row-wise partitioning, the complement, places an all-reduce on the output instead. In practice, a column-wise split followed immediately by a row-wise split eliminates the intermediate all-gather, leaving only a single all-reduce per transformer block. This is the Megatron-style tensor parallelism pattern that has become the standard for attention and feed-forward layers in large transformers.

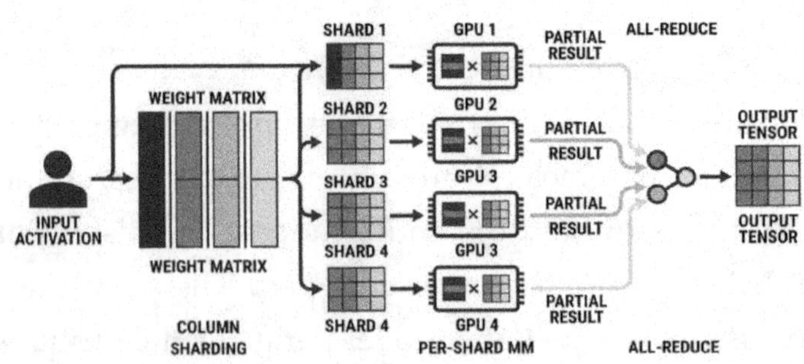

Layer parallelism, often called inter-layer model parallelism or vertical model parallelism, assigns entire layers of the network to different devices rather than splitting individual operations across devices. Device 0 holds layers 1 through L/K, device 1 holds layers L/K+1 through 2L/K, and so on. A forward pass requires each device to compute its layers and then pass the activations to the next device. A backward pass requires each device to compute its gradients and pass them back in reverse order. The communication volume is determined by the activation size at the layer boundary, not the gradient size, which is typically much smaller than the full model gradient. Layer parallelism is simple to implement and requires no changes to the internal operations of each layer. Still, it serializes execution unless combined with a scheduling strategy that keeps multiple micro-batches in flight simultaneously. Naive layer parallelism produces the pipeline bubble problem described in the next section.

The choice between tensor parallelism and layer parallelism depends on the model's computation graph and the cluster's network characteristics. Tensor parallelism introduces frequent, fine-grained all-reduce operations within each layer, which requires high-bandwidth, low-latency interconnects between the devices in the tensor-parallel group. NVLink between GPUs within a node, or InfiniBand at very high bandwidth within a rack, is sufficient. Tensor parallelism across nodes is rarely efficient. Layer parallelism, by contrast, requires only a single activation transfer at each layer boundary, which is a coarser and less frequent communication event. Layer parallelism tolerates higher inter-device latency and is the right choice when the only available interconnect is an Ethernet network. In practice, large model training uses both: tensor parallelism within a node and layer (pipeline) parallelism across nodes.

Pipeline Parallelism: Micro-Batching and Scheduling

Pipeline parallelism solves the idle-device problem that naive layer parallelism creates. The core idea is to split the training batch into multiple micro-batches and inject them into the pipeline in rapid succession so that multiple devices are always computing simultaneously. While device 1 is processing micro-batch 2 in its layers, device 0 is already processing micro-batch 3. The pipeline is never full because the forward and backward passes require devices to process micro-batches in a fixed order. Still, careful scheduling minimizes the fraction of time any device spends waiting. The wasted idle time at the start and end of each batch is called the pipeline bubble.

The bubble fraction is a function of the number of pipeline stages, P, and the number of micro-batches, M. With a naive schedule where all forward passes precede all backward passes, the bubble fraction approaches (P-1)/M. A pipeline with 8 stages requires at least 8 micro-batches to keep the bubble below 12.5 percent. The 1F1B (one forward, one backward) schedule, introduced with the GPipe and PipeDream frameworks and widely adopted since, interleaves forward and backward passes for individual micro-batches rather than separating them. In the 1F1B schedule, a device completes the backward pass for micro-batch K before beginning the forward pass for micro-batch K+1, thereby bounding the number of in-flight activations a device must store and reducing peak activation memory usage. For a pipeline with P stages and M micro-batches under the 1F1B schedule, the bubble fraction is approximately (P-1)/(M+P-1), which approaches zero as M grows large relative to P.

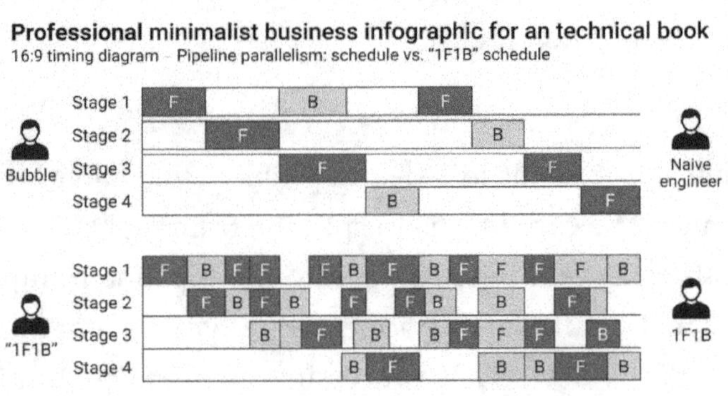

The interleaved pipeline schedule, sometimes called the virtual pipeline schedule, is a further refinement that divides each device's layers into multiple non-contiguous chunks. A device with 4 layers in a 4-stage pipeline might instead hold chunk 1 (layers 1-2) and chunk 5 (layers 9-10) in an 8-virtual-stage configuration. The micro-batch completes one virtual stage per pass through the pipeline, then circles back. The interleaved schedule reduces the bubble fraction by a factor of the number of chunks per device, at the cost of increased activation memory usage and more frequent pipeline communication. The choice between 1F1B and interleaved scheduling is a trade-off between bubble fraction and memory pressure, and the right answer depends on the available device memory and the number of micro-batches in each batch.

Pipeline depth has diminishing returns. Adding more pipeline stages increases potential parallelism but also increases bubble overhead and activation traffic at stage boundaries. The optimal pipeline depth for a given model and cluster is typically set to ensure that the pipeline stages have roughly equal computation times and that the stage-boundary activations fit comfortably in device memory without requiring rematerialization. Profiling the per-layer compute time before designing the pipeline split is a prerequisite, not an afterthought. Models in which some layers are much heavier than others, such as those with large MoE layers at specific depths, require careful load balancing across pipeline stages to avoid a single stage becoming the sustained bottleneck.

Sequence Parallelism for Long-Context Models

Sequence parallelism addresses a constraint that neither data parallelism nor standard tensor parallelism resolves: the activation memory cost of long input sequences. In a transformer model processing sequences of length S with hidden dimension D, the attention mechanism requires storing an activation tensor of shape [B, S, D] at each layer during the forward pass, where B is the batch size. For a model with 32 layers, a batch size of 4, a sequence length of 128,000 tokens, and a hidden dimension of 8,192, that activation tensor alone consumes tens of gigabytes per layer, far exceeding the device memory budget of any practical training setup even after applying standard gradient checkpointing.

Sequence parallelism partitions the activation tensor along the sequence dimension. Each device in the sequence-parallel group holds a contiguous slice of the input sequence, processes its slice through the non-attention components of each layer, and communicates with other devices for the operations that require a global view of the sequence, primarily the attention mechanism. The attention step typically requires an all-gather on the query, key, and value tensors before computing attention scores, followed by a reduce-scatter on the resulting attention scores. Ring attention is an optimization that computes attention in a ring topology, passing key-value slices between devices to avoid storing the full key-value matrix on any single device. Ring attention can reduce the per-device activation memory for the attention step from O(S) to O(S/K), where K is the sequence-parallel degree.

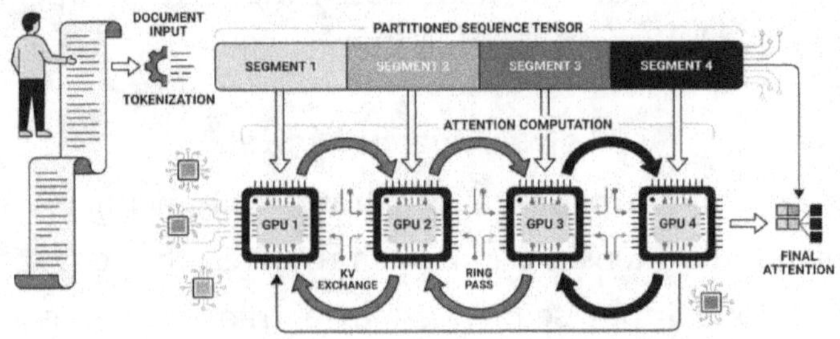

Sequence parallelism naturally composes with tensor parallelism when both activation and parameter memory are constrained. The two dimensions can be stacked: tensor parallelism reduces the per-device parameter memory for large weight matrices, and sequence parallelism reduces the per-device activation memory for long sequences. In the combined configuration, the set of devices is partitioned into tensor-parallel groups, and within each tensor-parallel group, the sequence dimension is additionally partitioned. The communication pattern includes both a tensor-parallel all-reduce and a sequence-parallel all-gather at each transformer layer, making high-bandwidth interconnects within the combined group a hard requirement. This combination is characteristic of long-context LLM training and of video or multimodal models where the input sequence is long by design.

The activation-to-parameter memory ratio drives the decision to use sequence parallelism. If the limiting factor in device memory is the weight tensors, tensor parallelism alone is the right tool. If the limiting factor is the activations, sequence parallelism is the right tool. If both are limiting, the combination applies. Practitioners should measure memory usage with a profiler before deciding: activation memory grows with sequence length and batch size, while parameter memory is fixed for a given model size. Models in the 1-billion to 10-billion range typically hit the parameter limit first at normal sequence lengths; models handling sequences above 32,000 tokens often hit the activation limit regardless of model size.

Expert Parallelism and Mixture-of-Experts Routing

Mixture-of-Experts (MoE) models replace the dense feed-forward layer at each transformer block with a collection of expert sub-networks, activating only a small subset of those experts for each input token. A model with 64 experts per layer and a top-2 routing policy activates 2 experts per token, keeping the per-token computation roughly constant regardless of the total number of experts. The total parameter count scales with the number of experts, but the effective compute per forward pass does not. This decoupling of model capacity from compute cost is the primary motivation for MoE architectures: the team can obtain a larger model within a given FLOP budget. The cost is a new class of infrastructure complexity that dense models do not face.

Expert parallelism is the pattern that distributes expert sub-networks across devices. In the simplest configuration, each device holds a subset of the experts and is responsible for computing the forward pass for any token routed to those experts. Tokens from all devices are routed to the appropriate expert devices via an all-to-all collective computation, performed locally, and returned to the originating devices via a second all-to-all. The all-to-all pattern is fundamentally different from the all-reduce used in data parallelism or the all-gather used in tensor parallelism. Each device sends a different amount of data to each other device, and the communication volume depends on the routing decisions made by the gating network at each step. This makes expert parallelism more sensitive to network topology irregularities than the other parallelism patterns.

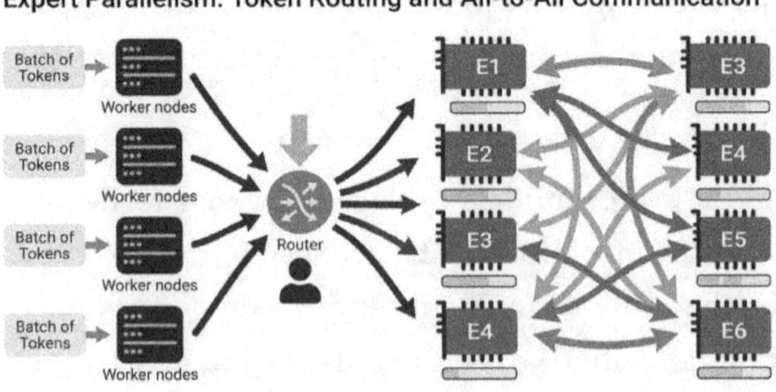

Load balancing is the central operational challenge of expert parallelism. A gating network that consistently routes most tokens to a small number of experts creates a hotspot: the overloaded expert devices become the bottleneck, while other expert devices sit idle. The pattern is called expert collapse, and it is common in early training before the router has learned a balanced distribution. The standard mitigation is a load-balancing auxiliary loss term that penalizes routing imbalance during training, combined with a capacity factor that hard-limits the number of tokens any expert can process per batch. Tokens that exceed the capacity limit are dropped or routed to a secondary expert. The capacity factor is a trade-off between load-balancing guarantees and the cost of dropped tokens.

Expert granularity is a second design dimension. Fine-grained experts with few parameters per expert require many experts to reach the target model capacity, increasing the routing communication volume and the number of all-to-all operations. Coarse-grained experts with many parameters per expert reduce communication but reduce specialization granularity and may concentrate the risk of expert collapse. The practical sweet spot in current large MoE models is expert feed-forward layers with roughly the same parameter count as a standard dense feed-forward layer, with 8 to 128 experts per layer depending on the capacity target.

Shared experts, a refinement used in some production MoE designs, designate a subset of experts to be always active for every token, alongside the routed top-K experts. Shared experts prevent the extreme specialization that can cause routed experts to ignore common linguistic patterns. The pattern trades some of the compute savings from sparse routing for improved stability and, often, better quality on tasks that require broad general knowledge. The decision to include shared experts falls within the scope of the model design discussion. Still, it has direct infrastructure implications: shared experts must be replicated across all devices in the expert-parallel group, which increases the memory requirements for those parameters.

Hybrid 3D Parallelism: Composing Data, Tensor, and Pipeline

No single parallelism primitive is sufficient for the largest model training workloads. Data parallelism reaches its limit when the model no longer fits on a single device. Tensor parallelism is limited by the bandwidth requirements of its fine-grained all-reduce operations and by the diminishing returns of splitting tensors into very many small shards. Pipeline parallelism is limited by the bubble overhead and the need for roughly equal stage compute times. The solution is to compose all three dimensions into a joint parallelism configuration, commonly called 3D parallelism, that exploits each dimension where it is most efficient and compensates for the others' weaknesses.

The standard 3D parallelism assignment works as follows. Tensor parallelism is applied within a single node, where the NVLink or equivalent high-bandwidth interconnect provides the bandwidth required by all-reduce operations. Pipeline parallelism is applied across nodes, grouping multiple consecutive layers onto each node and arranging nodes in a chain. Data parallelism is applied across pipeline groups, running multiple independent replicas of the full tensor-and-pipeline-parallel model on different subsets of the training data. The global batch is divided first by the data-parallel degree, then by the number of pipeline micro-batches. The combined structure forms a three-dimensional grid of devices, with tensor-parallel rank, pipeline stage, and data-parallel rank as the three axes.

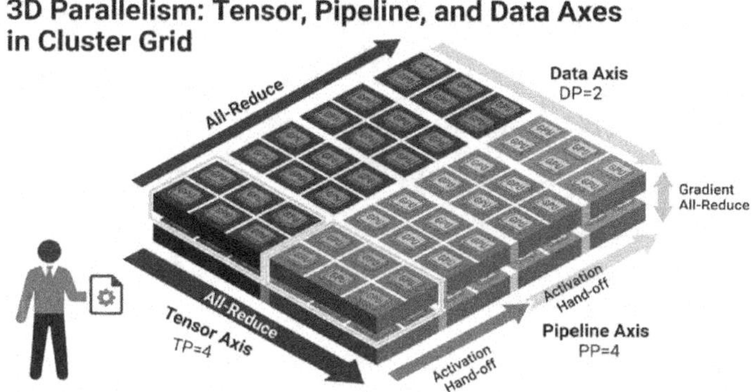

Sizing the three dimensions of a given cluster requires answering a sequence of questions in the right order. First, determine whether the model fits on a single node without tensor parallelism. If not, choose the smallest tensor-parallel degree that achieves memory feasibility, because the all-reduce cost within the tensor-parallel group grows with tensor-parallel degree. Second, determine how many nodes are available and set the pipeline-parallel degree so that the number of pipeline stages divides evenly into the node count. A pipeline-parallel degree of 4 on a 32-node cluster leaves 8 nodes per pipeline group. Third, set the data-parallel degree as the ratio of total nodes to nodes per pipeline group. The remaining design decision is the micro-batch size, which should be set to ensure the bubble fraction is acceptably small, typically requiring at least P times as many micro-batches as pipeline stages.

The failure modes of 3D parallelism are compositional. A tensor-parallel communication stall stops all pipeline stages simultaneously. A pipeline stall at a slow stage slows all data-parallel replicas sharing the same pipeline. A data-parallel gradient all-reduce stalls weight updates for the entire model. The interaction between these failure modes means that a 3D parallel job is more sensitive to hardware heterogeneity and network variability than a data-parallel-only job. Running 3D parallelism on a cluster with variable GPU performance, for example, due to mixed hardware generations or thermal throttling, tends to amplify the slowest-device effect rather than average over it. This makes 3D parallelism a pattern best deployed on homogeneous, high-bandwidth clusters with active hardware health monitoring.

ZeRO and FSDP-Style Sharding

ZeRO, which stands for Zero Redundancy Optimizer, is a family of sharding strategies that reduce the per-device memory footprint of data-parallel training by eliminating the redundancy imposed by conventional data parallelism. In standard data parallelism, each device holds a complete copy of the model parameters, gradients, and optimizer state, most notably the first- and second-moment estimates in Adam-style optimizers. For a model with P parameters in 16-bit floating point, the parameters alone consume 2P bytes, the gradients consume 2P bytes, and the Adam optimizer state consumes 8P bytes (two 32-bit tensors), for a total of 12P bytes per device. With 32 data-parallel replicas, the cluster holds 32 copies of those 12P bytes, the vast majority of which is pure redundancy.

ZeRO-1 shards the optimizer state across data-parallel workers. Each worker stores only 1/N of the optimizer state tensors, where N is the number of data-parallel replicas. After the gradient all-reduce, each worker applies the optimizer update to its parameter shard using its optimizer state shard. The updated parameter shards are then all-gathered so every worker has the full updated parameters for the next forward pass. ZeRO-1 reduces the optimizer state memory by a factor of N, with no change in communication volume relative to standard all-reduce data parallelism.

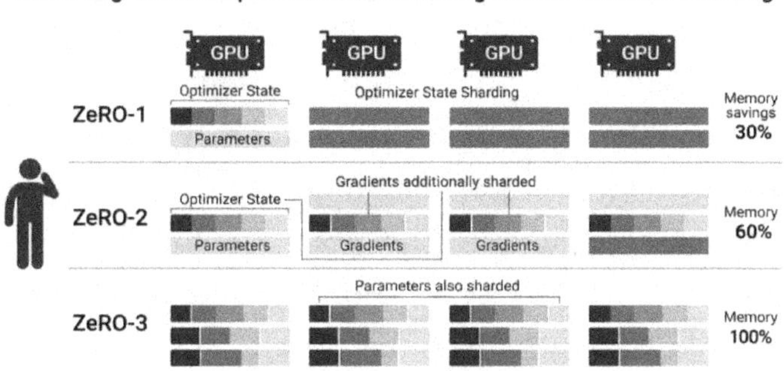

ZeRO-2 adds gradient sharding. Each worker accumulates gradients only for its shard of the parameter space, rather than the full gradient tensor. The communication pattern changes from a single all-reduce to a reduce-scatter, followed by a local optimizer step, then an all-gather for the parameter update. The total communication volume remains the same as ZeRO-1 (roughly 2 times the parameter size per step), but a factor of N reduces the peak memory usage for gradients.

ZeRO-3 extends sharding to the parameters themselves. No worker holds the full model at any time; each worker holds only 1/N of the parameters during steady-state computation. Before each forward pass, the necessary parameters are all gathered from their owners, used for computation, and then discarded. The trade-off is a dramatic reduction in peak parameter memory, which makes ZeRO-3 the only viable data-parallel strategy for models that do not fit on a single device without model sharding. Fully Sharded Data Parallel (FSDP), the PyTorch implementation of ZeRO-3 semantics, wraps each module in the model with a sharding boundary and manages all-gather and reduce-scatter operations automatically, so the engineer can apply sharding to any existing model without rewriting its forward pass.

The communication overhead of ZeRO-3 is approximately twice the parameter size per step, because the all-gather before the forward pass doubles the communication volume relative to standard all-reduce. The memory savings scale with the data-parallel degree, easily an order of magnitude or more in a large cluster, which more than justifies the overhead for most large model training scenarios. Practitioners should profile the all-gather latency at target scale to confirm that many small ZeRO-3 all-gather operations are not becoming individually latency-bound, but rather bandwidth-bound.

Gradient Accumulation and Communication-Computation Overlap

Gradient accumulation is a technique for simulating a larger effective batch size without increasing per-step memory usage. Rather than computing gradients over a full global batch in a single step, the training loop processes micro-batches sequentially, summing their gradients in a local accumulation buffer, and then applies a single optimizer step after all micro-batches have been processed. The effect is identical to training with a batch size A times larger, at the cost of A times as many forward and backward passes per optimizer step. Gradient accumulation is the canonical solution when the target batch size exceeds what can fit in device memory in a single step, and it composes naturally with pipeline parallelism because pipeline micro-batches are already a form of gradient accumulation.

The interaction between gradient accumulation and distributed data parallelism requires careful implementation. A naive implementation performs the all-reduce on the gradient after every micro-batch, which is correct but wasteful. The correct approach defers the all-reduce to the last accumulation step, using a no-sync context or equivalent, thereby suppressing intermediate communications. This reduces communication frequency by a factor of A and significantly improves hardware utilization when the network is a bottleneck relative to compute.

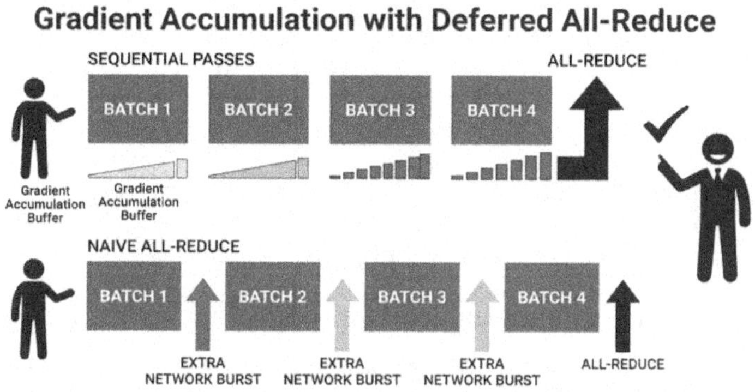

Communication-computation overlap is the broader architectural principle that motivates gradient accumulation deferral and extends to other parts of the training loop. The insight is that modern GPUs can execute computation and network communication simultaneously using separate hardware pipelines: the compute engines and the NIC DMA engine operate independently. Training frameworks exploit this by launching all-reduce or all-gather operations asynchronously while the GPU continues computing subsequent layers. The overlap is most effective when the communication tensor is large enough that the network transfer takes longer than a few layers of computation, and when the layers preceding the communication call have enough work to keep the compute engine busy during the transfer.

Bucket-based communication is the mechanism most implementations use to achieve computation-communication overlap. Rather than launching an all-reduce per-parameter tensor as gradients become available, the framework accumulates gradients into fixed-size buckets. It launches all-reduce for each bucket asynchronously while the backward pass continues. Small buckets produce many latency-sensitive communications that are hard to overlap; large buckets delay communication until late in the backward pass. A bucket size of 25-50 megabytes is a common starting point, tuned by profiling the compute-to-communication ratio on the target hardware.

Gradient Compression and Mixed Precision Training

Gradient compression reduces the volume of data communicated during gradient synchronization by transmitting a compressed approximation of the gradient rather than the full tensor. The two primary families of compression are quantization and sparsification. Gradient quantization represents each gradient value with fewer bits, for example, 8-bit integers rather than 32-bit floats, reducing the communication volume by a factor of 4. Gradient sparsification transmits only the K largest gradient values, setting the rest to zero, reducing communication volume by a factor of the sparsification ratio. Both approaches introduce error into the gradient estimate, which the training system must manage to avoid divergence.

Error feedback is the standard mechanism for managing compression error. When a gradient value is compressed, the quantization or sparsification error is added to a per-device error buffer. At the next step, the error buffer is added to the fresh gradients before compression, so errors omitted in one step are carried forward and eventually included in a subsequent step.

Error feedback prevents the compression error from accumulating indefinitely and, in practice, allows sparsification ratios of 99 percent and quantization to 8 bits with minimal degradation in convergence quality for many workloads. The convergence guarantee is empirical rather than theoretical for most sparsification schemes, and practitioners should validate convergence behavior with and without compression on a representative workload before deploying compression in production.

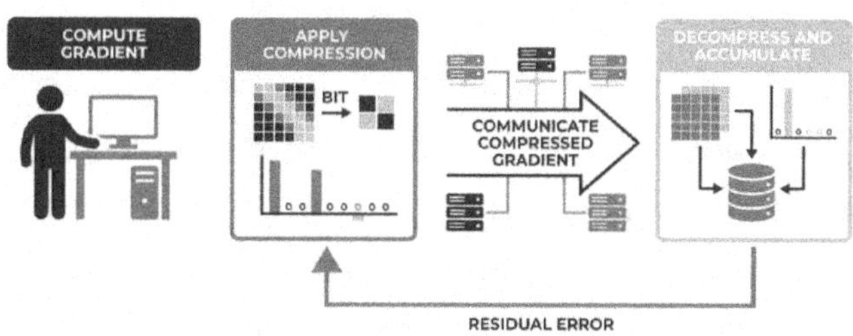

Mixed-precision training uses 16-bit floating-point (FP16 or BF16) for forward- and backward-pass computations, while maintaining a 32-bit master copy of the parameters for optimizer updates. The compute savings are significant: modern accelerators provide 2 to 4 times the throughput for 16-bit matrix multiplications compared to 32-bit, and the reduced tensor size halves the memory footprint for activations and working memory during the forward pass. The optimizer state (the master parameters plus Adam moments) is kept in 32-bit, preserving the numerical precision needed for stable gradient updates over millions of steps.

BF16 (Brain Float 16) has largely displaced FP16 for large model training because BF16 provides the same exponent range as FP32, which eliminates the gradient overflow and underflow problems that require loss scaling in FP16 training. FP16 training without dynamic loss scaling routinely produces NaN gradients when gradient magnitudes are large, which requires loss scaling: multiplying the loss by a large constant before backpropagation to amplify gradient values into the representable FP16 range, and dividing the scaled gradients by the same constant before the optimizer step. BF16 avoids this complexity at a small cost in mantissa precision, which is acceptable for most gradient-based optimization workloads. Loss scaling remains necessary on hardware that lacks native BF16 support.

FP8 training, available on the most recent generation of accelerators, takes mixed precision one step further by using 8-bit floating-point for matrix multiplications in the forward and backward passes, while maintaining FP32 or BF16 precision for gradient accumulation and optimizer state. The additional throughput gain is hardware-dependent, and the numerical stability requirements are more demanding than those of BF16, but FP8 represents the frontier of precision reduction for production training. The pattern-level principle is consistent across all mixed-precision variants: use the lowest precision the computation tolerates at each stage of the training loop, and maintain higher precision only where numerical stability requires it.

Communication Patterns and Collective Operations

Every parallelism strategy in this chapter is implemented, at the network level, by a small set of collective communication operations. A collective operation is one in which all members of a process group participate, and the result depends on the inputs from all members. Understanding each collective's cost model is essential for predicting whether a given parallelism configuration will be compute-bound or communication-bound at the target scale. The binding constraint determines where optimization effort is most valuable.

All-reduce is the collective most closely associated with data parallelism. Every worker contributes an input tensor; the tensors are reduced using an associative operation (typically summation for gradient aggregation), and every worker receives the result. The bandwidth cost of an all-reduce in-the-ring algorithm is 2(N-1)/N times the tensor size per worker, which approaches 2 times the tensor size as N grows. Latency is proportional to 2(N-1) rounds of point-to-point transfers, though the pipelined ring implementation reduces the effective latency. All-reduce is most efficient when the tensor is large relative to the latency overhead, making it suitable for dense gradient tensors but expensive for small, frequent updates.

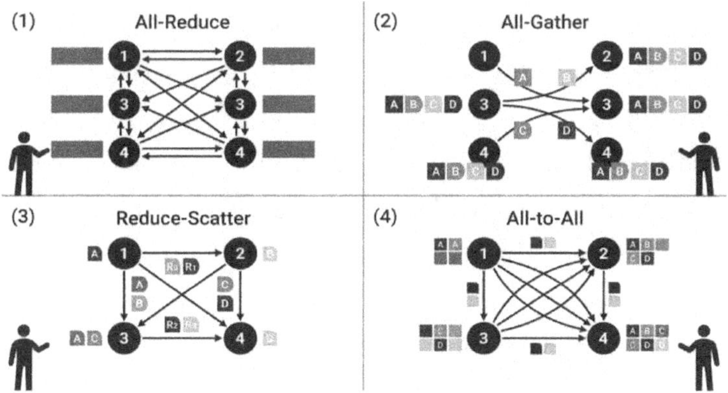

All-gather is the collective used at the beginning of tensor-parallel operations and in ZeRO-3 parameter fetching. Each worker holds a shard of a tensor and contributes it; each worker receives the full concatenated tensor. The bandwidth cost per worker is (N-1)/N times the full tensor size, which approaches 1 as N grows. All-gather is cheaper than all-reduce because it involves no reduction operation, only data movement. The output is N times larger than any individual input, so all-gather is typically followed immediately by a computation that consumes the full tensor before it is stored.

Reduce-scatter is the complement of all-gather. Every worker contributes a full tensor; every worker receives a reduced shard. The bandwidth cost per worker is (N-1)/N times the full tensor size, which is the same as for all-gather. All-reduce is equivalent to reduce-scatter followed by all-gather, which is why the ring all-reduce implementation achieves a 2(N-1)/N bandwidth cost: it performs both operations back-to-back. ZeRO-2 uses reduce-scatter for gradient synchronization and delays all-gather until the parameter update step, allowing gradient shards to be processed locally and discarded without ever materializing the full gradient tensor.

All-to-all is the collective used in expert parallelism for token routing. Each worker sends a distinct slice of its data to each other worker and receives a distinct slice from each. Unlike all-reduce and all-gather, the all-to-all communication volume depends on the routing pattern and can vary widely if the routing is unbalanced. The bandwidth cost in the balanced case is (N-1)/N times the total data volume per worker, similar to all-gather. Still, the access pattern involves N separate point-to-point transfers rather than a single scan, making it more sensitive to congestion at network switches in fat-tree topologies. Implementations on hierarchical networks often decompose all-to-all into a local all-to-all within a rack, followed by an inter-rack all-to-all, reducing pressure on the inter-rack links.

Anti-Patterns: Parallelism Choices That Look Right and Are Not

The most expensive anti-pattern in distributed training is naive model parallelism without a pipeline schedule. A team that splits model layers across devices and processes the full batch sequentially through each device will observe GPU utilization near 1/P, where P is the number of pipeline stages, because only one device computes at any given moment. This anti-pattern appears frequently because it is the obvious first attempt when data parallelism fails due to memory limits: the engineer splits the model, verifies that training runs without crashing, and does not immediately recognize that the throughput collapse is a consequence of the implementation strategy rather than an inherent property of model parallelism. The fix is always to introduce pipeline micro-batching. The diagnostic signal is GPU utilization well below 50 percent, with long compute-free intervals visible in a profiler trace.

Oversized data-parallel groups are a subtler anti-pattern. Increasing the data-parallel degree reduces the per-replica batch size for a fixed global batch, which can push the batch below the threshold at which the GPU's matrix multiplication units are well utilized. A batch size that is too small fills only a fraction of the GPU's SIMD width, and the hardware overhead per operation dominates the useful computation. The anti-pattern manifests as a system that scales linearly in device count on paper. Still, it shows sublinear throughput scaling in practice, often with GPU memory utilization comfortably low and compute utilization low as well, because the batches are too small to saturate either resource. The fix is to increase the micro-batch size or the number of accumulation steps to restore hardware utilization, even if it means reducing the degree of data parallelism.

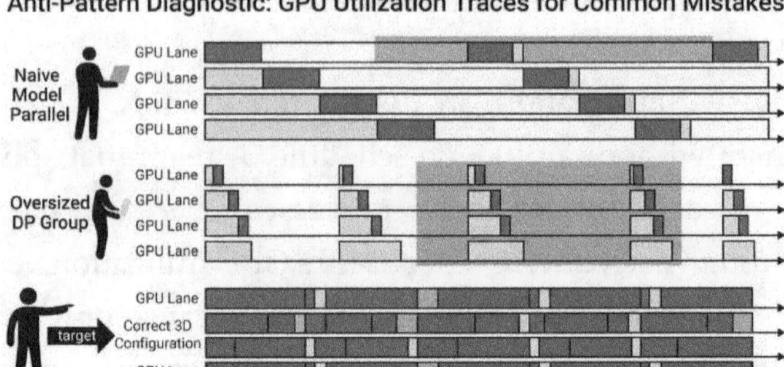

Micro-batch sizes that are too large for the pipeline schedule are the mirror anti-pattern to undersized batches. In pipeline parallelism, the activation tensors for all in-flight micro-batches must be stored simultaneously in device memory. A micro-batch size chosen without accounting for the number of in-flight activations can overflow device memory mid-training when the pipeline is filled, even if the first few steps succeed. The failure mode is a delayed out-of-memory error that occurs hundreds of iterations into training, and is particularly costly if the checkpointing frequency is low. The mitigation is to compute the expected peak activation memory analytically before starting a run: multiply the per-batch activation size by the pipeline depth and ensure the product fits within the device memory budget.

Copying parallelism configurations from papers without matching the cluster shape is a systemic anti-pattern driven by pressure to reproduce published results quickly. A configuration that achieves state-of-the-art results on a specific interconnect may perform poorly when inter-node bandwidth is lower or when the node count does not divide evenly. Treat published configurations as starting points and run a sweep over parallelism degrees, micro-batch sizes, and accumulation steps on the actual target cluster before committing to a long run. The sweep cost is always smaller than the cost of a failed or degraded run.

Ignoring the communication topology is a meta-level anti-pattern that causes all of the above mistakes. The cluster's network topology, the bandwidth between every device pair, and the latency structure of the interconnect are not incidental details. They are the primary inputs to the parallelism design. A team that does not document the cluster topology before designing the parallelism configuration is making decisions without the most important constraint in mind. The pilot-to-production pathway for any distributed training system should include a topology characterization step, using the cluster's built-in bandwidth measurement tools or a custom all-pairs benchmark, before finalizing the parallelism configuration.

A Decision Framework for Choosing a Parallelism Strategy

The parallelism decision framework presented here is a sequential checklist rather than a lookup table. It must be applied in order because each step's answer constrains the options available in subsequent steps. The framework takes four inputs: the model's memory profile (parameters, gradients, optimizer state, activations at target batch size and sequence length), the cluster's network topology (intra-node bandwidth, inter-node bandwidth, topology type), the available device count, and the target training throughput. Filling in these four inputs is a prerequisite before applying the framework.

Step one: Determine whether the model fits on a single device with the target optimizer state. For a model with P parameters in BF16, the minimum memory requirement is approximately 2P bytes for parameters and 8P bytes for the Adam optimizer state, totaling 10P bytes per device in data-parallel mode. A 7-billion-parameter model in BF16 with Adam requires approximately 70 gigabytes per device for parameters and optimizer state alone, before activations. If this exceeds the device memory budget, the team must use ZeRO-3, FSDP, or model parallelism, and data-only parallelism is ruled out.

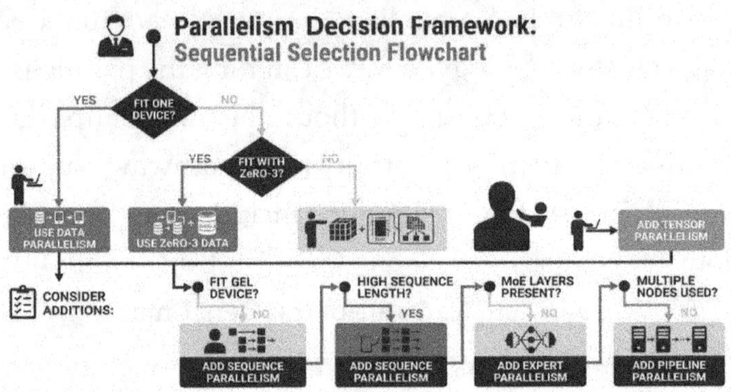

Step two: determine the tensor-parallel degree. If model parallelism is required and intra-node high-bandwidth interconnect is available, set the tensor-parallel degree to the smallest value that makes the model fit per device, up to the number of devices per node. A tensor-parallel degree larger than the number of GPUs per node forces tensor-parallel communications over the inter-node network, which is rarely efficient. If no intra-node high-bandwidth interconnect is available, skip tensor parallelism and proceed directly to step three.

Step three: Determine the pipeline-parallel degree. Set the pipeline-parallel degree to the number of nodes in the tensor-parallel group that can be arranged in a chain with the target stage compute balance. Verify that the number of micro-batches is at least P times the pipeline depth to keep the bubble fraction below 10 percent. If achieving this requires a micro-batch count that results in an effective batch size larger than the convergence-optimal batch size, reduce the pipeline depth.

Step four: fill the remaining devices with data parallelism. The data-parallel degree is the total device count divided by the product of the tensor-parallel and pipeline-parallel degrees. Apply ZeRO-1 or ZeRO-2 if the optimizer state is the binding memory constraint. Apply ZeRO-3 if the per-device parameter memory remains too large after tensor parallelism is applied. Step five is a sanity check: run the configuration on a scaled-down version of the workload, profile communication and compute utilization, and verify that the dominant bottleneck matches the expected one before committing to a full-scale run.

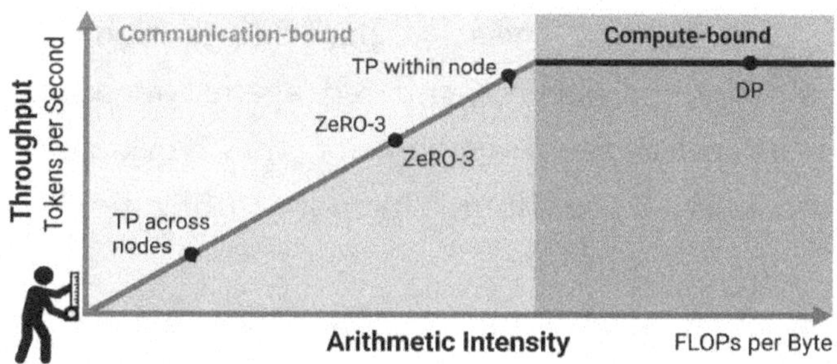

Technical Checklist

Use the following checklist as a release gate before declaring any distributed training configuration production-ready. Each item should be verifiable by a team member other than the person who made the original design decision.

- The model's memory profile has been measured: parameters, gradients, optimizer state, and peak activations at the target batch size and sequence length are all documented before any parallelism

decision.
- The cluster's network topology has been documented: intra-node bandwidth, inter-node bandwidth, topology type (fat-tree, rail-optimized, fully connected), and the bandwidths of all inter-node links used by the chosen parallelism configuration.
- The parallelism primitives in use are named explicitly: tensor-parallel degree, pipeline-parallel degree, data-parallel degree, and any sharding variant (ZeRO-1/2/3 or FSDP) are listed with the rationale for each choice.
- Communication collectives have been profiled at the target scale: the dominant collective, its measured bandwidth utilization, and its latency contribution per step are known.
- Micro-batch size, gradient accumulation steps, and pipeline schedule have been tuned rather than copied from a tutorial or paper: the target effective batch size was computed first, and the micro-batch schedule was derived from it.
- The chosen strategy has been compared against at least one alternative on a scaled-down version of the workload: the comparison includes throughput, memory utilization, and convergence stability.
- The peak activation memory in the filled pipeline has been computed analytically and verified against a profiler measurement: the system does not rely on the first few steps succeeding as evidence that it will not OOM later.
- Gradient accumulation uses deferred all-reduce (no-sync context or equivalent): intermediate micro-batch gradients are not synchronized over the network until the final accumulation step.
- Communication-computation overlap has been confirmed via a profiler: GPU compute and network DMA timelines show measurable overlap rather than sequential execution.
- Mixed precision configuration is documented: the precision of

each stage (forward activations, gradient computation, gradient communication, optimizer state, master parameters) is explicit and intentional.
- Anti-patterns from this chapter have been explicitly checked against the current configuration: naive model parallelism without a pipeline schedule, oversized data-parallel groups, and topology-mismatched tensor-parallel degrees have all been ruled out.
- A rollback plan exists: the team can revert to the previous parallelism configuration and restart from the last valid checkpoint within a defined time window if the current configuration fails to scale or diverges.

Team Conversation

The following questions are designed for use in team meetings to pressure-test alignment on compute parallelism. They do not have a single correct answer; their value is in surfacing hidden assumptions and gaps.

1. Which parallelism primitives described in this chapter are we using today, and which ones does the team silently avoid because no one has implemented them in a production system? What would it cost us to fill that gap?
2. Where in our current training pipeline is the network the binding constraint, and how do we know? Can we point to a profiler trace, or is the answer based on intuition?
3. Have we ever changed the parallelism strategy in a running system without a controlled experiment? What happened, and what would we do differently now?
4. Which of the anti-patterns described in this chapter have we shipped at least once? What was the detection path, and how long did the misconfiguration survive in production before we identified it?

5. If we could only invest engineering time in mastering one parallelism pattern next quarter, which one would have the largest payoff for our current model sizes and cluster shape, and why?
6. How would we explain our current parallelism configuration to a new team member without naming any specific framework? If we struggle to answer this question, what does that tell us about our understanding of the system?
7. What is the largest model we could train on our current cluster if we applied this chapter's decision framework end-to-end? How confident are we in that estimate, and what would it take to verify it?
8. If we needed to halve our training cost tomorrow without reducing model quality, which parallelism or communication optimization from this chapter would we apply first, and what risk would that decision carry?

Key Takeaway

Computing parallelism is not a single choice but a layered design problem with a specific decision order. The model's memory profile comes first. The cluster's network topology comes second. The parallelism dimensions follow from those two inputs, not from what a framework makes easy or what a paper happened to use. Teams that invert this order, starting from a familiar configuration and checking afterward whether the constraints are satisfied, pay the price in out-of-memory crashes, sublinear scaling, and the expensive reruns that follow. The patterns in this chapter exist to make the correct order of reasoning explicit.

The primitives are few and composable: data parallelism for replication, tensor parallelism for intra-layer sharding, pipeline parallelism for inter-layer staging, sequence parallelism for activation-heavy long-context workloads, and expert parallelism for sparse MoE routing. ZeRO and FSDP-style sharding are optimizations on top of data parallelism, not separate primitives. Gradient accumulation and communication-computation overlap are implementation techniques that improve the efficiency of any of the above. Mixed precision is a universal strategy for precision management. Knowing which category a given technique belongs to makes it far easier to reason about interactions and failure modes.

The communication profile, not the arithmetic, is usually the binding constraint in distributed training at scale. Every parallelism strategy produces a characteristic collective communication pattern: all-reduce for data parallelism, all-reduce within tensor-parallel groups, activation transfers across pipeline stages, and all-to-all for expert routing. The cost of those collectives on the specific cluster hardware determines whether a given parallelism configuration will achieve near-linear scaling or plateau well below it. Responsible by design means knowing these costs before the training run starts, not after it ends.

4 Data Sharding and I/O Optimization

Opening Scenario

A computer vision team at a mid-sized technology company has spent six months building a single-node training pipeline for a product defect detection model. The pipeline processes roughly four million labeled images per day, the model has converged reliably on an eight-GPU workstation, and leadership has approved a move to a thirty-two-GPU cluster to cut the training cycle from two days to six hours. The team copies their pipeline almost verbatim, adds a distributed data-parallel wrapper, and starts the first full-scale run with genuine optimism.

The results are deflating. Throughput improves by less than fourfold despite having eight times as many accelerators. Profiling reveals the cause: workers spend more than fifty percent of wall-clock time waiting on storage. The original pipeline was designed to saturate a single NVMe drive; with 32 workers, that drive is now a shared bottleneck. A second problem surfaces within the first epoch. One shard contains roughly twice as many samples as the others because the dataset grew after the initial split, and no one updated the shard boundaries. The worker assigned to that shard falls behind, and every other worker in the job idles at the synchronization barrier waiting for it to catch up. The training run is simultaneously I/O-starved and load-imbalanced.

The on-call engineer corrects the imbalance by hand, rebalancing the shards in an afternoon. Throughput climbs, and the team declares the problem solved. Three weeks later, after two more dataset increments, the same long-tail straggler pattern returns. The engineer rewrites the balancing script to run nightly. Still, that script does not account for semantic distribution: the nightly rebalance keeps shard sizes equal by byte count, but bunches all the images from a newly added product category into a single shard. The model trains faster on some classes than others, and the convergence curve develops an unexplained kink.

The fix that finally holds is not a single change. The team adopts a hash-based sharding scheme that distributes samples uniformly across shards regardless of dataset growth. They switch from raw JPEG files on a shared network file system to a sequential binary format that supports large contiguous reads. They build a prefetch pipeline that decouples I/O from compute, ensuring the GPU never idles while the CPU fetches the next batch. They add a tiered caching layer that keeps the hot working set on local NVMe and falls back to network storage only for cold shards. And they add a sharding health check to their CI pipeline so that imbalances are caught before a run starts, not halfway through a forty-hour job. This chapter teaches all five patterns together, along with the reasoning behind each choice.

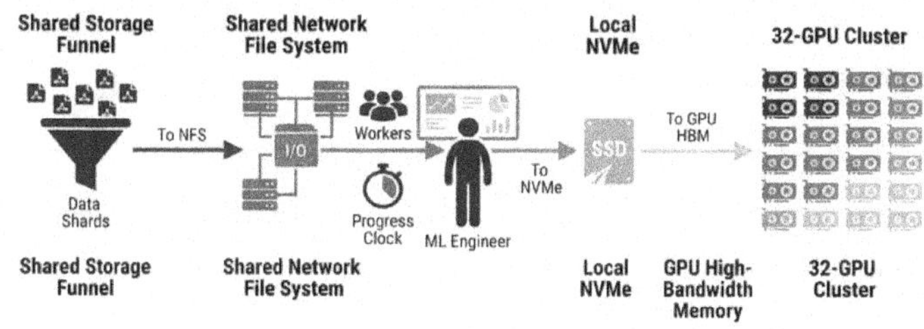

Why It Matters

Data pipeline inefficiency is one of the highest-leverage problems in distributed ML and one of the most under-measured. Most teams carefully instrument their model training loop by tracking loss curves, gradient norms, and per-epoch validation metrics. Far fewer teams track the fraction of wall-clock time their accelerators spend waiting on data. That fraction is the single most important number for understanding pipeline efficiency, and for many teams running at scale, it is above 30%. Every point above zero is a wasted computed budget. At the scale where training jobs run for days or weeks, the operational and financial impact is substantial.

The patterns in this chapter have workflow-level impact that extends beyond raw throughput. A sharding scheme that degrades gracefully as the dataset grows reduces the operational burden on the team that maintains it. A streaming pipeline that is deterministic across restarts makes training jobs schedulable on preemptible compute, which is where the cost savings in modern ML infrastructure are concentrated. A tiered caching layer that keeps the hot working set close to the accelerator reduces the pipeline's sensitivity to network congestion, which becomes a critical property as clusters grow larger. The network becomes a shared resource among many concurrent jobs.

This chapter identifies the canonical patterns for each stage of the data path, from cold storage to accelerator memory. The goal is a vocabulary that the team can carry to any framework and any infrastructure provider. The patterns are presented with their trade-offs, not as a ranked list of recommendations, because the right choice depends on the dataset shape, cluster topology, and the storage tier's cost model. A team running foundation-model pretraining on petabyte-scale data needs a different set of patterns than a team fine-tuning a vision model on a curated million-sample corpus. The chapter covers both ends of that range and the decision points between them.

Sharding Strategies for Training Data

Sharding is the process of dividing a dataset into units that workers consume independently. It is a first-class design decision with consequences that compound through every downstream stage of the pipeline. A sharding scheme that produces unequal shard sizes creates stragglers. A scheme that clusters semantically related samples into the same shard biases the loss landscape. A scheme that is not monotonically extendable as the dataset grows requires periodic manual rebalancing, which is both operational toil and a source of hard-to-reproduce training bugs. Choosing a sharding scheme is not a one-time configuration task; it is an architectural commitment.

Size-balanced sharding is the most common approach and the right default for homogeneous datasets. The dataset is divided into shards of equal byte count or equal sample count. An equal byte count is appropriate when samples are variable-length (e.g., text documents, audio clips) and the downstream bottleneck is I/O bandwidth. An equal sample count is appropriate when samples are fixed-length, and the downstream bottleneck is computational. The trade-off: size-balanced sharding is simple to implement and reason about. Still, it requires a rebalancing pass whenever the dataset grows, and that pass is typically a blocking operation that cannot run incrementally.

Hash-based sharding assigns each sample to a shard by computing a hash of the sample's identifier modulo the number of shards. The assignment is deterministic, stable across dataset versions, and requires no global scan of the dataset to produce. When the dataset grows with new samples that have new identifiers, those samples automatically hash into existing shards. The load variance across shards depends on the hash function's collision properties and the distribution of identifiers. Still, for most practical datasets with more than ten thousand samples per shard, that variance is small enough to be operationally negligible. Hash-based sharding is the recommended default for continuously growing datasets.

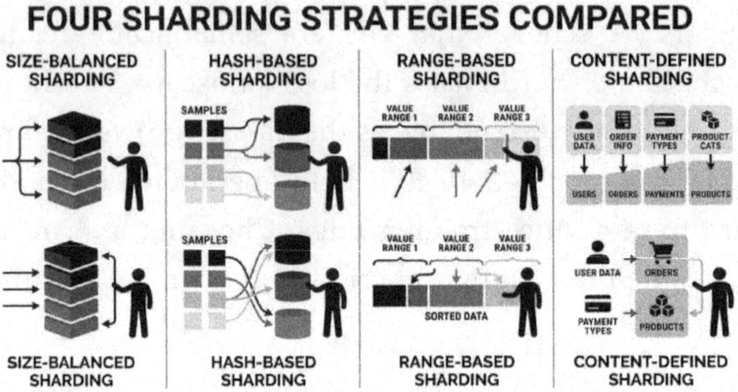

FOUR SHARDING STRATEGIES COMPARED

Range-based sharding divides the dataset by a sort key: samples with keys in a given numeric or lexicographic range are assigned to the same shard. Range sharding is the natural choice when the downstream consumer benefits from sorted order, for example, when training on time-series data where local temporal context matters, or when the dataset is stored in a columnar format that supports efficient range scans. The risk is hotspot formation: if the distribution of keys is skewed, some shards will be much larger than others, and if keys are monotonically increasing (as timestamps often are), new data always lands in the last shard until a rebalance splits it across shards.

Semantic sharding, sometimes called content-defined sharding, groups samples that share a meaningful attribute into the same shard. Use cases include curriculum learning, where early training epochs prefer simpler samples; multi-task training, where each shard corresponds to a task; and retrieval-augmented setups, where related context documents should be co-located for efficient batch construction. Semantic sharding requires domain knowledge to define the grouping criterion, and it introduces the risk of semantic imbalance. If one semantic group dominates the dataset, the shard that contains it becomes a hotspot by design. The pattern is powerful but should be paired with size monitoring.

Locality-aware sharding is an extension of any base strategy that accounts for the cluster's physical layout. Workers near the same storage rack or availability zone are assigned shards stored on that rack or zone, reducing cross-rack or cross-zone traffic. This pattern is most relevant in multi-rack clusters where cross-rack bandwidth is significantly lower than intra-rack bandwidth. The trade-off is that locality-aware shard assignment couples the sharding scheme to the cluster topology, making it harder to port to a different cluster without a reassignment pass.

The decision among these strategies reduces to four questions. First, does the dataset grow continuously? If yes, prefer hash-based sharding to avoid rebalancing. Second, does the workload benefit from sorted order? If yes, range sharding is the right starting point, paired with active hotspot monitoring. Third, does training logic require semantic grouping? If yes, semantic sharding is appropriate, paired with size balancing across groups. Fourth, is cross-rack or cross-zone bandwidth a measurable bottleneck? If yes, add locality awareness as an overlay to whichever base strategy best fits the dataset's shape.

Efficient Dataset Formats: TFRecord, Parquet, WebDataset, and Arrow

The choice of dataset format determines the shape of every read operation in the pipeline. A format designed for random access on a relational database is a poor fit for the sequential, high-throughput reads that ML training requires. A format that supports compression but not streaming is a poor fit for datasets too large to decompress in memory. Understanding the trade-offs among the dominant ML dataset formats is a prerequisite for building a pipeline that is fast by design rather than fast by accident.

TFRecord is a row-based binary format built around length-delimited protocol buffer records. Its primary strength is sequential read performance: a single linear pass through a TFRecord file maps to a single sequential read from the underlying storage, achieving close to the device's theoretical bandwidth. TFRecord files are splittable at the record boundary, which makes them straightforward to shard. They support transparent compression at the record level. The limitation is that TFRecord files require a schema definition and the protocol buffer serialization toolchain, which adds friction for teams outside the ecosystem that popularized the format. Random access is expensive because the format has no index.

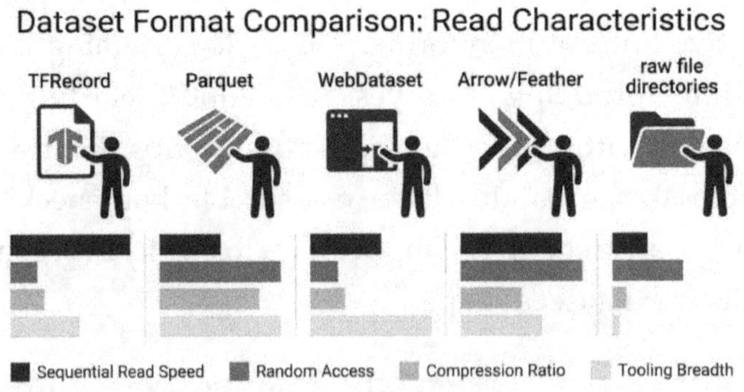

Parquet is a columnar binary format designed for analytical workloads that read a subset of columns across a large number of rows. For ML training, the columnar layout is advantageous when the model consumes a small number of features from a wide feature table, because only the relevant columns need to be read from disk. Parquet files include a footer with row-group statistics and column chunk offsets, which enables pushdown filtering and coarse random access without a full file scan. The limitation of ML training is that columnar layouts require reassembling rows before samples can be passed to the model, and that reassembly is not free. Parquet is the right choice when a data warehouse manages the dataset and ML training is one consumer among many, not when raw per-sample throughput is the only constraint.

WebDataset is a tar-based streaming format that packages heterogeneous per-sample files (image, label, metadata) into a single sequential archive. A WebDataset shard is a standard tar file, which means it is readable with commodity tools and can be stored in any object store that supports byte-range reads. The format was designed specifically for cases where samples contain multiple modalities (an image file, a caption file, and a metadata JSON file), and it bundles them into a single contiguous read. WebDataset shards are naturally ordered and support shuffle buffers for approximate shuffling without a full sort. The limitation is that the tar format offers no built-in random access, which requires a fixed held-out set for evaluation, making it slightly more involved than indexed formats.

Apache Arrow's Feather format (and its IPC streaming variant) is the format of choice when the training pipeline is co-located with a data processing framework that already uses the Arrow memory layout. Arrow's zero-copy interoperability eliminates the deserialization cost between data loading and training when both use Arrow. For pipelines where the bottleneck is CPU deserialization rather than I/O bandwidth, replacing a slower format with Arrow can recover significant time without changing any other aspect of the pipeline. Arrow files are not compressed by default, which means they can be substantially larger than compressed alternatives; this trade-off is acceptable when local storage is abundant and deserialization time is the measured bottleneck.

Raw file directories, in which each sample is an individual file, are common in early-stage datasets because they require no packaging step. The pattern fails at scale for a straightforward reason: object stores and distributed file systems impose a per-file overhead on list and open operations that grows linearly with the number of files. A dataset of ten million images, each a ten-million-item list, generates ten million lists and open calls per epoch. For object stores that charge per API call, this translates directly into billing costs. On distributed file systems, the metadata server becomes a bottleneck under concurrent list load. Migrating from raw file directories to any of the packaged formats above is one of the highest-return I/O optimizations available to teams in early pipeline development.

Prefetching and Overlapping I/O with Compute

The fundamental challenge of an ML data pipeline is that I/O and compute operate at different rates and on different hardware. Storage delivers data at a rate determined by the device type, the network, and the concurrency of other requests. The accelerator consumes data at a rate determined by the model architecture and the batch size. If the pipeline is synchronous, the accelerator waits whenever the I/O rate falls below the compute rate. The producer-consumer pipeline pattern breaks this coupling by placing an asynchronous I/O stage between storage and the accelerator, so that the next batch is being fetched. In contrast, the current batch is being processed.

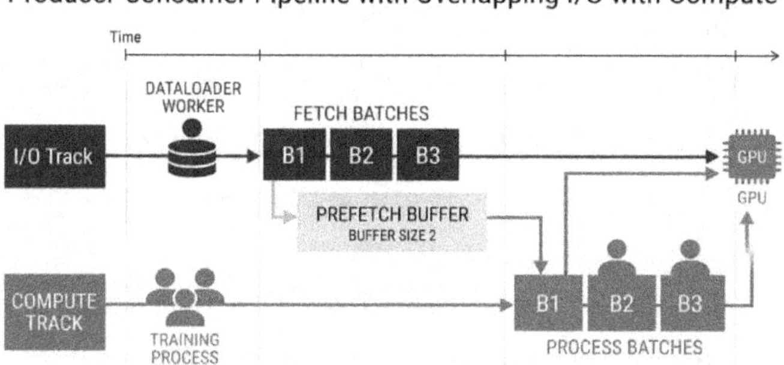

The critical parameter of a prefetch pipeline is prefetch depth: the number of batches held in the prefetch buffer at any time. A depth of 1 means the pipeline begins fetching the next batch as soon as the current batch is consumed, not before. A depth of two means two batches are queued ahead of the current one, which tolerates moderate I/O jitter without stalling the accelerator. In practice, the optimal prefetch depth is the smallest value that maintains accelerator utilization above a target threshold (typically 95%). Higher depths consume host memory and can cause cache thrashing if the buffer is too large relative to available memory.

The degree of parallelism is the second tuning parameter: the number of concurrent I/O threads or processes that load data simultaneously. Increasing the degree of parallelism improves throughput until the underlying storage device or network is saturated. Beyond that saturation point, additional parallelism adds latency without improving throughput, because requests queue at the storage layer. A common mistake is increasing the degree of parallelism to compensate for a slow format or a cold cache; the correct intervention is to fix the underlying cause (switch to a sequential format or warm the cache) and then tune the degree of parallelism against the resulting baseline.

Memory pressure is the hidden risk of deep prefetch pipelines. A prefetch buffer holding large batches of high-resolution images or long text sequences can consume tens of gigabytes of host memory. When the buffer grows to compete with other system memory consumers (the OS page cache, the training process's own tensors), the result is increased swapping, reduced cache hit rates, and counterproductive latency. The responsible-by-design approach is to set an explicit memory budget for the data pipeline and derive the prefetch depth and batch size from that budget, rather than tuning them independently.

Overlapping I/O with compute requires that the dataloader run in a separate execution context from the training loop. The standard pattern is to use background worker processes (not threads, for languages where a global interpreter lock prevents true CPU parallelism) that fetch and preprocess data concurrently with the backward pass. The training loop requests batches from a queue rather than blocking on I/O directly. Profiling this pattern requires instrumentation at the queue level: if the queue is empty when the training loop requests a batch, the pipeline is I/O-bound and requires more workers or faster storage. If the queue is consistently full, the pipeline has spare I/O capacity that could be redirected to other jobs.

Caching, Locality, and the Tiered Storage Pattern

The tiered storage pattern organizes the storage hierarchy used by the data pipeline into three levels: a fast local tier (NVMe SSD attached to the training node), a warm shared tier (a distributed file system or high-throughput object store accessible to all workers in the cluster), and a cold archival tier (a low-cost object store with higher latency and per-request cost). Data migrates up the hierarchy as it becomes hot and down as it cools. The goal is to make the hierarchy appear flat to the training loop: the loop requests a batch and receives it, regardless of which tier actually served it.

Tiered Storage Architecture

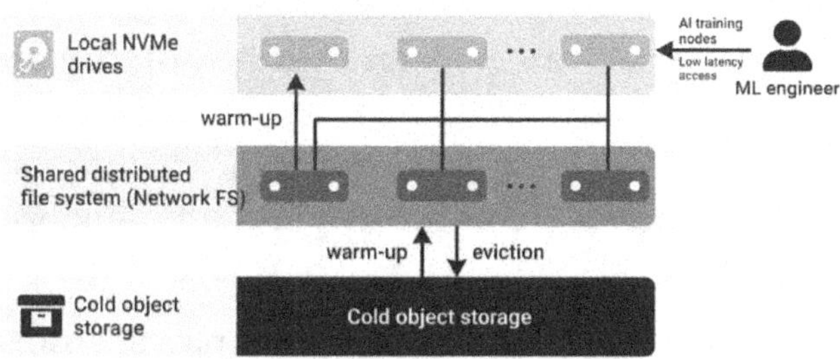

The cache warm-up strategy determines how quickly the first epoch achieves full throughput. A cold cache means that the first epoch is served entirely from the warm or cold tier, which may be significantly slower than subsequent epochs served from local NVMe. Two warm-up strategies are common. The eager warm-up strategy pre-fetches the entire working set to the local tier before training begins, ensuring consistent throughput from the first batch. The lazy warm-up strategy begins training immediately and copies shards to the local tier as they are accessed for the first time. Eager warm-up is preferred for jobs running on preemptible compute where the warm-up cost is a known one-time overhead; lazy warm-up is preferred for interactive exploration where time-to-first-batch matters more than steady-state throughput.

The eviction policy governs which shards are removed from the local tier when the tier is full. The least-recently-used policy is the standard default: the shard accessed most recently is evicted first. For training workloads that sweep the dataset in a fixed order, LRU correctly evicts shards that will not be needed until the next epoch.

For workloads with non-uniform access patterns (curriculum learning, importance sampling, online hard example mining), the access frequency pattern is more complex, and a frequency-aware policy that tracks hit rates per shard may perform better. The key operational requirement is that eviction decisions are logged and observable, so that unexpected eviction storms (caused by a sudden expansion of the working set) can be diagnosed before they affect training throughput.

Distributed cache coherence is a concern when multiple workers share the same warm tier. If two workers try to warm the same shard to their local tiers simultaneously, they both pay the transfer cost, and the shard is stored twice. A shard-level lock or a coordinator-mediated claim protocol prevents this duplication. For most clusters, a simple cooperative protocol is sufficient: a worker that wants to warm a shard checks a shared manifest, claims the shard if unclaimed, and releases the claim once the transfer is complete. Other workers that need the same shard wait for the claim to be released, then read from the local copy produced by the first worker.

Choosing the right storage tier for a given workload starts with characterizing the access pattern. If the working set (the shards accessed over one epoch) fits entirely in the local NVMe tier, then tiered storage reduces to a cache-everything strategy: warm all shards at job start and read entirely from local NVMe throughout the job. If the working set exceeds local NVMe capacity, the eviction policy and warm-up strategy determine whether the pipeline can sustain the target throughput. If the working set exceeds even the warm-tier capacity, the pipeline is inherently cold, and the optimization effort should focus on maximizing sequential read throughput from the cold tier rather than on cache management.

Streaming Data for Large-Scale and Foundation-Model Training

Foundation-model pretraining operates on datasets measured in trillions of tokens or hundreds of terabytes of images. Materializing the entire dataset on a single storage tier before training begins is impractical at this scale: the capacity cost is too high, and the latency to start training in a full materialize-then-train pipeline is unacceptable. Streaming data patterns address this by serving samples directly from the source as training proceeds, without materializing the full dataset. Streaming is now the default operational posture for any dataset that cannot fit comfortably in the warm tier.

Checkpointable Streaming Pipeline for Foundation-Model Training

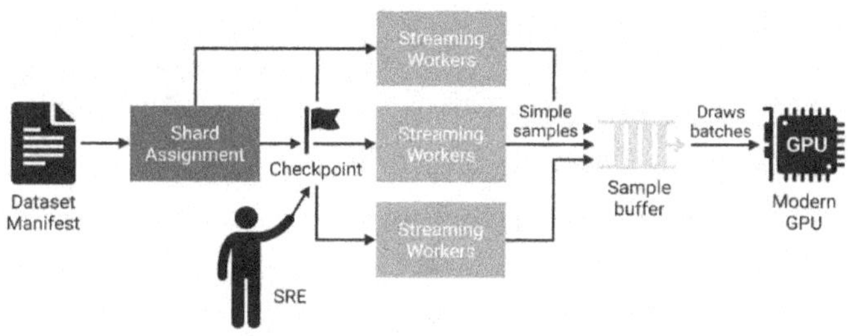

Deterministic streaming requires that the pipeline can reproduce the exact sequence of samples seen by the training loop from a given checkpoint, even after a restart. This property is non-negotiable for fault-tolerant training on preemptible compute, where restarts are routine. The key mechanism is a shard-level iterator state that records which shards have been fully consumed and which position within the current shard was last processed. When the job resumes from a checkpoint, the streaming pipeline restores this iterator state. It resumes from the correct position, producing the same sample sequence the training loop would have seen if the job had not been interrupted.

The streaming architecture constrains the shuffle strategy for streaming data. A global shuffle, where every sample in the dataset can be paired with every other sample in a batch, requires random access to the full dataset, which is incompatible with sequential streaming. The standard compromise is a multi-level shuffle: shards are shuffled in a shard-level shuffle buffer before being assigned to workers (coarse-grained), and samples within a shard are shuffled in a sample-level shuffle buffer before being batched (fine-grained). The statistical quality of this shuffle depends on the buffer sizes and the number of shards. For most training workloads, a shard buffer of a few hundred shards and a sample buffer of a few thousand samples are sufficient to break within-shard ordering, which would otherwise bias the batch distribution.

Checkpoint-friendly streaming requires that the iterator state be serializable and compact. A naive implementation that stores the full sequence of sample identifiers consumed since the beginning of training is not compact: for a trillion-token dataset, that state grows without bound. The correct abstraction is a shard-cursor representation: for each worker, store the index of the current shard and the offset within that shard. This representation has constant size regardless of how many samples have been consumed, and it supports exact restart semantics as long as the shard assignment is deterministic (which hash-based shard assignment guarantees by construction).

Multi-source joins under a shard are required when the training data comes from multiple independently sharded datasets that must be combined into a single batch. A common example is retrieval-augmented training, where each sample from a primary corpus is paired with one or more retrieved context documents from a secondary corpus. If the two corpora are sharded independently, joining them at batch construction time requires either a co-sharding step (re-sharding both corpora with the same key so that matching samples land in the same shard) or a cross-shard lookup (fetching context documents from a separate index at training time). Co-sharding is the more efficient pattern, but it requires coordination during dataset preparation. Cross-shard lookup is more flexible but adds latency proportional to the lookup cost.

Data Deduplication and Shuffling at Scale

Deduplication and shuffling are preprocessing concerns that interact with sharding in ways that are easy to overlook. A dataset that contains duplicate samples is not merely redundant; it biases the training distribution toward the duplicated content and can cause the model to memorize specific samples rather than generalize from them. At the foundation-model scale, where training corpora are assembled from web crawls and multiple institutional sources, the duplication rate can exceed thirty percent without explicit deduplication. The deduplication step belongs in the data preparation pipeline, upstream of sharding. Still, its output (a deduplicated manifest or a deduplicated set of shard files) must be designed for the downstream sharding scheme.

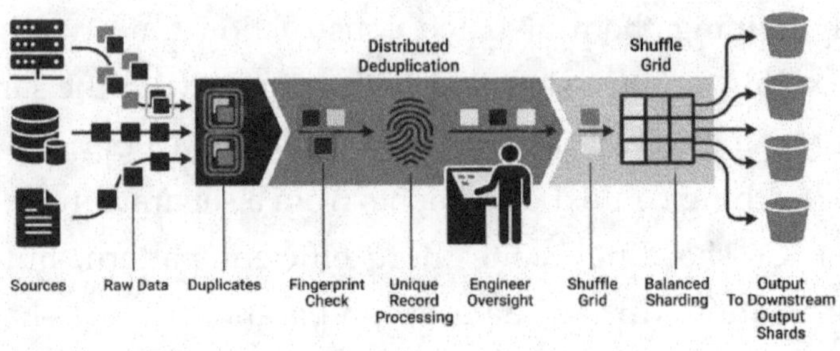

Exact deduplication identifies and removes samples that are byte-for-byte identical. The standard technique is to compute a cryptographic hash (SHA-256 or xxHash for speed) of each sample and collect all hashes into a deduplication index. Samples whose hashes appear more than once are marked as duplicates, and all but one representative is removed. Exact deduplication is efficient and parallelizable: each shard can be hashed independently, and the index merge is a distributed sort-and-dedup operation. The limitation is that exact deduplication does not catch near-duplicates: two images that differ by a single pixel, or two text documents that differ by a single punctuation mark, are treated as distinct samples even if they carry identical semantic content.

Near-duplicate detection uses locality-sensitive hashing or MinHash to identify samples that are similar but not identical. MinHash represents each document as a compact sketch of its n-gram set; two documents with a high Jaccard similarity in their n-gram sets will have a high probability of producing identical sketches. LSH groups samples with similar sketches into the same hash bucket, making near-duplicate detection a linear-time operation (linear in the number of samples) rather than a quadratic comparison. Near-duplicate detection is computationally more expensive than exact deduplication. It introduces a similarity threshold that requires domain judgment: a threshold that is too low removes semantically distinct samples that happen to share common phrases, and a threshold that is too high misses paraphrased duplicates, which would bias training.

Shuffling at scale must balance statistical quality against operational cost. A full Fisher-Yates shuffle of a trillion-sample dataset requires either loading the entire dataset into memory or performing multiple passes over it on disk, both of which are infeasible at foundation-model scale. The practical alternative is the approximate multi-pass shuffle: the dataset is divided into chunks, chunks are shuffled independently (an embarrassingly parallel operation), and then chunks are interleaved randomly. The resulting shuffle is not perfectly uniform, but the bias introduced by the chunk boundaries is small relative to the batch-level variance from other sources (label noise, augmentation stochasticity). For most training workloads, a two-pass chunk shuffle produces shuffle quality indistinguishable from a global shuffle in validation loss trajectories.

Sharding Across Heterogeneous Clusters

Homogeneous clusters, in which every worker has the same CPU speed, memory capacity, NVMe bandwidth, and network connectivity, are the underlying assumption in most introductory treatments of data sharding. Production clusters are rarely homogeneous. A cluster assembled across multiple purchasing cycles may mix older and newer GPU generations, each with different host memory configurations. A cloud cluster running on spot instances may mix instance types with different local storage capacities. A multi-region cluster may mix nodes with different network latencies to the shared storage tier. Sharding strategies designed for homogeneous clusters produce stragglers in heterogeneous ones, because equal shard sizes do not translate to equal processing times when workers have different throughput capacities.

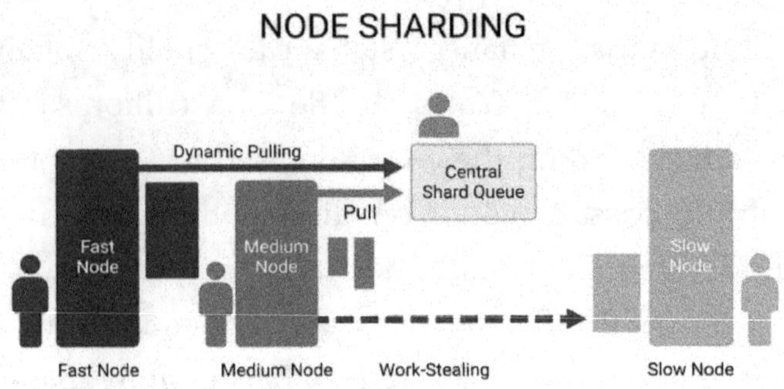

Dynamic sharding addresses heterogeneity by decoupling shard assignment from worker identity. Rather than pre-assigning a fixed set of shards to each worker before training begins, the dynamic sharding pattern maintains a global shard queue that workers pull from on demand. A fast worker drains its current shard quickly and immediately pulls the next one from the queue. A slow worker takes longer to drain its shard, but does not hold back the fast workers, because the fast workers are not waiting for it. The cluster's total throughput is determined by the aggregate capacity of all workers, not by the slowest worker's capacity.

Work-stealing is an extension of dynamic sharding that addresses the case where a worker has been assigned a large shard (or a shard stored on slow storage) and is falling behind while other workers are idle. In the work-stealing pattern, an idle worker actively looks for a peer that is mid-shard and proposes to take over the remaining portion of that shard. The shard is split at the current cursor position, and the stealing worker begins processing from that position. In contrast, the original worker continues to the end of the portion it has already loaded. Work-stealing requires that shards be splittable at arbitrary positions, which is a property of record-oriented binary formats (TFRecord, WebDataset) but not of formats with header dependencies (some compressed Parquet variants).

Locality-aware work assignment is a complementary pattern that routes shards to the workers closest to the storage that holds them, reducing network traffic in the common case. In a cluster with multiple racks or availability zones, the assignment algorithm prefers to give each worker shards stored in the same rack or zone. When a worker's local shards are exhausted, it falls back to remote shards. This pattern reduces peak cross-rack bandwidth and improves overall cluster network utilization. The interaction with dynamic sharding is straightforward: the global shard queue can be annotated with the storage location of each shard, and workers that request a new shard specify their location so that the assignment logic can prefer local shards.

Dataloader Fault Tolerance and Checkpoint-Friendly Streaming

Training jobs running on preemptible compute or across many nodes face a non-trivial probability of worker failure during any given run. A training run that restarts from the most recent model checkpoint and replays the data pipeline from the beginning introduces a subtle but consequential bug: the model continues training from a mid-training parameter state but sees data samples in a different order than it would have seen if the failure had not occurred. If the data pipeline includes any randomized components (e.g., shuffle buffers or augmentation stochasticity seeded by the global step), the replay diverges from the original trajectory. At a minimum, this produces a slightly different model. At worst, it produces a model that has seen some samples many more times than others, which biases the training distribution in ways that are difficult to detect from loss curves alone.

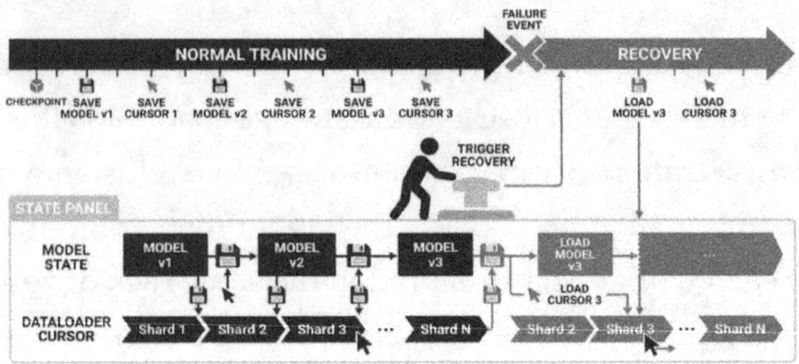

The solution is joint checkpointing: the dataloader cursor state is saved in the same atomic checkpoint operation that saves the model parameters. When the job resumes, both the model and the dataloader restore from the same checkpoint, and the training trajectory continues exactly as it would have without the interruption. Joint checkpointing requires that the dataloader expose a serializable state object (the shard index and intra-shard offset for each worker) and that the training loop pass this state to the checkpoint manager alongside the model parameters. This is an operational clarity requirement, not a research concern: the engineering work is straightforward, but it must be done deliberately and tested with synthetic restarts before the pipeline goes to production.

Graceful degradation handles the case where a single dataloader worker fails mid-run without causing the entire job to fail. The standard pattern is a worker pool with automatic replacement: the training coordinator detects that a worker has stopped producing batches, marks the shards the worker was responsible for as unassigned, and launches a replacement worker that picks up those shards from the last checkpoint. The training loop continues without interruption during the replacement window by drawing batches from the other workers' prefetch buffers. This pattern requires that the prefetch buffer be deep enough to absorb the replacement latency, which is a function of worker startup time and the time to warm the replacement worker's local cache.

End-to-end reproducibility from a checkpoint is a stronger guarantee than fault tolerance alone. It means that two runs starting from the same checkpoint, even if they use different numbers of workers or different hardware, produce identical loss trajectories. Achieving this requires that the shard assignment be deterministic given the checkpoint state (which hash-based sharding provides), that the intra-shard sample order be deterministic given the shard seed (which requires a seeded PRNG for shuffle), and that the augmentation pipeline be seeded consistently from the sample identifier (not from the global training step, which changes across restarts). End-to-end reproducibility is the benchmark for a production-grade data pipeline and forms part of the pilot-to-production pathway for any new training infrastructure.

Data Quality Gates in the I/O Path

Data quality problems that reach the training loop are expensive to diagnose. A corrupted shard may not produce an error; instead, it may produce syntactically valid but semantically wrong samples, such as images paired with incorrect labels, or text documents truncated mid-sentence. The model trains on these samples, the loss does not diverge catastrophically, and the quality problem surfaces only when downstream evaluation reveals degraded performance on specific subsets of the evaluation set. By that point, the corrupted training run may have consumed days of accelerator time. A data quality gate placed in the I/O path catches these problems at minimal cost: the gate runs on every batch before it reaches the model, adding microseconds of CPU time in exchange for hours of debugging saved.

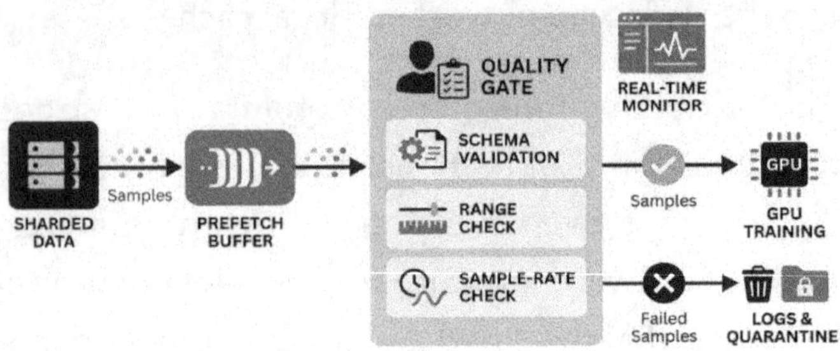

Schema enforcement validates that each sample conforms to the expected structure: the correct keys are present, the correct data types are used, and required fields are not null. For typed binary formats, schema enforcement can be embedded in the deserialization step at near-zero cost, because the deserializer already validates the wire format. For less-structured formats (JSON lines, raw file directories), an explicit schema check adds a small per-sample overhead that is worth the cost. Schema violations should be logged with the offending shard identifier and byte offset, enabling the data engineering team to locate and repair the source shard without reprocessing the entire dataset.

Range checks validate that numerical fields fall within expected bounds: pixel values between 0 and 1, audio amplitudes between -1 and 1, and label indices within the vocabulary size. A tensor containing out-of-range values often indicates a preprocessing bug (e.g., a normalization step applied twice or not at all), and catching it at batch time prevents it from silently affecting model parameters. Range checks are inexpensive when vectorized across the batch dimension, and they can be implemented as assertions that are enabled in development and optionally disabled (or sampled at 1%) in production to reduce per-batch overhead.

Sample-rate sanity checks monitor the distribution of labels, languages, or other categorical attributes across batches and flag anomalies where the distribution deviates significantly from the expected training distribution. A sudden shift in label distribution within a batch is a strong signal that the sharding scheme has produced a semantic hotspot, or that a data source contributing to the training corpus has changed its output format. Sample-rate checks can be implemented as exponential moving averages over the batch-level class histogram, with alerts triggered when any class frequency deviates from its rolling average by more than a configurable threshold.

The trade-off between training-time validation cost and the cost of a corrupted training run depends on the run duration and cost. For short fine-tuning jobs (hours, thousands of dollars), full validation on every batch may be appropriate. For foundation-model pretraining (weeks, millions of dollars), full per-batch validation on every sample is operationally prohibitive, and the right policy is a sampled validator that applies the full check suite to a random 1% of batches, with a fast anomaly detector (range check only) on the remaining 99%. The sampling rate can be tuned dynamically: increase it when the anomaly detector fires, decrease it when the pipeline has been clean for a configurable window.

RDMA, GPUDirect, and the Hardware Path from Storage to Accelerator

The conventional data path from storage to accelerator traverses three copies: the data is read from the storage device into kernel page cache (first copy), copied from page cache into a user-space buffer in the dataloader process (second copy), and then transferred from the CPU-side buffer across the PCIe bus into GPU high-bandwidth memory (third copy). Each copy consumes CPU cycles, memory bandwidth, and PCIe bandwidth. For workloads where the I/O bottleneck is measured at the PCIe interface or in CPU copy overhead rather than in raw storage bandwidth, hardware-accelerated I/O paths that eliminate one or more of these copies can recover measurable training throughput.

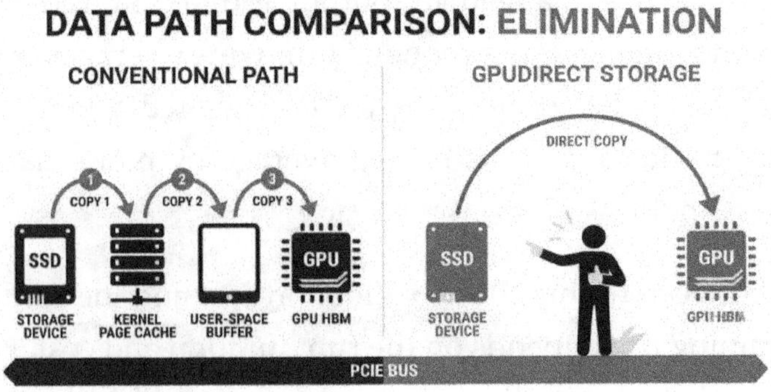

Remote Direct Memory Access (RDMA) allows a network adapter to transfer data directly between the memory of two systems without involving the host CPU of either system. In a distributed ML cluster, RDMA is used for two distinct purposes: fast inter-worker communication (covered in the compute parallelism chapter) and fast data loading from a remote storage server whose memory is exposed over an RDMA fabric. When the training cluster is connected to a storage cluster over an InfiniBand or RoCE fabric, RDMA-capable storage clients can fetch data directly into host memory with substantially lower latency and CPU overhead than TCP-based storage protocols. The operational cost is a network fabric that supports RDMA (typically InfiniBand or

Ethernet with RoCE configuration) and storage software that exposes an RDMA interface.

GPUDirect Storage is a hardware capability that enables a compatible storage controller to transfer data directly to the GPU's high-bandwidth memory, bypassing the host CPU and system memory entirely. For I/O-bound training workloads where the bottleneck is precisely the CPU-mediated copy from host memory to GPU memory, GPUDirect Storage eliminates that copy. It can reduce data transfer latency by a factor that depends on the PCIe topology and the host memory bandwidth. The practical requirement is a GPU model and a storage controller both certified for GPUDirect Storage, a supported operating system kernel, and a compatible storage driver. Not all storage backends support GPUDirect Storage, and the operational overhead of maintaining a GPUDirect-enabled stack is non-trivial.

The decision to invest in hardware-accelerated I/O should be driven by profiling, not by intuition. The first question is whether I/O is the bottleneck being measured. If the accelerator is at 95% utilization and the prefetch buffer is consistently non-empty, the pipeline is compute-bound, and hardware I/O acceleration will not improve throughput. The second question is whether the bottleneck is in copy overhead or in raw bandwidth. If the storage device can deliver data faster than the CPU can copy it, RDMA or GPUDirect Storage is the right intervention. If the device itself is the bottleneck, the intervention is a faster device or more devices in a striped configuration. Hardware-accelerated I/O is a specialized optimization for mature, profiled pipelines, not a first response to any I/O complaint.

The operational cost of hardware-accelerated I/O must be amortized against the throughput gain. RDMA and GPUDirect

Storage require specific hardware, kernel configurations, and storage software, all of which must be maintained through kernel upgrades, hardware refreshes, and storage backend migrations. A pipeline that achieves 92% accelerator utilization through careful format selection, prefetching, and tiered caching may not require the additional complexity of a hardware-accelerated data path. The responsible-by-design guideline is to reach the point of diminishing returns on software optimization before investing in hardware specialization, and to document the profiling evidence that justified the investment.

Technical Checklist

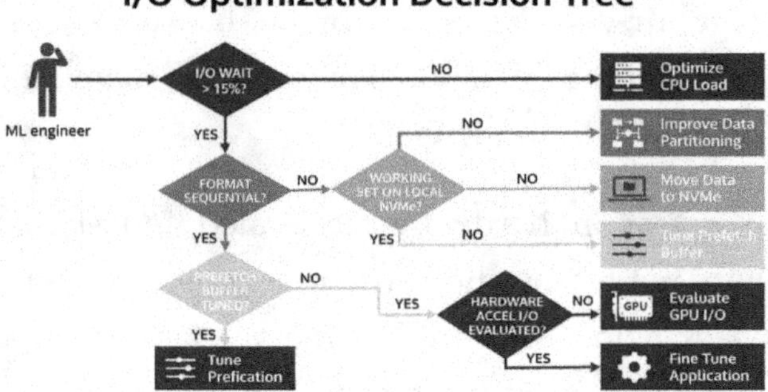

Apply this checklist as a release gate before declaring any data pipeline production-ready. Each item corresponds to a pattern covered in this chapter. A pipeline that passes all items is operationally sound; a pipeline that fails any item poses a known risk that should be documented with a mitigation plan if it cannot be resolved immediately.

- Workers spend less than fifteen percent of wall-clock time waiting on I/O at the target scale. Verify this by profiling at least one full epoch, logging accelerator utilization, and I/O wait time concurrently.

- The sharding scheme is documented, has an assigned owner, and rebalances automatically or semi-automatically as the dataset grows. A shard imbalance greater than twenty percent by sample count or byte count should trigger an automated alert.
- The prefetch pipeline has been profiled at target batch size and target worker count. Prefetch depth and worker parallelism are set to values that achieve the target accelerator utilization, rather than being left at framework defaults.
- The dataset format supports efficient sequential reads at the target batch shape. Raw file directories have been replaced with a packaged binary format (TFRecord, WebDataset, Parquet, Arrow, or equivalent).
- The storage class is chosen for the workload's access pattern: object store for cold archival, distributed file system for warm shared access, local NVMe for hot working set. The choice is documented with a rationale.
- Streaming pipelines are deterministic and checkpointable. A restart from any checkpoint produces the same sample sequence that the training loop would have seen without interruption.
- The dataloader cursor state is saved in the same atomic operation as the model checkpoint. Joint checkpointing has been tested with at least one synthetic restart before the pipeline was promoted to production.
- Data quality gates (schema enforcement, range checks, sample-rate sanity checks) are active in the I/O path. Failures are logged with the shard identifier and byte offset, not silently dropped.
- Deduplication has been run on the training corpus. Exact deduplication is the minimum; near-duplicate detection has been evaluated and applied if the corpus was assembled from multiple sources or web crawls.

- Shuffle quality has been validated. The shuffling strategy (multi-pass chunk shuffle or shard-plus-sample-buffer shuffle) produces batches whose label and attribute distribution match the global dataset distribution to within a configurable tolerance.
- Hardware-accelerated I/O paths (RDMA, GPUDirect Storage) have been evaluated. If adopted, the adoption is documented with profiling evidence of the throughput gain. If declined, the reason is documented.
- The I/O checklist has been run and signed off by a second reviewer before each production training run of more than four hours duration.
- Tiered storage eviction and warm-up policies are configured, monitored, and logged. Unexpected eviction storms trigger an automated alert before they affect training throughput.
- Multi-source joins under the shard have been co-sharded where possible. Cross-shard lookup latency has been measured and is below the acceptable threshold for the target training throughput.

Team Conversation

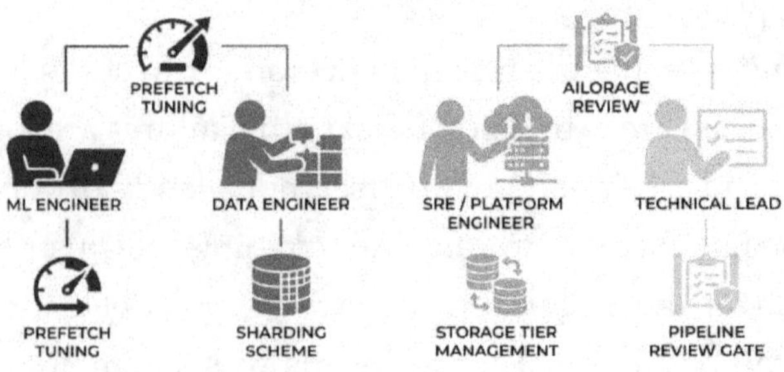

The following questions are intended for use in team meetings, architecture reviews, and sprint retrospectives. They are open-ended by design: the goal is to surface alignment gaps and untested assumptions about the data pipeline, not to produce definitive answers in a single session.

- Which of our current training jobs would benefit most from a redesigned sharding scheme? What evidence do we have, and what profiling would we need to run to confirm it?
- Have we measured I/O wait time on our most expensive training job in the last quarter? If not, what is preventing us from doing so, and who owns the action to add that instrumentation?
- Are we paying object-store list costs we could avoid with a shard manifest or a packaged binary format? How much of our storage budget is consumed by per-file API calls rather than by actual data transfer?
- Do our streaming pipelines survive a restart without subtle non-determinism? Have we tested this with a synthetic restart in the last release cycle, or are we relying on the assumption that the framework handles it correctly?
- Where does host memory pressure show up in our pipeline today? Is the prefetch buffer competing with the page cache, and if so, what is the next pattern that would relieve that pressure?
- If we standardized on one dataset format across the team, which format would serve the broadest range of our workloads? What would we lose compared to the format each team currently uses, and is that loss worth the operational simplicity of standardization?
- Do we have a data quality gate inside the training I/O path, or do we discover corruption only when a model fails to converge or underperforms on evaluation? What is the cost in accelerator

hours of the last quality problem we could have caught at the I/O layer?

- Have we evaluated hardware-accelerated I/O paths for our highest-throughput training jobs? If not, is the reason a hardware constraint, a software complexity concern, or simply that we have not yet measured the copy overhead?
- Is our sharding scheme documented and owned? Does the person responsible for it have the authority and tooling to automatically rebalance as the dataset grows?
- What is our pilot-to-production pathway for a new data pipeline pattern? Who reviews the I/O checklist, and at what point in the development cycle is that review required?

Key Takeaway

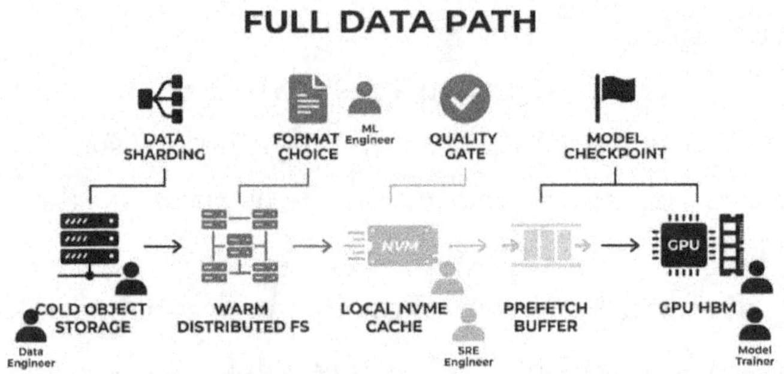

The data pipeline is where distributed ML performance is most often lost, and most often recovered. The patterns in this chapter follow a consistent logic: measure the actual bottleneck before choosing the remedy, design for the dataset's growth trajectory rather than its current size, and make every stage of the pipeline observable so that regressions are caught by instrumentation rather than by loss curves. A team that applies this logic consistently will spend less time debugging unexplained training slowdowns and more time on the model and evaluation work that actually advances the mission.

Sharding is a first-class design decision that deserves an owner and a documented policy. The choice among hash-based, range-based, size-balanced, and semantic sharding strategies is not a configuration detail; it determines the cluster's load balance, the statistical properties of the training distribution, and the operational burden of maintaining the pipeline as the dataset grows. Storage class choice has a greater impact on training costs than most teams realize: an object store that charges per API call for directory listings can consume a significant fraction of the infrastructure budget for a dataset stored as individual files, and that cost disappears entirely with a format migration. These are workflow-level impact decisions that belong in architecture reviews, not in sprint tickets.

Streaming and tiered storage are now the default shape of large-scale ML data pipelines. Foundation-model pretraining has established the pattern, and the operational patterns that make streaming pipelines reliable (deterministic iterators, joint checkpointing, shard-cursor state) are now mature and available in the tooling ecosystem across multiple frameworks. A team that has not yet adopted streaming for its largest training jobs should treat the migration as a strategic investment in operational resilience rather than an optional performance optimization. The pilot-to-production pathway for any new data infrastructure should include a synthetic restart test, a profiling session to measure the I/O wait fraction, and a review of the checklist in this chapter, all of which should be signed off on before the first production run.

5 Orchestration and Scheduling

Opening Scenario

On a Friday afternoon, a seventy-two-hour training run is humming along at reasonable throughput when the first preemption signal arrives. The scheduler evicts the entire pod group from the spot node pool, triggering a 30-minute checkpoint reload and a 20-minute warm-up phase. The team accepts the setback and requeues the job. Ninety minutes later, the second preemption arrives. Then a third. By midnight, the job has been preempted four times in six hours; the wall-clock loss now exceeds the originally estimated remaining training time; and the model that was supposed to clear staging review before Monday stands no chance of hitting that gate.

The on-call engineer opens the scheduler dashboard and finds nothing illuminating. GPU utilization is low on the on-demand pool. The queue shows the job as high priority. But the scheduler continues to place each restarted pod group on the same spot node pool that just evicted it. No affinity rule prevents this. No policy caps preemption frequency. No fallback pool is configured. The placement decision is fully automatic, fully opaque, and fully wrong.

The team patches the immediate issue by adding a node-affinity rule that directs the job to on-demand capacity. The job finishes Sunday evening, twelve hours late. A brief retrospective follows. The team identifies three root causes: the training job used no gang scheduling primitive so that the scheduler could place individual pods on separate node pools with different preemption risk profiles; the checkpoint policy was calibrated for occasional short interruptions, not for the burst-preemption pattern that tightened spot markets produce; and the queue configuration had never been updated from cluster defaults, so a lower-priority batch job was able to occupy the on-demand pool and crowd out the training run.

Each of those three root causes is a solved problem at the orchestration layer. Gang scheduling with topology-aware placement keeps pod groups together on capacity that shares a risk profile. Preemption-aware checkpoint policies cap the cost of an eviction to a bounded window. Queue management with fair-share policies and priority preemption lets the team declare their actual priorities, and the scheduler honors them. The team had the primitives available. They had not been wired together into a coherent orchestration strategy.

This chapter is the playbook that closes that gap. It covers the full orchestration stack from the base container primitives through the ML-aware scheduler extensions, the fault-tolerance machinery, the queue management patterns, and the declarative job definitions that make the whole system auditable. Every pattern is presented with the failure mode that motivates it, the trade-offs it makes, and the metrics that prove it is working.

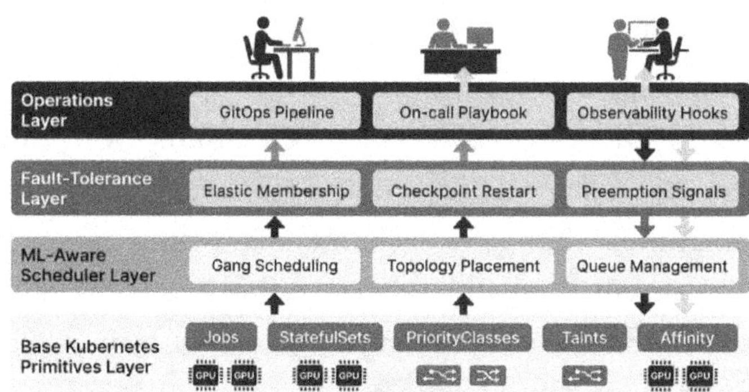

Why It Matters

Orchestration failures are invisible until they are catastrophic. A job that runs on the wrong network topology does not fail; it runs slowly, and the team spends days ruling out model bugs and data pipeline issues before someone checks switch locality. A job without a preemption policy does not fail on the first eviction; it degrades gradually as evictions compound, and the effective training throughput falls below the threshold at which the deadline remains achievable. By the time the failure is unmistakable, the blast radius has already been set.

The operational clarity that a mature orchestration layer provides is mission-aligned in a direct sense: it enables the team to make promises to stakeholders and keep them. A scheduler that honors topology, respects priorities, and absorbs preemptions predictably is not a luxury. It is the mechanism by which a distributed ML platform becomes a system that other teams can depend on. Without it, every training run is a probabilistic event, and the team's role shifts from engineering to risk management.

The patterns in this chapter also have a compounding effect on cost. Gang scheduling reduces wasted compute time when partial pod groups sit idle while waiting for missing members that the scheduler placed elsewhere. Topology-aware placement lifts effective communication bandwidth and shortens wall-clock training time for the same dollar spent. Spot-aware checkpoint policies unlock significant capacity discounts that would otherwise be operationally unsafe. Together, these patterns are often the largest single lever available to a team trying to reduce training cost without changing the model or the data.

Kubernetes for ML Workloads

Kubernetes was designed for long-lived stateless services. Distributed ML training is the opposite: tightly coupled, stateful, communication-heavy, and often finite in duration. The gap between what the base Kubernetes primitives offer and what distributed training requires is not a flaw in Kubernetes; it is an accurate signal of where each abstraction layer belongs. The base primitives are necessary but not sufficient. Understanding where they are sufficient and where they must be extended is the first decision every ML platform team makes.

The most relevant Kubernetes primitives for distributed training are the Job and the PodGroup. A Job declares a finite workload with a completion criterion. For single-node training, a Job with a single Pod is the right primitive. For distributed training, the Job remains the right outer wrapper. Still, it needs a coordination layer that the base Kubernetes Job controller does not provide: a guarantee that all Pods in the group start together, or none start at all. That guarantee is gang scheduling, and it requires either a custom scheduler extension or a dedicated scheduler that supports a minimum gang size.

PriorityClasses assign a numeric priority to Pods and determine which Pods are evicted when the cluster runs out of capacity. A training job with a high PriorityClass will not be preempted to make room for a lower-priority batch job. A training job without a PriorityClass will inherit the default, which is often 0, making it a candidate for preemption by any higher-priority workload. Setting explicit PriorityClasses for every ML workload class is one of the highest-leverage, lowest-effort changes a platform team can make.

Taints and Tolerations give the team a coarse-grained mechanism for dedicating node pools to specific workload classes. A GPU node pool can be tainted so that only Pods with the matching toleration will schedule onto it. This prevents CPU-only batch jobs from occupying accelerator capacity. Affinity and Anti-Affinity rules provide a finer-grained mechanism: a training job can declare affinity for nodes with a specific GPU type, within a specific availability zone, or that share a network switch with the other members of the job.

The Kubernetes operator pattern is what makes ML-specific primitives composable with the base scheduler. An operator defines a custom resource (for example, a PyTorchJob or an MPIJob), monitors instances of that resource, and translates them into the underlying Kubernetes primitives required to run the workload. The operator handles the lifecycle complexity that would otherwise fall to the engineer: waiting for all workers to be ready before starting training, restarting failed workers within the retry budget, and cleaning up resources when the job completes. The operator is the pilot-to-production pathway for workloads that are too complex for the base Job primitive to handle.

The StatefulSet is appropriate for serving deployments in which each replica has a stable network identity and stable storage. A multi-replica model server that requires each instance to load a specific model shard uses StatefulSet semantics. Training jobs generally do not use StatefulSets directly because the stable-identity guarantee adds overhead, and shard assignment is usually handled by the training framework rather than the scheduler.

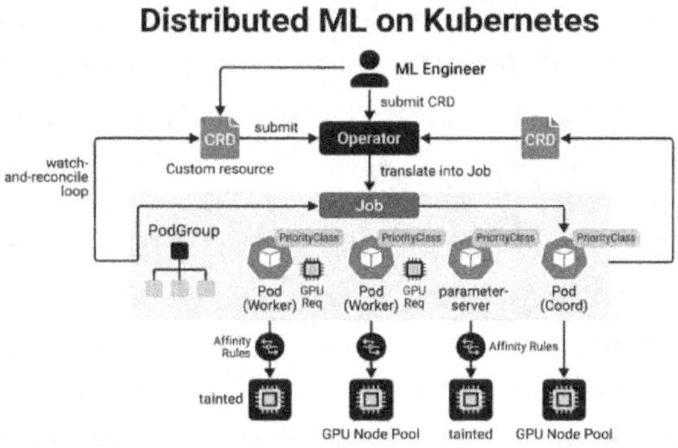

MPI-Based Orchestration Patterns

MPI (Message Passing Interface) predates Kubernetes by two decades. Still, it remains the right default for tightly coupled collective communication workloads, where the application manages its own communication topology and the scheduler's only job is to start the process group and keep it running. HPC clusters have run MPI workloads for decades. The challenge for modern ML platforms is to make MPI workloads cloud-native without sacrificing the performance properties that make MPI worth using.

The canonical cloud-native MPI pattern uses an operator that maps MPI concepts onto Kubernetes primitives. The launcher process runs as a single Pod and manages the job lifecycle. Workers run as replicas in a headless service so they have stable DNS names. The operator waits for all worker Pods to reach the Ready state before it starts the launcher, which is a primitive form of gang scheduling. SSH-based communication between the launcher and workers is replaced by an operator-managed credential-injection mechanism, removing the need to bake SSH keys into container images.

The failure mode introduced by MPI's rigid process-group model is all-or-nothing recovery. If one worker fails, the entire MPI job must restart from the last checkpoint. There is no mechanism for elastic membership in the standard MPI model. For jobs with low failure rates and frequent checkpoints, this is an acceptable trade-off. For jobs where node failure rates are high or checkpoints are expensive, the all-or-nothing model becomes the primary bottleneck, and the team should evaluate frameworks that support elastic worker membership.

Sidecar logging is the pattern that makes MPI worker output observable in a cloud-native environment. Each worker Pod carries a logging sidecar container that streams stdout and stderr to a centralized log aggregation system. Without this, per-worker output is only accessible while the Pod is running, and a failed worker takes its logs with it when it is evicted. The sidecar survives the main container's exit and flushes buffered log data before terminating, preserving the evidence needed for post-mortem analysis.

The bandwidth-pinning pattern is equally important for MPI workloads. MPI collectives saturate the network, and the slowest link in the communication path determines the effective bandwidth available to the collective. If some workers are co-located on a high-bandwidth switch island and others are on a different switch, the slow cross-switch links dominate the overall performance. The network-topology-aware placement patterns covered in the next section apply directly to MPI workloads and are often the difference between a job that trains efficiently and one that trains at a fraction of the theoretical peak.

MPI Orchestration Pattern on Kubernetes

Specialized ML Schedulers: Volcano, Kueue, and Their Patterns

The base Kubernetes scheduler does not understand the concept of a job as a unit. It places Pods one at a time, optimizing each placement independently. For distributed training, this produces a class of deadlock called the all-or-nothing problem: the scheduler places half the Pods of a training job, then finds no capacity for the remaining Pods, leaving the partial group idle and consuming resources that block other jobs from running. The ML-aware scheduler solves this by treating the PodGroup as the atomic unit of scheduling.

Volcano is an illustrative example of a batch scheduler built for ML and HPC workloads on Kubernetes. Its core concept is the Queue, which holds PodGroups waiting for capacity. When capacity becomes available, the queue's policy determines which PodGroup is admitted next. Volcano's gang scheduling plugin enforces the minimum-available constraint: a PodGroup will not be admitted unless the number of Pods that can be placed is at least as large as the minimum gang size. This eliminates the partial-placement deadlock.

Kueue is a more recent scheduler extension that focuses on resource quota management across teams and namespaces. Its core concepts are the ClusterQueue, which holds a pool of resource quotas, and the LocalQueue, which is the team-facing view of that pool. A job is admitted only when its resource request fits within the quota available to its team's LocalQueue. Kueue's borrowing mechanism allows a team to temporarily exceed its quota by borrowing from another team's unused quota, up to a cluster-level cap. This gives teams burstability without permanently reallocating capacity.

The patterns that both illustrative schedulers implement are more important than their specific APIs. The gang scheduling pattern says: admit a PodGroup atomically or not at all. The queue management pattern recommends assigning each job to a queue and managing admission at the queue level rather than at the Pod level. The borrowing pattern says: allow temporary quota exceedance within a bounded ceiling. The preemption pattern says: when a higher-priority job arrives, identify the lowest-priority running job that would free enough capacity, and evict it gracefully. These four patterns combine to form a scheduling policy that is both utilization-efficient and predictably fair.

The Slurm-style batch scheduler, widely used in HPC environments, implements the same four patterns using different primitives. Jobs are submitted to partitions (analogous to queues). The scheduler backfills small jobs around large jobs to maximize utilization while preserving the large jobs' expected start times. Priority is numeric and explicit. Preemption is configured at the partition level. Teams migrating from HPC environments will recognize these concepts and should expect to map them directly onto the Kubernetes-native scheduler's vocabulary.

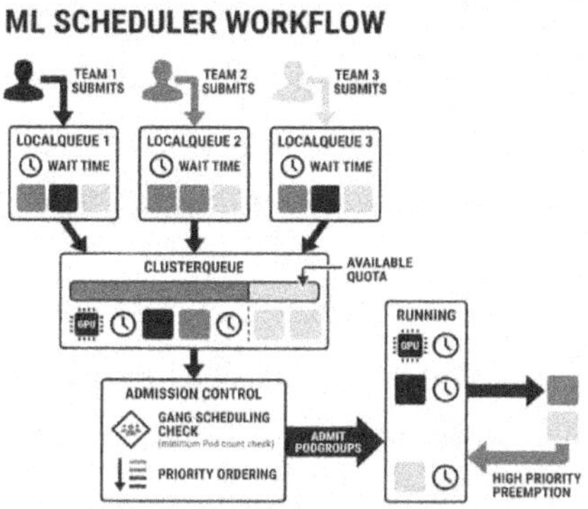

Gang Scheduling and Topology-Aware Placement

Gang scheduling is the constraint that a group of Pods must start together or not at all. It is non-negotiable for any training job that uses collective communication, because a collective operation like AllReduce requires all participants to call it simultaneously. A Pod that starts late will block every other Pod waiting for it. A missing Pod means no collective operation can make progress, and the entire group idles.

The minimum gang size is the tunable that determines the constraint's strictness. Setting the minimum gang size to the total job size enforces strict all-or-nothing semantics: the job runs at full size or not at all. Setting it to a value less than the total job size enables elastic scheduling: the job can start with fewer workers and expand as capacity becomes available. Elastic scheduling improves cluster utilization but adds complexity to the training code, which must handle a changing number of workers at runtime.

Topology-aware placement extends gang scheduling by imposing constraints on Pod placement. A training job that uses RDMA for gradient communication requires all its workers to be reachable within the same RDMA fabric, which typically maps to a single rack or a single network-switch island. A job that uses standard

TCP but is sensitive to latency variance needs workers on nodes that share a low-latency network path. A job running on a cloud cluster needs workers within the same availability zone to avoid cross-zone bandwidth charges and latency penalties.

The topology constraint is expressed as a hierarchy of domains. At the finest granularity, the domain is the NUMA node within a single host. NUMA-aware placement ensures that a worker's CPU and GPU are in the same NUMA domain, avoiding the latency and bandwidth penalties of crossing NUMA boundaries for PCIe transfers. At a coarser granularity, the domain is the network switch island: a group of hosts all connected to the same top-of-rack switch. At the coarsest granularity, the domain is the availability zone or cloud region.

The placement policy declares a preference ordering over these topology domains. The scheduler selects the domain that minimizes the worst-case communication latency across the entire gang. For a two-level fat-tree network, this means packing workers onto as few top-of-rack switches as possible before spilling to adjacent switches. For a cloud cluster, this means keeping all workers within the same availability zone and, within that zone, on hosts that share a placement group with dedicated high-bandwidth interconnect.

The metric that proves topology-aware placement is working is the all-reduce bandwidth achieved by the job expressed as a percentage of the theoretical peak for the interconnect type. A job that achieves 85 percent of peak all-reduce bandwidth on an InfiniBand cluster is well-placed. A job that achieves 30 percent of peak on the same cluster is almost certainly spanning switch boundaries. This metric is cheap to collect and should be part of the telemetry for every distributed training job.

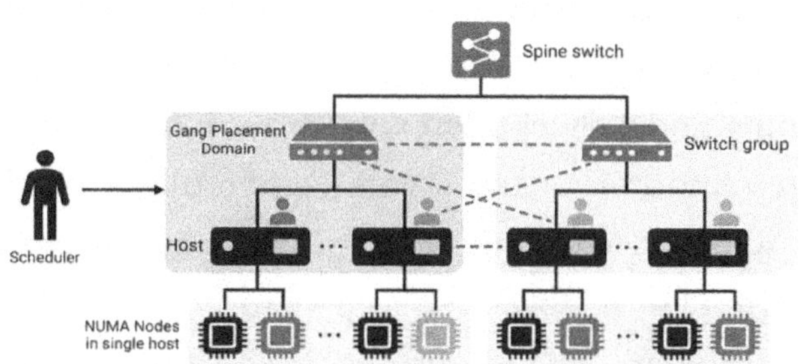

Spot Instance Strategies and Preemption Handling

Spot and preemptible instances offer significant unit cost reductions relative to on-demand capacity, typically 60 to 80 percent lower. The reduction is real and meaningful at scale. A training run that costs $10,000 on on-demand capacity might cost $2,000 on a spot basis. But spot instances can be reclaimed by the cloud provider with minimal notice, usually two minutes or less, and the reclamation can affect multiple nodes simultaneously during periods of high demand. Safely using spot capacity requires an orchestration strategy, not just a flag in the instance type configuration.

The foundational pattern for spot-safe training is the checkpoint-before-eviction protocol. When the platform receives a preemption signal (a two-minute warning from the cloud provider, or a preemption event from the cluster scheduler), it triggers an immediate checkpoint write. If the checkpoint completes before the instance is reclaimed, the job loses at most the progress made since the previous scheduled checkpoint. If the checkpoint does not complete in time, the job falls back to the previous scheduled checkpoint. The checkpoint-before-eviction protocol converts a random loss window into a bounded one.

Checkpoint cadence is the parameter that controls the worst-case loss. Frequent checkpoints reduce the loss window but incur I/O overhead and storage costs. For a job with a two-minute preemption warning and a checkpoint write time of thirty seconds, a cadence of fifteen minutes means a worst-case replay window of fifteen minutes, with a two-minute margin for the checkpoint-before-eviction write. For a job with a four-hour checkpoint write time, no checkpoint-before-eviction protocol can save it on a two-minute warning. Reducing checkpoint time (through incremental checkpointing, which saves only the weight deltas since the last full checkpoint) is therefore a prerequisite for spot-safe training at scale.

The multi-pool fallback pattern complements the checkpoint strategy. The scheduler maintains an ordered list of node pools: the primary pool is spot; the fallback pool is on-demand. When a spot preemption occurs, the job is requeued to the same queue it came from, but its requeue priority is elevated. If the spot pool has enough capacity within a configurable wait window, the job restarts there. If the spot pool remains constrained beyond the wait window, the scheduler promotes the job to the on-demand fallback pool. The fallback pool is more expensive, but the job keeps running, and the deadline is protected.

The spot-and-on-demand mix pattern is a refinement of the fallback approach. Instead of running entirely on spot and falling back to on-demand, the job intentionally places a small fraction of its workers on on-demand nodes. These stable anchors hold the checkpoint state and can act as a nucleus for recovery if spot workers are evicted. The all-reduce communication still runs across the full worker set, but the recovery path is faster because the anchor workers have already loaded the model and optimizer state.

Quantifying the operational cost of spot is as important as quantifying the savings. Track preemption frequency, the average checkpoint-reload overhead per preemption, the deadline miss rate for jobs running on spot, and the fraction of spot-targeted jobs that had to fall back to on-demand. If a job class has a high deadline miss rate or a high fallback rate, the actual cost of running it on the spot may exceed the on-demand cost once time costs and engineering overhead are included.

Spot Instance Preemption and Fallback Pattern

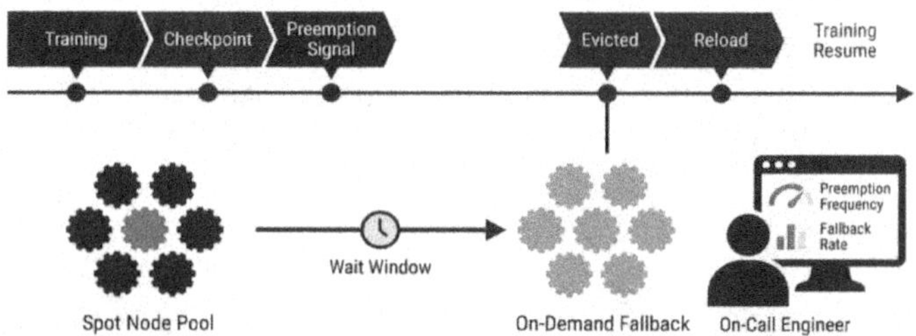

Fault Tolerance and Automatic Recovery

A node failure in a distributed training job is not the same event as a node failure in a stateless microservice deployment. For a stateless service, the scheduler replaces the failed instance, and the system recovers in seconds. In a distributed training job, the failed worker participated in every collective communication operation, and their absence stalled the other workers in the gang. The recovery pattern must restart the failed worker, reload its state, and re-synchronize it with the rest of the gang, all while keeping the healthy workers from idling or diverging.

The simplest recovery pattern is a full-job restart from the last checkpoint. When any worker fails, the scheduler terminates the entire gang, waits for all Pods to exit, and then restarts the job from the most recent checkpoint. A full job restart is correct and simple to implement, but it is expensive. If the job has 128 workers and one fails, 127 workers are terminated and must reload their state. For a large model, the checkpoint reload time can be tens of minutes.

Elastic worker membership is the recovery pattern that avoids the full-job restart cost. In this pattern, the training job is written to tolerate a changing set of workers at runtime. When a worker fails, the remaining workers continue training with a reduced batch size (or adjust the learning rate to compensate) while the scheduler provisions and initializes a replacement worker. The replacement worker joins the gang, downloads its shard of the model state from the shared checkpoint store, and participates in the next forward pass. The job never fully stops.

Elastic membership requires the training framework to support dynamic re-initialization of the communication group. The rendezvous mechanism (the protocol by which workers discover each other and agree on their ranks) must be robust enough to handle a worker joining mid-run without stalling the rest of the group. The heartbeat mechanism must detect failures quickly enough that the group does not stall waiting for a dead worker to respond. And the optimizer state must be re-sharded to account for the new worker count, which is a non-trivial operation for certain optimizer variants.

The graceful degradation pattern is a simplified form of elastic membership. Instead of immediately bringing in a replacement worker, the job continues training with the reduced worker set permanently. The batch size is adjusted proportionally. The learning rate schedule may be adjusted. The job finishes with lower throughput but without the overhead of provisioning and re-synchronizing a replacement. Graceful degradation is appropriate for jobs where the failure rate is low, and the throughput reduction from a missing worker is acceptable relative to the cost of a full job restart.

The checkpoint-driven restart pattern applies to both full-job restart and elastic membership—every restart, whether due to a node failure or a preemption, begins at a checkpoint. The checkpoint must be consistent: all workers must have written their state as of the same training step before any worker is restarted. Achieving this consistency requires a checkpoint barrier: a point in training where all workers pause, flush their state, confirm to a coordinator that the write is complete, and then resume. The coordinator writes a manifest that records the checkpoint step and the list of workers who contributed to it. A restart reads the manifest and rejects any checkpoint that is not fully consistent.

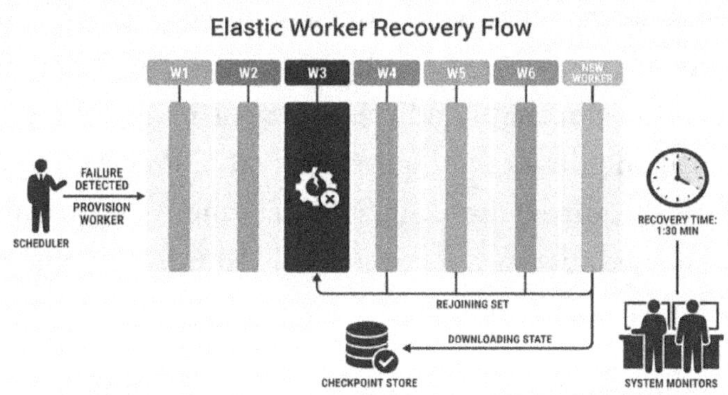

Queues, Priorities, and Backfill

A shared GPU cluster without a queue policy is one in which the longest-running job wins. The team that submits first accumulates resources; teams that submit later wait indefinitely. Priority inversions are common: a low-value exploratory job submitted before a high-value deadline-driven job will finish first simply because it arrived earlier. Queue management is the mechanism by which the platform enforces the team's actual priorities rather than the order of submission.

The priority queue pattern assigns a numeric priority to each job. The scheduler always admits the highest-priority job that fits within available capacity. Jobs of equal priority are ordered by submission time. When no job fits within available capacity, the scheduler waits rather than fragmenting capacity across partial allocations. This simple pattern eliminates priority inversion and makes the scheduler's decisions transparent: the team can see the queue, see each job's priority, and understand exactly why their job is waiting.

The preemption-at-queue-level pattern extends priority queues to running jobs. When a high-priority job arrives and finds no available capacity, the scheduler identifies the lowest-priority running job that, if terminated, would free enough capacity for the high-priority job. It preempts that job gracefully (triggering its checkpoint-before-preemption handler), waits for its resources to be released, and then admits the high-priority job. Preemption at the queue level requires that all jobs in the queue have checkpointing enabled, because a preempted job must be able to resume from where it was terminated.

Fair-share scheduling is an alternative to strict priority ordering for multi-tenant clusters. In a fair-share system, each team is assigned a share of the cluster's total capacity. A team's effective priority is inversely proportional to how much of its share it has used recently. Teams that have used less than their share get priority over teams that have used more. This prevents any single team from monopolizing the cluster for an extended period while still allowing short-term bursts when other teams are idle.

The backfill pattern improves cluster utilization within a priority or fair-share system. After the scheduler admits all jobs that fit within the available capacity and cannot find room for the next job in the queue, it searches the queue for smaller jobs that could run in the available gaps without delaying the waiting job's expected start time. Backfill is a utilization optimizer, not a priority mechanism: it never delays a higher-priority job to make room for a smaller one, but it prevents capacity from sitting idle. At the same time, the higher-priority job waits for a future preemption to free up a large block.

The hierarchical queue pattern is appropriate for organizations with multiple teams and multiple workload classes. Cluster-level capacity is divided into team-level queues. Each team's queue is divided into workload-class queues (for example, production, development, and batch). Priority and fair-share policies are applied within each level of the hierarchy. This structure makes the cluster's priority model legible to stakeholders: a team lead can see exactly how much capacity their team has, how it is distributed across workload classes, and how it compares to other teams.

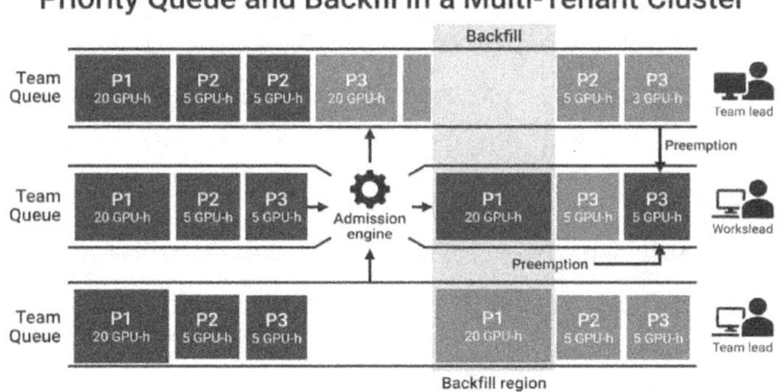

Multi-Cluster and Multi-Cloud Orchestration

A single cluster imposes a hard ceiling on the scale of any individual job and a single point of failure for every job. Multi-cluster orchestration lifts both constraints by federating multiple independent clusters into a unified scheduling domain. The patterns are more complex, and the operational overhead is real. Still, for organizations with scale requirements that exceed a single cluster or with regulatory requirements that mandate geographic distribution, the investment is justified.

The queue-of-queues pattern is the foundational multi-cluster orchestration primitive. A global scheduler sits above the individual cluster schedulers. Each cluster scheduler manages its own local queue. The global scheduler receives job submissions, selects the target cluster based on capacity, policy, and topology, and forwards the job to that cluster's local queue. From the job's perspective, submission is always to the global scheduler; the cluster selection is transparent.

Cluster selection at the global level uses the same criteria as placement at the local level, but with coarser granularity. The global scheduler prefers clusters that have sufficient idle capacity to absorb the job without preempting lower-priority jobs. It respects geographic affinity constraints: a job with a data residency requirement must run on a cluster in the compliant region. It respects cost constraints: a job flagged as cost-sensitive is steered toward the cluster with the lowest unit cost at the current moment, which may be a spot-heavy cluster in a less popular region.

Data locality is a complicating factor in multi-cluster scheduling. A training job that reads its dataset from a storage system co-located with cluster A should not be placed on cluster B without accounting for the cross-cluster data transfer cost. The global scheduler must model data placement alongside compute placement. For large datasets, the cost of transferring data across cluster boundaries often dominates the difference in compute costs between clusters, making data locality the primary scheduling constraint.

The federation pattern for serving is simpler than for training because individual inference requests are stateless and replicas are interchangeable. A global load balancer distributes requests across replicas that may span multiple clusters and cloud regions. Capacity scaling at the serving layer is handled per-cluster, but the global scheduler can rebalance capacity across clusters if one region becomes saturated. The primary concern in multi-cluster deployments is model version consistency: all active replicas across all clusters should serve the same model version during normal operation.

Crossing cluster boundaries introduces latency and bandwidth costs that do not exist within a single cluster. For training jobs that use collective communication, these cross-cluster costs are prohibitive: no multi-cluster deployment should attempt to run a tightly coupled all-reduce across a wide-area network. Pipeline parallelism across clusters is feasible if the pipeline stages are sufficiently large that inter-stage activation transfers are small relative to the per-stage compute time. Still, the design requires careful analysis of the communication-to-compute ratio before deployment.

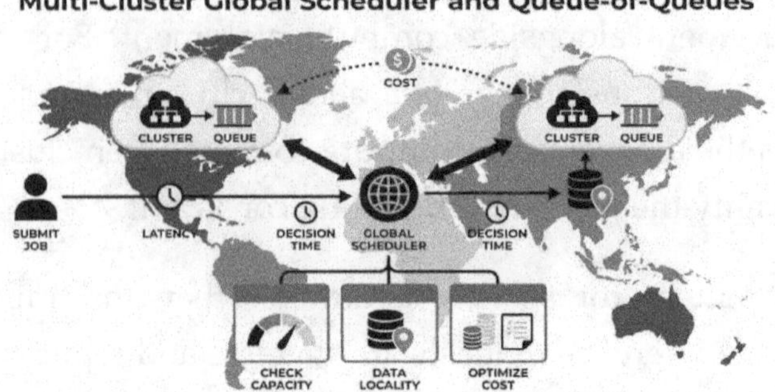

GitOps and Declarative Job Definitions

Ad hoc job submission, where an engineer constructs a YAML manifest on a local machine and applies it directly to a cluster, is the operational equivalent of running database migrations by hand in production. It is fast in the short term and a reliability risk in the long term. The manifest is not reviewed, tested, version-controlled, or auditable after the fact. When a job fails because it depends on a configuration detail, the engineer who submitted it may not remember what they submitted.

Declarative job definitions stored in version control directly solve these problems. Every job, environment, and policy configuration is expressed as a file in a repository. Changes to job definitions go through the same review process as changes to application code. The audit trail is the repository's commit history. Rolling back to a previous configuration involves running git revert and triggering a pipeline. The team always knows exactly what is running in the cluster and why.

The GitOps pattern applies the declarative model to the deployment pipeline. A change to a job definition triggers an automated pipeline that validates the manifest against the cluster's resource quotas and policy constraints, runs a dry-run admission through the scheduler to catch placement conflicts before they reach production, and applies the change to the cluster if validation passes. The pipeline can be configured to require manual approval for high-cost or high-priority jobs, giving the team lead a review gate without adding operational friction to routine changes.

Job templating is the pattern that makes declarative definitions scalable across a large job library. Instead of maintaining a single manifest per job, the team defines a parameterized template for each job class (e.g., distributed training, hyperparameter sweep, serving deployment). It generates concrete manifests by instantiating the template with job-specific parameters. The template encodes the cluster's best practices: the correct PriorityClass, topology-aware affinity rules, and the checkpoint configuration for spot safety. A new job that uses the template automatically inherits all of these defaults.

Environment promotion is the pattern that enforces the pilot-to-production pathway. A job definition progresses through a sequence of environments: development (small cluster, no production data, spot capacity), staging (production-scale cluster, anonymized data, on-demand capacity), and production (full cluster, live data, full fault tolerance). A pipeline stage triggers promotion from one environment to the next and requires passing the validation gates for the target environment. The declarative definition is the same across all environments; only the environment-specific parameters change.

Policy-as-code is the pattern that makes the team's orchestration rules auditable and enforceable at the source rather than at runtime. Admission policies (for example, every training job must declare a PriorityClass, every spot job must declare a checkpoint cadence, no job may request more than 256 GPUs without a team lead approval annotation) are expressed as code in the same repository as the job definitions. The pipeline evaluates every manifest against the policy library before applying it. Jobs that violate a policy are rejected at the pipeline stage, not at cluster admission, giving the engineer a faster feedback loop.

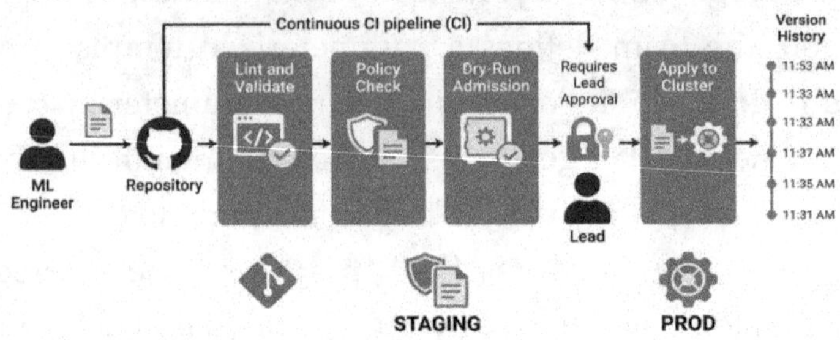

Custom Resources and Accelerator-Aware Scheduling

GPUs and TPUs are not interchangeable with CPU cores. They have finite counts per node, specific NUMA topology relationships to the host's CPU and memory, and inter-device communication capabilities (NVLink, NVSwitch, TPU interconnect) that the base Kubernetes resource model does not capture. The extended resource model, which allows cluster administrators to advertise arbitrary countable resources on nodes, is the mechanism by which accelerator-aware scheduling is expressed in a Kubernetes-compatible environment.

The extended resource pattern works as follows. The cluster administrator registers a resource name (for example, nvidia.com/gpu) and publishes the available count on each node via a device plugin. A Pod that requests that resource will be scheduled only on a node with at least that many units available. The scheduler tracks allocation and prevents double-booking.

From the Pod's perspective, the resource appears as an environment variable set to the device identifiers it has been allocated.

GPU topology is the aspect of accelerator-aware scheduling that the extended resource model alone cannot capture. Each node may have eight GPUs. On one node, all eight GPUs are connected via NVLink, providing 600 GB/s of bidirectional bandwidth. On the other node, the GPUs are arranged in two groups of four, and cross-group communication traverses the CPU interconnect at 64 GB/s. For a training job that does inter-GPU gradient communication, these two nodes have very different effective performance even though their extended resource counts are identical.

The topology hint mechanism extends the extended resource model with placement affinity information. A device plugin that understands GPU topology can advertise topology hints alongside device counts. The scheduler uses these hints to co-locate a job's GPU allocations within a single NVLink domain rather than spanning multiple domains. For jobs that saturate NVLink during gradient aggregation, the difference between a topologically optimal allocation and a naive allocation can be a factor of five in effective all-reduce bandwidth.

TPU pods introduce a different scheduling constraint. A TPU pod is a fixed-size rectangular slice of a larger TPU fabric, and jobs must request an entire pod rather than individual TPU cores. The scheduler must understand the pod topology and ensure that the requested pod dimensions are available as a contiguous allocation within the TPU fabric. If prior allocations fragment the fabric, the scheduler must either compact existing allocations (which may require preemption) or wait for them to complete before admitting the new job.

NUMA-aware CPU allocation is the third dimension of accelerator-aware scheduling. Many ML workloads use CPU threads for data loading and preprocessing, while GPUs run forward and backward passes. If the CPU threads are on a NUMA node that is topologically distant from the GPU (on the other side of the CPU interconnect), data transfers between host memory and GPU memory traverse the interconnect, incurring latency and bandwidth penalties. The NUMA-aware scheduling pattern pins a Pod's CPU allocation to the NUMA node that is directly connected to the allocated GPU, eliminating this penalty.

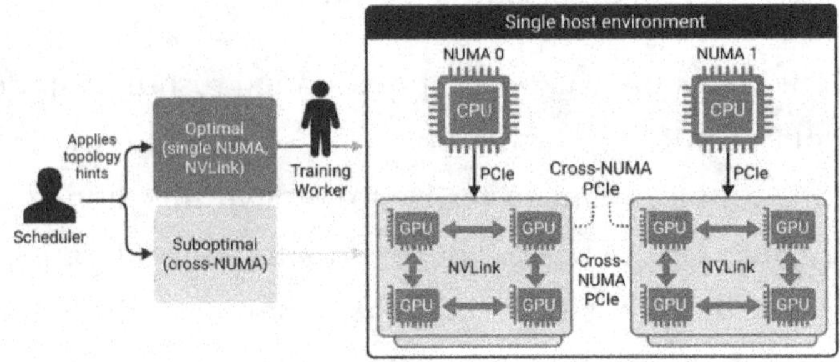

The Orchestration Playbook

The patterns covered in this chapter are most valuable when wired together into an operational playbook that the team can follow under time pressure. A pattern documented in a design review has limited operational clarity. A pattern encoded in a runbook with specific diagnostic steps, decision trees, and escalation paths becomes the team's default response to incidents. The difference between a controlled incident and a weekend-long fire usually comes down to the existence and quality of the playbook.

The triage sequence for an orchestration incident begins with three questions. First: Is the job running? If the job has no running Pods, the problem is at the scheduling or admission layer. Check the queue state, the job's priority, the resource request against available capacity, and the gang scheduling constraint. A job that is stuck at admission will have a queued event with a reason; read the reason before doing anything else. Second: Is the job making progress? If the job has running Pods but metrics show zero throughput, the problem lies in the communication or fault-tolerance layer. Check for a stalled collective (all workers blocked, waiting for a missing participant), a checkpoint write blocking training, or a worker in a crash loop. Third: Will the job finish on time? If the job is running and making progress but is behind schedule, the problem lies in the placement or capacity layer. Check whether topology-aware placement is being honored, whether spot preemptions are adding overhead, and whether the job can be rescheduled to a faster node pool.

The scheduler-placement check is a specific diagnostic step that every team should run after each job submission. It reads the node assignments of all Pods in the gang and evaluates whether they satisfy the topology constraint. An automated script that reads Pod node assignments, resolves the network topology of those nodes, and reports whether the assignment is optimal takes an hour to write and saves hours of debugging time over the life of the cluster.

The checkpoint health check is the second specific diagnostic step. Before relying on spot capacity for a high-value job, verify that the checkpoint write completes within the preemption warning window. A synthetic test that triggers a checkpoint write and measures elapsed time is sufficient. If the write time exceeds the warning window minus a safety margin, the job is not spot-safe and should be run on on-demand capacity until the checkpoint is optimized.

The queue audit is the third specific diagnostic step. Before each planning cycle, review the queue configuration against the team's current priorities. Priorities set six months ago may no longer reflect the team's actual work. A queue configured for a previous project structure may route jobs to the wrong capacity pool. The queue audit is a fifteen-minute review that prevents a class of scheduling bugs that are entirely human in origin.

The escalation boundary is the decision that belongs to the team lead, not the on-call engineer. When an incident requires rescheduling a high-priority production job by preempting another team's high-priority job, that decision requires human judgment about organizational priorities. The playbook should explicitly mark this boundary: the on-call engineer gathers the facts, presents the options, and escalates. The team lead makes the call. Encoding this boundary in the playbook prevents the on-call engineer from making a decision that carries organizational consequences.

Orchestration Incident Triage Playbook

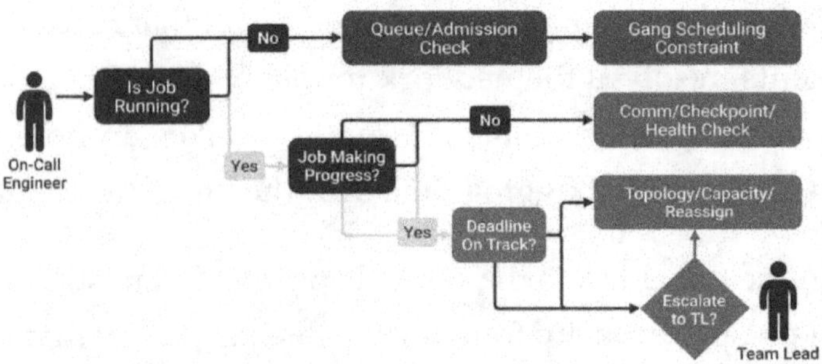

The escalation boundary is the decision that belongs to the team lead, not the on-call engineer. When an incident requires rescheduling a high-priority production job by preempting another team's high-priority job, that decision requires human judgment about organizational priorities. The playbook should explicitly mark this boundary: the on-call engineer gathers the facts, presents the options, and escalates. The team lead makes the call. Encoding this boundary in the playbook prevents the on-call engineer from making decisions with organizational consequences and preserves trust between teams that depend on a shared cluster.

Checkpoint Cadence and Spot Preemption Window Sizing

PREEMPTION SIGNAL
EVICTION
FREQUENT CADENCE
SAFE REPLAY WINDOW
2 min
MODERATE CADENCE
2 min
RARE CADENCE
PREEMPTION SIGNAL
EVICTION
COST-VERSUS-SAFETY TRADE-OFF
CHECKPOINT OVERHEAD %
WORST-CASE REPLAY TIME
WORST-CASE REPLAY TIME
PLATFORM ENGINEER

Technical Checklist

Use this checklist as a release gate before declaring an orchestration configuration production-ready. Each item should be verifiable by a specific command or metric, not by inspection alone.

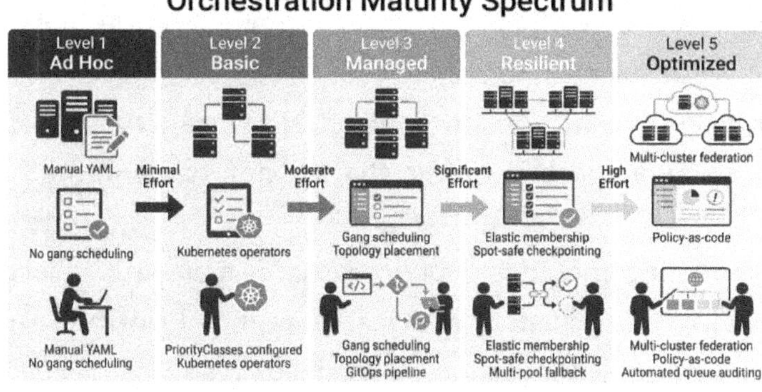

- Every distributed training job specifies a PodGroup or equivalent gang scheduling primitive with the correct minimum gang size.

Verify by checking job manifests in version control for the gang annotation.

- Topology-aware placement is configured for every training job that uses collective communication. Verify by running the placement check script after job admission and confirming that all Pods are within the same topology domain.
- Spot capacity is used only for jobs that have a documented checkpoint cadence and a measured checkpoint write time that fits within the preemption warning window minus a twenty-percent safety margin.
- The checkpoint-before-eviction handler is deployed and tested for every job that runs on spot or preemptible capacity. Verify by triggering a synthetic preemption signal and confirming that the checkpoint completes within the target window.
- Every job has an explicit PriorityClass assigned. No job relies on the cluster default priority.
- Queue configuration is reviewed at the start of each planning cycle and reflects the team's current priorities. The review is documented in the team's operational log.
- Elastic worker membership or graceful degradation is configured for training jobs longer than four hours. A single-node failure should not require a full-job restart.
- All job manifests live in version control. No production job is submitted from a local manifest that is not tracked in the repository.
- The GitOps pipeline includes a policy-as-code gate that rejects manifests violating cluster policies (missing PriorityClass, missing checkpoint annotation, oversized resource request without approval).

- An on-call orchestration playbook exists, covers the three primary triage questions, and has been exercised in at least one real or synthetic incident.
- Multi-cluster or multi-cloud strategy has been formally evaluated. If adopted, the cluster selection policy is documented and tested. If declined, the reason is documented.
- GPU and NUMA topology hints are configured via a device plugin and verified by comparing the achieved all-reduce bandwidth against the theoretical peak for the interconnect type.
- Backfill policy is enabled and configured to maximize utilization without delaying priority jobs. Utilization reports are reviewed weekly.
- The team has a documented process for escalating preemption decisions that affect other teams' jobs.

Team Conversation

These questions are designed for use in team meetings and planning sessions to pressure-test alignment on orchestration strategy. They have no single right answer; the value is in the discussion they generate.

1. What percentage of our distributed training jobs were preempted in the last quarter? How much total wall-clock time did we lose to preemption, and how does that compare to the cost savings from spot capacity? Do we know whether spot is a net positive?
2. Does our scheduler honor topology constraints, or do we discover after the fact that a training job ran across multiple network domains? When did we last run the placement check script, and what did it find?
3. If our primary cluster failed for an hour right now, how would our three most important jobs respond? Is the answer "they

would resume from the last checkpoint within thirty minutes" or "we would be paging people all night"?

4. Are our job priorities and queue configurations aligned with what we tell stakeholders we prioritize? When did we last audit the queue, and what did the audit find?
5. Which orchestration runbooks have been exercised in a real incident? Which ones are aspirational? What would it take to exercise the aspirational ones in a controlled simulation before a real incident forces the test?
6. Where in our orchestration stack are we still submitting jobs by hand from local manifests? What is the plan for moving those jobs into the GitOps pipeline?
7. What is the single largest orchestration change we could make in the next quarter to reduce on-call load? Is that change blocked by a technical dependency or by a team-process dependency?
8. Are we tracking the all-reduce bandwidth achieved by our training jobs relative to the theoretical peak? If not, how confident are we that our topology-aware placement is actually working?
9. Does our preemption policy reflect our actual organizational priorities, or did we configure it once and never revisit it? Who owns the queue configuration, and how often is it reviewed?
10. If a new team joined the cluster tomorrow, how long would it take them to understand our queue structure, priority model, and spot policies? Is that understanding written down anywhere?

Key Takeaway

Orchestration is a layered concern, and each layer has its own failure modes and its own patterns. The base container orchestrator provides the primitives: Jobs, Pods, PriorityClasses, and affinity rules. The ML-aware scheduler extends those primitives with gang scheduling, topology-aware placement, and queue management. The fault-tolerance layer absorbs the failures that neither of the lower layers can prevent: node failures, preemptions, and network partitions. The operations layer, the playbook, the GitOps pipeline, and the queue audits are what make the lower layers function reliably over time. Skipping any layer in this stack means accepting the failure mode that layer was designed to prevent.

The patterns that matter most operationally are also the simplest to state. Every tightly coupled training job needs gang scheduling. Every gang-scheduled job needs topology-aware placement. Every job on preemptible capacity needs a checkpoint-before-eviction protocol with a measured write time. Every shared cluster needs a queue configuration that reflects the team's actual priorities, not the cluster defaults. These are not advanced optimizations; they are the baseline that separates a reliable distributed ML platform from a probabilistic one.

The orchestration playbook is where all of these patterns converge into a system that the team can operate. A pattern that is understood but not operationalized provides no protection when a Friday-afternoon preemption cascade starts. A pattern that is encoded in a runbook, exercised in practice, and reviewed regularly becomes muscle memory. The goal of this chapter is to give the team enough of the underlying pattern vocabulary that writing and maintaining that playbook is a natural act rather than a scramble. Responsible by design means the orchestration layer was built to absorb failures, not to hope they do not happen.

6 Distributed Training at Scale

Opening Scenario

The call came at 2:47 a.m. A foundation-model team had just watched the most expensive training run in the company's history die on hour thirty of a seventy-two-hour job. A node failure on rack seven stalled the ring-allreduce, and the watchdog timer killed the job forty minutes later. When the on-call engineer opened the checkpoint directory, she found that the newest checkpoint was eight hours stale. The team had been told, by no one in particular, that checkpointing too frequently was wasteful. Eight hours of gradient updates across 400 accelerators were lost to a single node failure.

The CTO's first question the next morning was not about the hardware. It was about the checkpoint cadence: who decided it? No one had decided it. The cadence had been copied from a public tutorial written for a single-node run. At the cluster scale, with eight hundred gigabytes of optimizer state distributed across hundreds of workers, the cost calculation is qualitatively different. The team had never done the math.

The post-mortem surfaced three compounding failures. First, no checkpoint policy existed: cadence was inherited from tutorial defaults rather than derived from a cost-of-loss calculation. Second, the cluster had no elastic training configuration, so a single-node death was a run death. Third, the run showed no reproducibility, so resuming from the stale checkpoint meant that data ordering and optimizer state were now partially inconsistent with the lost work. The team could restart, but could not meaningfully resume. This chapter is the playbook the team should have had before launch.

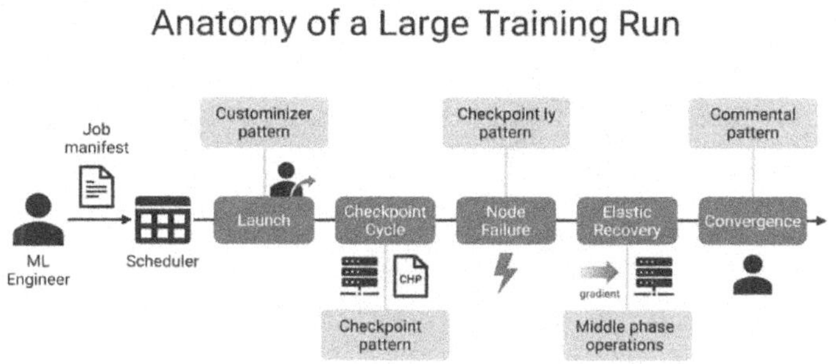

Anatomy of a Large Training Run

Why It Matters

Training at scale is an operational discipline before it is a research one. The research problem of minimizing a loss function is well understood. The operational problem, doing that on hundreds of accelerators over days or weeks while hardware fails and capacity fluctuates, is a systems engineering problem with its own failure modes and its own patterns. Teams that treat scale as merely 'more of the same' tend to learn the difference the hard way, usually at the worst possible moment.

The workflow-level impact is direct. A team without a codified checkpoint policy burns compute whenever hardware fails, and hardware eventually fails. A team without elastic training configured treats every node loss as catastrophic. A team without a reproducibility discipline finds that its most important training results cannot be audited or replicated when a compliance question arises. These are not hypothetical risks; they are the incident shapes that recur across organizations at scale.

The patterns in this chapter address risk at every layer of the training stack. Checkpoint strategies address the cost of failure. Elastic training addresses the probability of failure. Reproducibility patterns address auditability. Numerical stability patterns address silent correctness failures that appear to be model quality issues but are precision artifacts. Taken together, they form a mission-aligned foundation for any training program that intends to survive contact with production infrastructure.

Gradient Synchronization Under Jitter

The defining operation in data-parallel distributed training is gradient synchronization. After each forward and backward pass, each worker communicates its local gradients to the collective, so that all workers update their parameters in the same way. In practice, the communication fabric of a real cluster introduces jitter: variable latency from network congestion, collective scheduling, OS noise, and thermal throttling. Synchronous gradient aggregation blocks the next step until every worker has contributed, so a single slow worker stalls all the fast ones. On a large cluster, the tail of the latency distribution, not the mean, determines throughput. This is the straggler problem.

Ring-allreduce remains the dominant synchronous pattern because it distributes communication load evenly: each worker sends and receives exactly one chunk per step, regardless of cluster size. The pattern is bandwidth-optimal but latency-bound. Hierarchical allreduce improves on this by grouping workers into intra-node and inter-node collectives that match the physical topology: GPUs within a node communicate over high-bandwidth interconnects. In contrast, only one representative per node crosses the slower inter-node network. For clusters with fast intra-node fabric and slower inter-node links, the throughput improvement is substantial.

Straggler mitigation without full asynchrony uses a backup-worker pattern: launch more workers than required, use the gradient from whichever N workers finish first, and ignore the rest for that step.

A 10 percent redundancy budget (110 workers doing the work of 100) eliminates most tail-latency events at the cost of 10 percent more accelerator hours. Gradient compression (top-K sparsification, PowerSGD) reduces the volume transmitted per step by one to two orders of magnitude, with modest degradation in convergence. The error feedback mechanism, which accumulates the compression residual and adds it to the next step's gradient, is essential for convergence and must be included in checkpoints; otherwise, resuming resets the residual, causing the loss to spike.

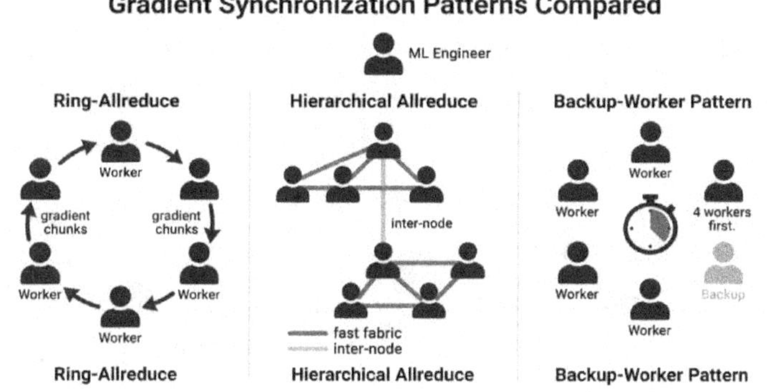

Large-Batch Training: LARS, LAMB, Warmup, and Batch-Size Schedules

Distributed training naturally produces large effective batch sizes. The linear scaling rule is the starting point: multiply the batch size by K, multiply the learning rate by K as well. This works in the moderate-batch regime but breaks down for tens of thousands of samples because the optimal learning rate grows more slowly than linearly once gradients become nearly deterministic. Using the linear rule with very large batch sizes can cause training instability during the early epochs.

Learning-rate warmup addresses early instability by starting with a low rate and ramping it linearly to the target over the first few hundred to thousands of steps. Warmup is now mandatory for large-batch training and for transformer-based models. A useful heuristic: warmup spans at least one full pass through the dataset at the initial small rate. LARS (Layer-wise Adaptive Rate Scaling) adapts the learning rate per layer based on the ratio of the weight norm to the gradient norm, enabling stable training with batch sizes of 32,000 or more for convolutional networks. LAMB extends LARS to Adam, adding adaptive moment estimates while preserving the layer-wise trust-ratio correction. LAMB was used to train BERT in 76 minutes by scaling to 65,536 sequences per batch on 1,024 TPUs.

Batch-size scheduling adds a temporal dimension: start small and grow the batch during the run. In early training, a small batch size introduces gradient noise that serves as implicit regularization. As training progresses and the loss landscape smoothes, a larger batch reduces noise and accelerates convergence. Batch-size schedules have been shown to achieve the same final accuracy as fixed-small-batch training in fewer total steps. Operationally, they align naturally with elastic training configurations in which cluster size grows during the run. The generalization gap at large batch sizes is real: models trained at very large batch sizes tend to converge to sharper minima. Warmup, LARS/LAMB, and explicit regularization help close the gap, but teams should validate generalization metrics, not just loss curves, when scaling batch size.

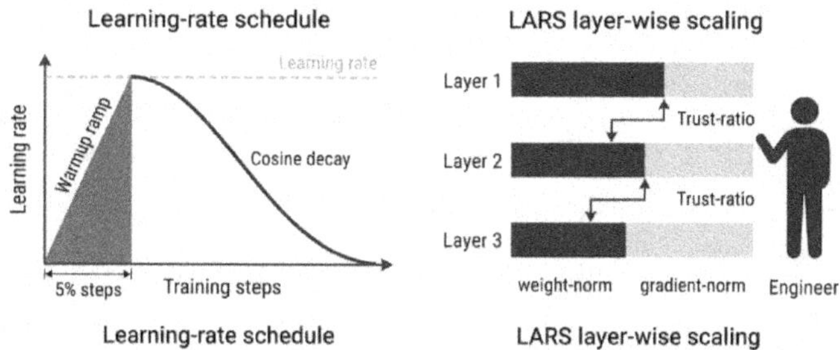

Checkpoint Strategies: Synchronous, Asynchronous, Sharded, and In-Memory

A checkpoint is a snapshot of all states required to resume training: model weights, optimizer state, data loader position, random number generator state, and step counter. At scale, checkpointing costs are non-trivial. A 70-billion-parameter model with Adam requires roughly 1 terabyte of data to be fully checkpointed in FP32. The cost-of-loss calculation frames the cadence decision correctly: let C denote the cost of a single checkpoint write, and let L(T) denote the expected cost of losing T hours of training. The optimal interval T* minimizes C/T + L(T)/2. For typical large runs, the optimal interval is often 15 minutes to 2 hours, not the 4-to-8-hour defaults inherited from tutorial code.

Synchronous checkpointing pauses all workers, writes the checkpoint, and resumes. The guarantee is global consistency. The cost is a training pause proportional to write latency, which can exceed ten minutes for very large models. Asynchronous checkpointing overlaps the write with training: each worker copies local state to a CPU-memory write buffer, resumes immediately, and a background thread flushes the buffer to storage. The pause reduces to a GPU-to-CPU copy, typically a few seconds. Atomic commit (writing to a temporary path, then renaming on completion) ensures that an interrupted write leaves the previous checkpoint intact.

Sharded checkpointing distributes the write across all workers, with each worker responsible for its own slice of the model and optimizer state. The checkpoint is a collection of per-rank files plus a metadata index. Writes are fast and parallel; restoring with a different worker count requires resharding, which must be tested before it is needed in production. In-memory checkpointing holds state in host RAM rather than storage. A ring of two to four in-memory checkpoints provides fast rollback to a recent step without any I/O, handling software faults and training divergence. In-memory checkpoints complement rather than replace persistent ones: the typical pattern combines a fast in-memory checkpoint every few minutes with a persistent checkpoint every thirty minutes to two hours. Checkpoint verification (test-loading every checkpoint immediately after writing and asserting round-trip correctness) is the pattern most consistently skipped with the most serious consequences.

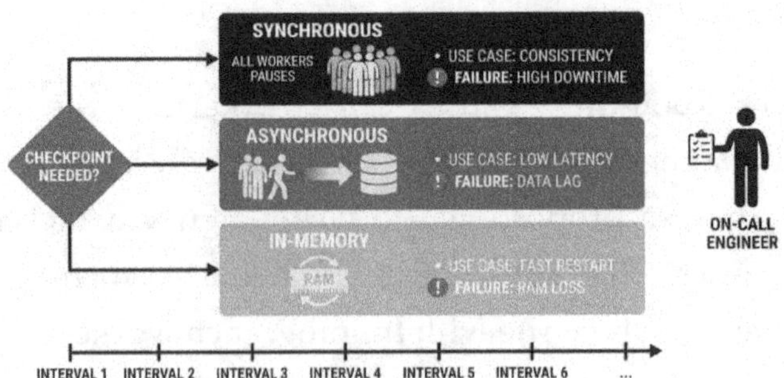

Elastic Training: Preemption Recovery and Dynamic World Size

Elastic training allows a distributed job to continue when the number of workers changes. Workers may be added (when capacity becomes available) or removed (due to node failure, spot-instance preemption, or policy reclaim). A job without elastic configuration treats every node loss as a catastrophic event. A job configured for elastic execution treats it as a managed one. The rendezvous protocol is the foundation: workers discover one another and agree on a collective identity. In a static job, rendezvous happens once at startup. In an elastic job, it may happen multiple times, after each membership change.

Preemption recovery is the elastic pattern for spot instances and priority-scheduled clusters. Upon receipt of a preemption signal (typically with 90 seconds to 2 minutes to complete), the worker immediately writes an asynchronous checkpoint, deregisters from the rendezvous backend, and exits. Surviving workers detect the departure through a heartbeat timeout, trigger a new rendezvous with reduced world size, reshard the model to the new rank count, and continue from the most recent checkpoint. The total interruption is bounded by checkpoint write time plus rendezvous timeout.

Resharding after a world-size change is operationally expensive for sharded checkpoints: redistributing shards across the new rank count may require both reading and writing checkpoint files. The sharded case is faster to write but slower to reshard; teams must test the reshard path before relying on it in production. The dynamic world size also affects the effective global batch size. An elastic configuration that automatically adjusts the learning rate as world size changes prevents silent miscalibration. Testing the elastic path on a scheduled basis (at a minimum, quarterly) in an environment that mirrors the production cluster, covering the full path from node removal through training resumption, with a loss check, is a mandatory operational practice.

Elastic Training: Preemption and Recovery Sequence

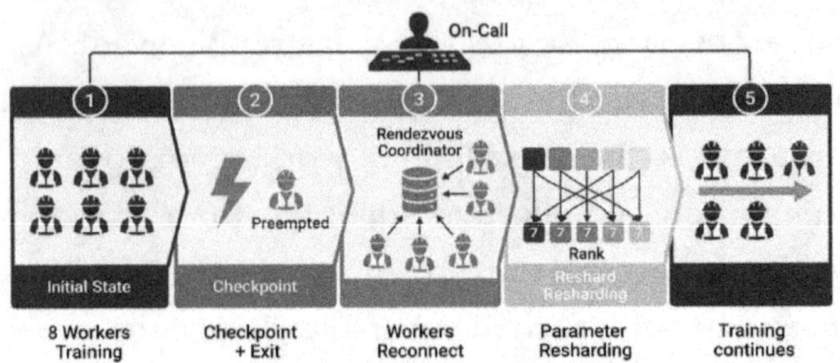

Fault Tolerance: Byzantine Actors, Stragglers, and Silent Failures

Hardware failures in large clusters are steady-state operating conditions, not rare events. A cluster of 1,000 accelerators, each with an individual mean time between failures of 3 years, will experience roughly one failure every 11 days. A ninety-day training run should expect multiple node failures as a matter of probability. Straggler detection is the first line of defense: monitoring per-worker step time and flagging workers whose step time exceeds the median by more than two to three standard deviations allows the orchestrator to replace or exclude the straggler before it becomes a full failure.

Byzantine fault tolerance addresses a qualitatively different failure mode: a worker that produces incorrect gradient values rather than simply being slow or absent. Byzantine failures can arise from memory corruption, silent data errors on interconnects, or software bugs that produce numerically plausible but wrong values. In a standard allreduce, a single Byzantine worker can poison the gradient average across all workers. Byzantine-robust aggregation algorithms (Krum, coordinate-wise median, trimmed mean) replace the standard average with an estimator that down-weights statistical outliers, at the cost of computational overhead and reduced convergence speed.

Silent training divergence is the hardest fault to detect. The job appears to run normally, but the model converges to a bad solution. Common causes: data pipeline bugs that produce repeated batches, gradient vanishing that averages to zero during allreduce, and precision issues that produce consistent but incorrect updates. Detection requires monitoring beyond loss curves: layer-wise gradient norms, weight update magnitudes, attention entropy for transformer models, and comparison against reference runs. Watchdog patterns that alert and optionally pause the job when any monitored metric leaves its expected range for more than a configurable number of consecutive steps prevent runaway compute burn. Gradient clipping helps prevent explosion: monitoring the fraction of steps where clipping occurs provides a diagnostic signal (near zero in a healthy post-warmup run).

Fault Taxonomy and Detection Signals

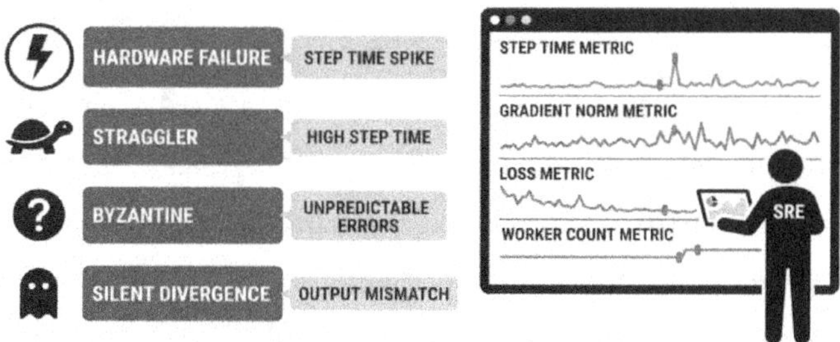

Mixed Precision and Numerical Stability

Mixed-precision training uses lower-precision data types for both the forward and backward passes while maintaining higher-precision master copies for parameter updates. Modern accelerators deliver two to four times higher throughput in FP16 or BF16 than in FP32, and the memory reduction enables larger batch sizes or larger models. The pattern is now standard for any training run on contemporary hardware.

FP16 has a dynamic range of approximately 6e-8 to 65,504. Gradients frequently fall outside this range: small gradients underflow to zero, and large ones overflow. Loss scaling addresses this by multiplying the scalar loss by a large constant (512 to 65,536) before the backward pass, shifting gradient values into the FP16 range. Dynamic loss scaling adjusts automatically: it doubles every N clean steps, halves, and skips the step on overflow. BF16 has the same exponent range as FP32, making overflow essentially impossible and eliminating the need for loss scaling. BF16 is now the default precision for training large models on hardware that supports it. FP8 (E4M3/E5M2) is the emerging frontier, providing additional throughput gains at the cost of per-tensor scaling factors and increased implementation complexity.

Master-copy weights in FP32 are maintained alongside FP16 compute copies. The optimizer operates on the master copies, applying gradient updates in full precision. FP16 copies are derived from master copies at the start of each forward pass, preventing precision loss from accumulating across optimizer updates. Mixed-precision regression (loss NaN or a sudden loss explosion after a code or data change) is diagnosed by temporarily disabling mixed precision and rerunning the affected batch in FP32. If the loss is finite in FP32 but NaN in FP16, the problem is a precision artifact.

The fix is additional loss scaling, explicit FP32 casting for the offending operation, or switching to BF16.

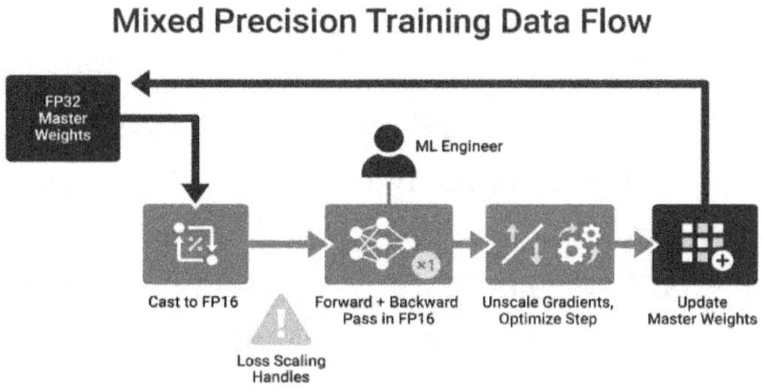

Mixed Precision Training Data Flow

Gradient Accumulation and Communication Optimization

Gradient accumulation decouples the global batch size from the number of workers and per-step memory capacity. Each worker accumulates gradients over K micro-steps before a single synchronization and update, thereby multiplying the effective global batch size by K without requiring K times as many workers. The primary use case is memory-constrained training, where the model or sequence length exceeds the device's capacity at the natural batch size. A correctness concern: batch normalization computes statistics over the micro-batch rather than the accumulated batch, producing different effective statistics. Layer normalization does not have this problem and is preferred in distributed training for this reason.

Communication overlaps with pipelines and computation. In the naive implementation, the backward pass completes before the allreduce begins, leaving the network idle during computation and the accelerators idle during communication. Overlap stages the gradient for each layer so it can be communicated as soon as it is computed, while the backward pass continues through earlier layers. On bandwidth-sufficient clusters, communication cost is fully hidden behind backward-pass compute. Gradient fusion reduces overhead by combining small gradient tensors into larger messages before submitting to the collective backend. Bucket size (commonly one to twenty-five megabytes) requires profiling on the target cluster: too small and kernel launch overhead dominates, too large and the first bucket is not ready until the backward pass is nearly complete, negating overlap. Gradient compression (top-K sparsification, PowerSGD, quantized communication) reduces the volume by a factor of 10 to 100 for bandwidth-constrained topologies such as cross-datacenter training.

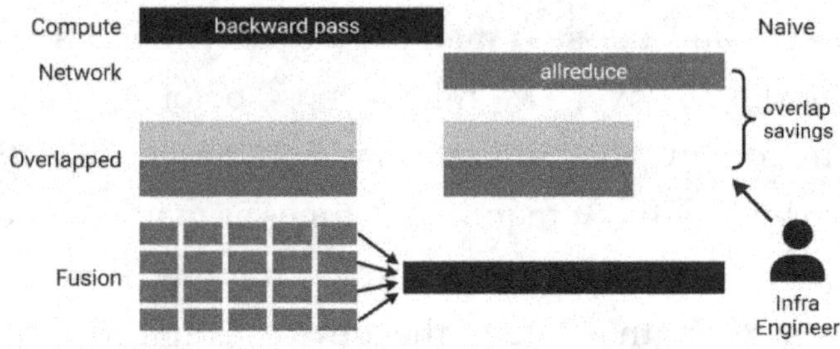

Optimizer Tuning for Large-Scale Training

Optimizer choice is an architectural decision that shapes the entire training trajectory. At scale, it interacts with batch size, learning rate schedule, gradient noise, and memory budget in ways that are not apparent at small scale. Adam and AdamW are the dominant optimizers for training large models. AdamW decouples weight decay from the adaptive learning rate by applying it directly to the weights, ensuring uniform regularization across parameters. AdamW consistently outperforms Adam with L2 regularization for transformer models. The beta values (beta1 = 0.9, beta2 = 0.95-0.999) are relatively scale-invariant; learning rate and weight decay require recalibration when batch size changes significantly.

Lion (Evolved Sign Momentum) uses sign updates with a uniform step size per parameter and maintains only one moment estimate, halving optimizer state memory compared to Adam. For large models where the optimizer state occupies a significant fraction of memory, this is practically useful. Lion's sign-based updates behave differently at large batch sizes: the learning rate need not increase linearly with batch size. Adafactor factors the second-moment matrix rather than storing it explicitly, reducing the optimizer state from O(mn) to O(m+n) for weight matrices. This is the optimizer of choice when device memory is the binding constraint. Shampoo, a second-order optimizer using Kronecker-factored preconditioners, outperforms Adam in convergence speed (fewer steps to reach the same loss) despite higher per-step compute and memory costs, making it strong for moderate-scale runs where step count is the binding constraint.

Optimizer state sharding makes large optimizers feasible at scale. In standard data-parallel training, each worker holds a full copy of the optimizer state, which can equal or exceed the model's memory footprint. State sharding (ZeRO stage 1 and 2) partitions the optimizer state across workers, so each worker holds only its shard, reducing per-device memory consumption in proportion to the number of workers. The trade-off is a communication step to redistribute parameters after each update; this cost is typically smaller than the memory savings justify.

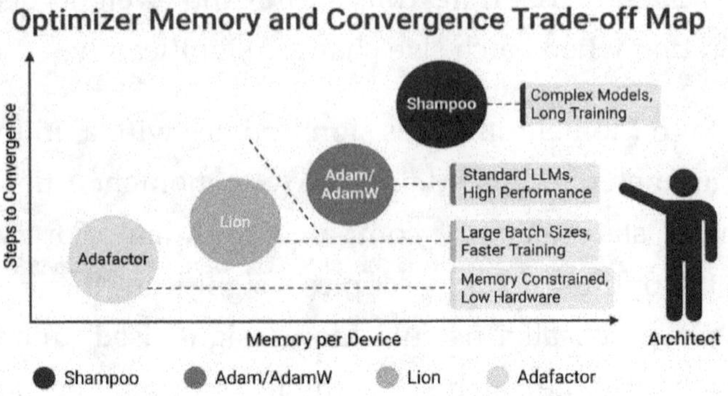

Learning-Rate Schedules and Curriculum at Scale

The learning rate schedule is one of the most consequential hyperparameters in a training run. Warmup is non-negotiable for large-batch training and for transformer models at any batch size. In transformers, attention logits in early training can grow large, pushing softmax into a saturated regime with tiny gradients. Warmup allows the model to establish a reasonable internal representation before the rate reaches full strength. Duration heuristic: warmup spans at least one full pass through the dataset at the initial small rate, or 1 to 5 percent of total training steps.

Cosine decay is the dominant schedule for the post-warmup phase. The learning rate decays from peak to a small final value following a cosine curve, avoiding the discontinuous drops of step decay and the slow mid-schedule annealing of linear decay. A critical operational constraint: cosine decay requires knowing the total number of training steps in advance, because the decay rate is calibrated to reach the final value exactly at the last step. Extending or truncating a cosine schedule mid-run can cause misaligned decay, potentially harming final performance. WSD (Warmup-Stable-Decay) addresses this by decoupling the schedule into three phases: warmup, a long, stable phase at peak rate, and a short final decay phase. The stable phase allows the run to be extended cheaply or branched into multiple continued pretraining or fine-tuning runs, each with its own independent decay. WSD is increasingly used for foundation model pretraining where the compute budget is uncertain at launch.

The sequence-length curriculum begins with short sequences and progressively introduces longer ones. Short sequences pack more samples per step at the same memory cost, increasing throughput in early training when long-range context is less critical. Curriculum approaches require a data pipeline that can filter or sort by sequence length without introducing selection bias, and checkpoints that capture the current curriculum stage to ensure correct resumption. Diagnostic signals for schedule health: gradient norm over time (should decrease as the schedule anneals), improvement rate in loss per step (should track changes in the learning rate), and validation loss relative to training loss. Teams that do not monitor these signals cannot distinguish a schedule problem from a data problem from a precision problem when all three produce similar surface behavior.

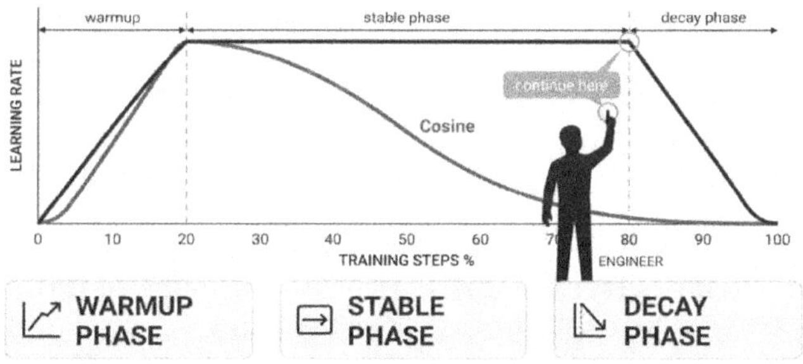

Continued Pretraining, Fine-Tuning, and PEFT at Distributed Scale

A foundation model that has completed pretraining is the starting point for continued pretraining on domain-specific data, supervised fine-tuning on instruction datasets, and parameter-efficient fine-tuning for resource-constrained teams. Each has distinct distributed requirements. The principal risk of continued pretraining is catastrophic forgetting: performance on the original pretraining distribution degrades as the model adapts to the new one. Data mixing is the primary mitigation: including ten to thirty percent original pretraining data in the continued pretraining batch preserves performance while allowing adaptation. The mixing ratio must be tuned for each domain and evaluated against benchmarks on both distributions.

The learning rate for continued pretraining must be lower than the peak rate used during pretraining. A common starting point is one-tenth of the peak rate, with a brief warmup and a WSD or cosine schedule calibrated to the continued pretraining token budget. Supervised fine-tuning on instruction data is typically one to three epochs over a dataset orders of magnitude smaller than the pretraining corpus. At this scale, the concerns shift to data quality and overfitting: closely monitoring validation loss and stopping early are more important than the checkpointing strategy. The distributed setup is often simpler and shorter than pretraining.

Parameter-Efficient Fine-Tuning (PEFT), with LoRA as the most widely adopted variant, freezes most pretrained parameters and trains only a small set of additional parameters (typically less than 1% of the original count). LoRA inserts low-rank matrix pairs alongside each attention weight. Memory savings are substantial: gradients and optimizer state for frozen parameters are not stored.

At a distributed scale, PEFT enables fine-tuning on hardware that cannot store the full model's optimizer state, and multiple fine-tuning jobs can run concurrently, sharing the same base model weights in memory. Frozen-layer strategies reduce allreduce cost in proportion to the frozen fraction; the optimal frozen/unfrozen boundary is dataset- and task-dependent and must be validated on held-out benchmarks.

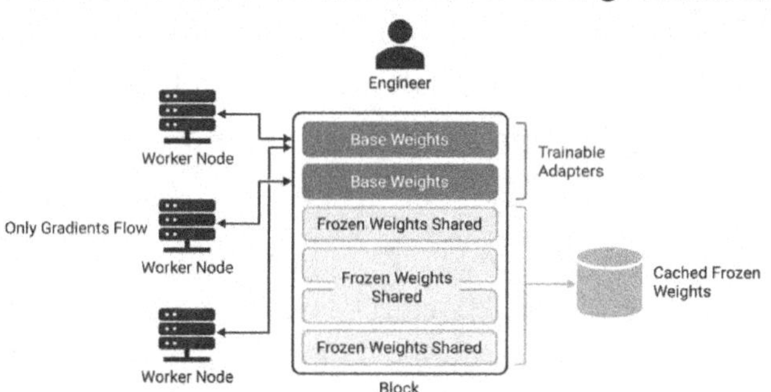

Hyperparameter Search at Scale

Hyperparameter tuning is the part of large-scale training most consistently left unbudgeted. Teams allocate compute for the final run but treat the search as overhead to minimize. The result is that production models frequently run on hyperparameter configurations inherited from a smaller pilot, which may be substantially suboptimal at the target scale. Random search consistently outperforms grid search at the same budget by exploring the hyperparameter space more efficiently: most hyperparameters have low effective dimensionality, and random sampling covers important dimensions better than a regular grid. Population-Based Training (PBT) maintains a population of parallel training runs, periodically copying weights of high-performing runs into low-performing ones and perturbing the hyperparameters of offspring. PBT searches for good hyperparameter schedules rather than fixed values, adapting learning rate and regularization throughout training. It requires that all population members be able to checkpoint and resume, linking this directly to the checkpoint patterns in this chapter.

ASHA (Asynchronous Successive Halving Algorithm) is a bandit-based early-stopping strategy. Trials begin simultaneously; after a configurable number of steps (the first rung), the bottom fraction by the validation metric is stopped, and computation is redirected to the survivors. The process repeats at each rung. ASHA is asynchronous: a new trial begins as soon as a slot opens, keeping the cluster saturated. It is the recommended approach for hyperparameter search on shared clusters. Bayesian optimization with a surrogate model predicts expected objective value across the hyperparameter space, requiring fewer total trials than random search for low-dimensional problems. Its sequential nature limits parallelism; batch acquisition, with asynchronous updates, partially addresses this. Budget governance is the pattern that makes search responsible-by-design: before any search, specify the trial count, maximum steps per trial, cluster quota, and success criterion, and have them reviewed by a technical lead and recorded in the run manifest.

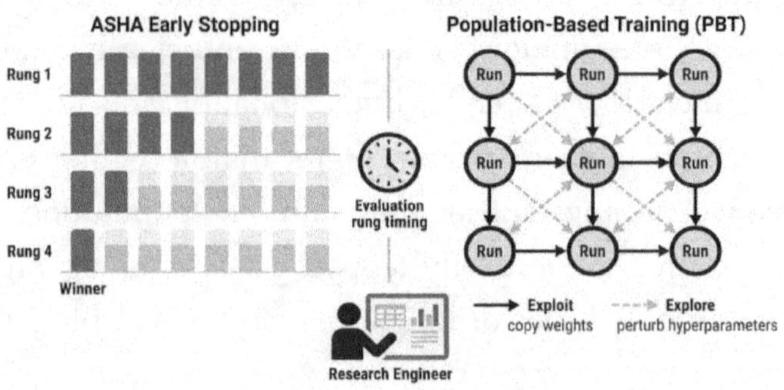

Reproducibility Under Nondeterminism

Reproducibility in distributed training is not binary. The practically useful target for most teams is run-manifest reproducibility: enough state is captured that the run can be closely replicated (with the same model quality within measurement noise) by any team member on any equivalent cluster. The sources of nondeterminism must be categorized before they can be controlled. Floating-point nondeterminism arises from the non-associativity of floating-point addition: the order in which gradient contributions are summed in an allreduce affects the result, and the order is not guaranteed to be consistent across runs or hardware generations. Deterministic allreduce algorithms exist, but may be slower; the alternative is to accept FP nondeterminism and target statistical reproducibility.

Data-ordering nondeterminism arises from shuffle operations. If the data is shuffled with an unrecorded seed, rerunning produces a different ordering. Deterministic data ordering requires a fixed shuffle seed per epoch, the same seed propagated identically to all workers, and the ability to reconstruct the exact data sequence up to any step. The typical implementation seeds the data-loader's random state with a function of epoch number and worker rank. Random-seed discipline across workers prevents silent model divergence: initialize weights on rank zero from a fixed seed, broadcast it to all other ranks before training begins, and scope data augmentation seeds to rank plus epoch. Deterministic operations (deterministic convolution mode, stable-sort flags for indexing operations, fixed scan order for attention) reduce nondeterminism without fully eliminating variability in floating-point accumulation.

The run manifest is the document that makes reproducibility actionable. A complete manifest records: the exact code revision (commit hash), dataset version and checksum, initial weight checkpoint or initialization seed, full hyperparameter configuration, cluster configuration, and software environment (framework versions, CUDA version, library versions). The manifest is committed to the same version control system as the model, associated with the output checkpoint, and retrievable by artifact identifier. Without a run manifest, a training result exists only as an artifact without provenance. A result that cannot be attributed to a specific configuration and dataset version cannot be trusted, reproduced, or defended.

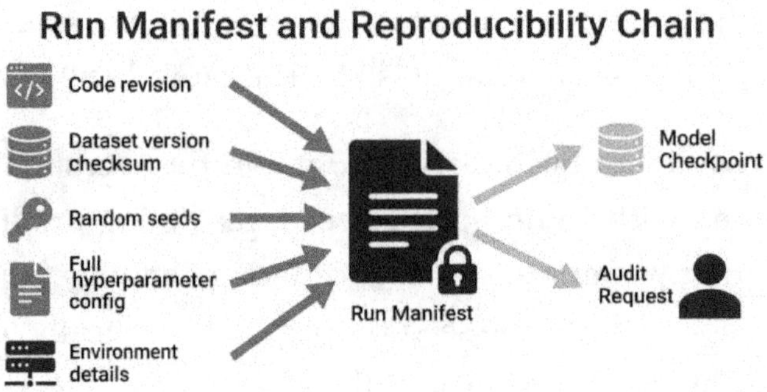

Run Manifest and Reproducibility Chain

Code revision
Dataset version checksum
Random seeds
Full hyperparameter config
Environment details
Run Manifest
Model Checkpoint
Audit Request

Sustainability and the Carbon Profile of a Training Run

The carbon and energy cost of large training runs is no longer a peripheral concern. A foundation-model pretraining run of several months on hundreds of accelerators can consume millions of kilowatt-hours and emit hundreds of tonnes of CO_2, depending on the region's energy mix. These numbers are consequential for sustainability commitments, regulatory disclosure, and procurement decisions. Treating carbon as a first-class engineering metric is now a mission-aligned requirement for any team operating at scale.

Measuring the energy footprint requires instrumentation that most teams lack. The components to measure: accelerator power draw (exposed via management libraries), host CPU and memory power, network switch power (estimated as a fixed fraction of rack power), and storage I/O power. The sum is the facility load, divided by the data center's Power Usage Effectiveness (PUE) to obtain the IT load, and then multiplied by the grid's carbon intensity to obtain the CO_2 estimate. Cloud providers are increasingly publishing per-region hourly carbon-intensity data. Carbon-aware scheduling chooses launch times and regions to minimize carbon intensity: many grids exhibit strong diurnal patterns, and scheduling a flexible job during low-intensity windows can reduce the effective carbon footprint by 20 to 50 percent.

Compute efficiency is the most direct lever: every optimization that reduces wasted compute also directly reduces carbon. A training run at 60 percent accelerator utilization, due to communication stalls, emits 40 percent more CO_2 per model than one at full utilization. Teams should record kilowatt-hours and estimated CO_2 for each training run in the same run manifest that records hyperparameters and dataset versions, and review the carbon profile alongside the cost profile in every post-run review. When the carbon profile of a large run is presented alongside its dollar cost, it consistently shifts the conversation about what constitutes waste and what earns its place in the compute budget.

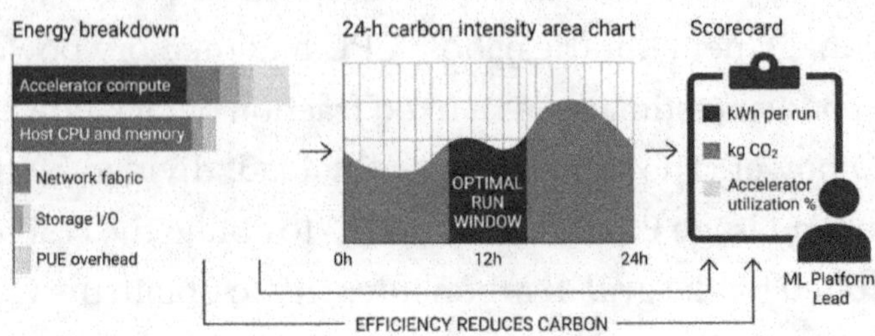

Technical Checklist

Apply this checklist before declaring a distributed training run production-ready at scale. Each item is a gate, not a suggestion.

- Checkpoint cadence is justified by an explicit cost-of-loss calculation that accounts for checkpoint write time, expected node failure rate, and rollback cost at the current cluster size.
- Asynchronous checkpointing is configured with atomic commit (write-to-temp, rename on completion) so an interrupted write does not corrupt the previous checkpoint.
- Sharded checkpoints include a metadata index mapping rank to shard, and the reshard path for a different world size has been tested end-to-end.
- Checkpoint verification is automated: every checkpoint is test-loaded on a separate process immediately after writing, asserting model parameters, optimizer state, and step counter.
- Elastic training rendezvous has been exercised with a simulated node loss in the last quarter, and the full recovery path (deregister, reshard, resume) has been timed and documented.
- The training job handles preemption signals gracefully: on signal receipt, it writes an asynchronous checkpoint and exits cleanly within the preemption window.

- World-size changes during elastic recovery automatically adjust the effective learning rate under the linear scaling rule or an equivalent calibrated policy.
- Straggler detection monitors per-worker step time and alerts when any worker exceeds the median by more than a configurable threshold for more than N consecutive steps.
- Gradient clipping is configured, and the fraction of steps where clipping activates is logged; near zero is expected in a healthy post-warmup run.
- Mixed precision configuration is documented: the precision format (FP16 or BF16), loss scaling strategy (dynamic or static), and master-weight copy policy are all explicit.
- A mixed precision regression test confirms all operations produce finite gradients in the configured precision on the target hardware.
- Communication overlap and gradient fusion bucket sizes have been profiled on the target cluster and are set explicitly rather than defaulted.
- The optimizer and schedule are calibrated for the global batch size, including warmup duration, peak learning rate, and decay schedule.
- The run manifest captures code revision, dataset version, and checksum, random seeds, full hyperparameter configuration, cluster configuration, and software environment.
- Data ordering is deterministic: shuffle seeds are fixed per epoch, scoped by worker rank, and included in the run manifest.
- Hyperparameter search budget (trial count, steps per trial, cluster quota) is agreed in writing before the search starts and reviewed by a technical lead.

- The carbon and energy profile (kWh, estimated CO_2, accelerator utilization) is recorded in the run manifest and reviewed alongside the cost profile.
- Carbon-aware scheduling has been evaluated: if the cluster and workload support flexible launch timing, the lowest-carbon-intensity window has been identified.

Team Conversation

Use these questions in your next team meeting to pressure-test how well the chapter's patterns are embedded in your current training program.

- When did we last simulate a node failure on our most expensive training run? What did we learn, and what changed as a result?
- Is our checkpoint cadence the result of an explicit cost-of-loss calculation, or did we inherit it from tutorial defaults? Can we show the math?
- Which of our most-cited training results could we reproduce exactly today, starting only from what is in our run manifests?
- Where has mixed precision quietly failed us, and how would we recognize it next time before the loss curve makes it obvious?
- Are our optimizer and learning rate schedule choices calibrated to the global batch size we actually train with, or were they set at a smaller scale and never revisited?
- What is our process for validating that elastic training works on the production cluster configuration, and when did we last run it?
- Do we have a carbon budget for our training program? Would our most senior leadership recognize the number if we presented it?
- What is the largest scaling-related risk in our current training pipeline that we have not written down?

- If our cluster lost 20 percent of its workers mid-run tomorrow, what would happen in the first ten minutes, and who would be responsible for each action?
- How do we currently distinguish between a schedule problem, a data problem, and a precision problem when the loss curve starts behaving unexpectedly?

Key Takeaway

Training at scale is an operational discipline before it is a research one. The patterns in this chapter exist because the failure modes of large-scale training are qualitatively different from those of single-node training. A node failure is not a nuisance but a probabilistic certainty; a checkpoint cadence is not a preference but a quantitative trade-off; a reproducibility discipline is not bureaucracy but the minimum condition for scientific credibility. Teams that treat these patterns as optional overhead will rediscover their necessity through incidents. Teams that internalize them will ship training runs that survive hardware reality, produce auditable results, and consume compute responsibly.

The pilot-to-production pathway for large training runs runs through every pattern in this chapter. A pilot run that trains successfully on eight nodes does not automatically scale to eight hundred. Checkpoint I/O becomes a bottleneck. The learning rate calibrated for a small batch is wrong at scale. The rendezvous configuration that worked in a stable environment fails under real cluster churn. The reproducibility discipline that was never implemented becomes visible only when an audit arrives. Every pattern described here should be validated at each scale milestone, not assumed to transfer automatically.

Carbon and energy are now first-class scaling concerns, not footnotes. The teams and organizations that build the instrumentation, scheduling discipline, and manifest practices now will have a structural advantage as carbon reporting becomes a universal requirement. Responsible-by-design training is not a constraint on ambition; it is the foundation on which ambitious training programs earn their compute budgets and maintain operational clarity with the rest of the organization.

7 Serving at Scale and Model Parallelism for Inference

Opening Scenario

The search and ranking team at a mid-sized retail platform had spent three months retraining their homepage ranking model on a richer feature set. Offline evaluation was unambiguous: the new model lifted normalized discounted cumulative gain by 11% and improved click-through rate by 8 points in a held-out replay experiment. The team was confident. They skipped the canary stage, promoted the new model directly to production, and watched the dashboards.

Within forty minutes of the rollout, P99 latency on the homepage rose from 40 milliseconds to 290 milliseconds. Conversion dropped by four percent. Revenue-per-hour fell. The on-call engineer triggered a rollback at minute 53, but it was not automated. It required a manual image swap, a configuration push, and a rolling restart of twelve serving pods. The window of impact lasted for 110 minutes before the old model was fully restored.

The post-mortem revealed four independent failure modes that had compounded. First, the new model had twice the parameter count of its predecessor and had never been profiled under production-shaped traffic; its P99 GPU kernel time was four times that of the P50. Second, the load balancer used a simple round-robin policy that ignored accelerator memory pressure; one replica received three times the steady-state request rate during the post-deploy traffic spike and queued requests rather than shedding them. Third, the serving batch size had been tuned for average concurrency, not peak concurrency, causing the dynamic batcher to degrade into single-sample inference at the tail of the load curve.

Fourth, there was no shadow deployment or live latency gate in the release pipeline; the only quality gate was the offline experiment.

None of the four failure modes was novel. Each one has a well-understood pattern that addresses it: model-aware load balancing, continuous batching, canary deployment with automated latency guardrails, and shadow deployment for online-offline comparison. The team knew about these patterns in the abstract, but had no systematic way to apply them. This chapter is the catalog of those patterns that came from. It treats serving architecture as a first-class discipline with the same rigor that the rest of the book applies to training.

The patterns in this chapter span the full serving lifecycle: how models are parallelized across accelerators at inference time, how batching strategies shape the latency-throughput curve, how caches are managed for large models, how multiple models share a platform, how requests are routed and queued, how deployments are made safe, and how autoscalers are tuned for accelerator-backed workloads. The goal is a serving architecture that is responsible by design, not one that is patched after the first incident.

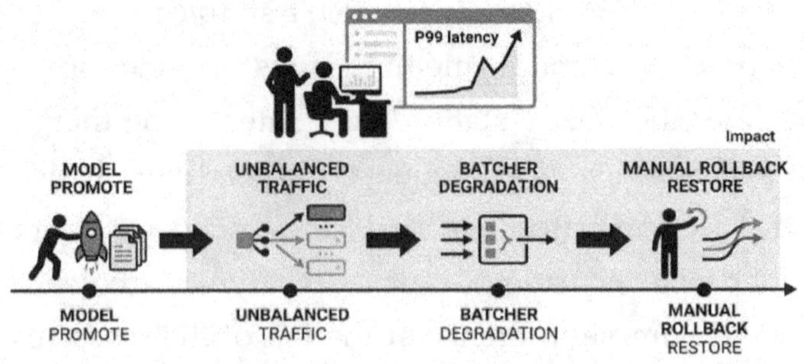

Why It Matters

Training is judged by model quality. Serving is judged by everything else: latency at every percentile that a human or downstream system depends on, throughput that scales with traffic without proportional cost, reliability that survives peak load without manual intervention, and deployment safety that lets the team ship changes without rehearsing a rollback under pressure. These constraints are not ML constraints. They are product and platform constraints that place the ML engineer directly on the critical path of the broader engineering organization.

The workflow-level impact of poor serving architecture is disproportionate to its engineering cost. A model that performs well offline but is slow or unreliable in production destroys user trust faster than a mediocre model served reliably. The serving layer is the interface between all the work done in earlier chapters, from data sharding and training optimization to orchestration and fault tolerance, and the end users who generate the revenue that funds that work. Getting serving right is not optional polish; it is the pilot-to-production pathway that determines whether any of the upstream investment pays off.

The patterns in this chapter also carry direct cost implications. An over-provisioned serving fleet wastes GPU budget that Chapter 7 will cover in detail. An under-provisioned fleet creates latency violations that trigger SLO penalties and manual on-call intervention. Batching strategies, cache hit rates, and autoscaling policies together determine whether a serving cluster runs at 40% or 80% utilization under a constant workload. Operational clarity about each of these levers is the difference between a cost-controlled serving platform and a cluster that grows faster than traffic.

Multi-Model Serving Architectures

Most production ML platforms serve many models simultaneously. A recommendation stack might include a retrieval model, a ranking model, a diversity filter, and a fallback model for cold-start users. A content platform might run separate models for video, image, and text classification alongside a user-intent model that routes traffic between them. The question of how to host all of these on shared infrastructure is the first architectural decision in serving, and it has a larger impact on operational complexity and cost than most teams realize when they start.

The canonical topologies are four. Model-per-replica assigns one model to each process and to a single set of accelerator devices. This is the simplest topology to reason about, the easiest to scale independently, and the most wasteful when models are small or traffic is sparse. Multi-tenant servers assign multiple models to a single process and share the accelerator memory pool between them. This is more efficient for small models but creates noisy-neighbor effects when one model's traffic spike forces another model's requests to wait for memory. Model multiplexing keeps a fixed number of model slots in memory and swaps models in and out of those slots based on traffic demand. Dynamic loading takes multiplexing further by loading models on demand from a model store and evicting them when demand falls below a threshold.

The decision between topologies is driven by three variables: the number of distinct models, the traffic shape for each model, and the latency budget. A platform with fewer than ten models and sustained traffic on each can run model-per-replica without waste. A platform with hundreds of models and heavy-tail traffic distributions, in which 90% of models receive fewer than 1 request per minute, is a strong candidate for dynamic loading with a result cache in front. Model multiplexing fits the middle ground: enough models that per-replica hosting is expensive, but enough traffic per model that the cold-start cost of dynamic loading would violate latency budgets.

GPU sharing adds another axis. Multi-Instance GPU (MIG) partitioning divides a physical accelerator into isolated logical devices, each with guaranteed memory and compute bandwidth, making it well-suited for small models and tenant isolation. Time-slicing shares a GPU without hard partitioning, which is cheaper to configure but unsuitable for tight tail-latency SLOs.

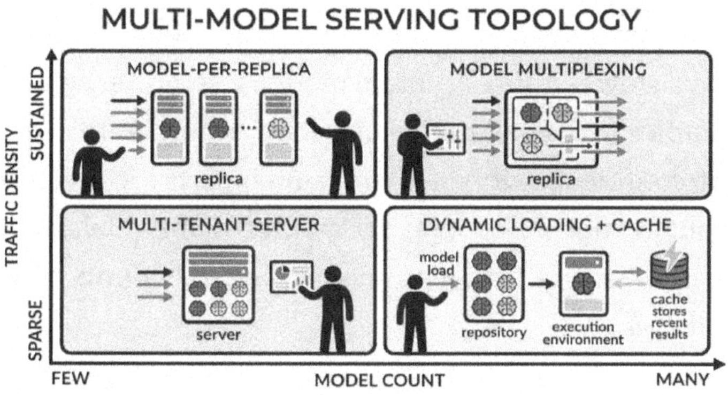

Swap-in and swap-out mechanics require careful design when multiplexing or dynamic loading is in use. A swap involves deserializing model weights from a model store into accelerator memory, which can take seconds for large models. The serving system must hold requests in a queue during the swap, ensure the swap does not cause a memory allocation failure on the device, and report accurate latency to the load balancer so the load balancer does not route additional traffic to the replica during the warm-up period. Serving systems that implement swap-in without these three controls produce latency spikes indistinguishable from hardware failures.

Model cohabitation introduces interference between models sharing the same device. Two models running concurrent inference compete for compute units, cache bandwidth, and memory bandwidth. The interference is not linear: a model that uses 90% of a GPU's compute when running alone can stall at 60% throughput when a second model consumes a disproportionate share of cache bandwidth. Profiling models together, not independently, before assigning them to a shared device is the canonical mitigation.

Ensemble Strategies and Model Composition

A single model rarely covers all the requirements of a production prediction pipeline. Feature-and-rank pipelines retrieve a candidate set using a fast approximate model, then score the candidates with a slower, more accurate model. Cascade ensembles route easy examples to a small model and hard examples to a large model based on confidence scores. Mixture-of-experts architectures route each input to one of several specialized submodels. Each composition pattern has a different latency profile, failure mode, and operational cost.

The cascade pattern reduces average latency and cost: most traffic is handled by the smaller model, and only a minority reaches the larger one. The failure mode is that the confidence threshold is tuned on a distribution that may not match production traffic. When the distribution shifts, the cascade either routes too many requests to the large model (cost spike) or passes too many low-confidence predictions through the small model (quality degradation). Monitoring both the routing rate and per-stage quality is a structural requirement.

Feature-and-rank pipelines introduce serial latency that compounds across stages. A retrieval stage at fifteen milliseconds and a ranking stage at twenty milliseconds produce a P50 of roughly thirty-five milliseconds, but P99 compounds worse because tail latency in each stage adds to the head of the next. Each stage needs its own SLO, its own circuit breaker, and a fallback path that returns a degraded result rather than propagating a timeout.

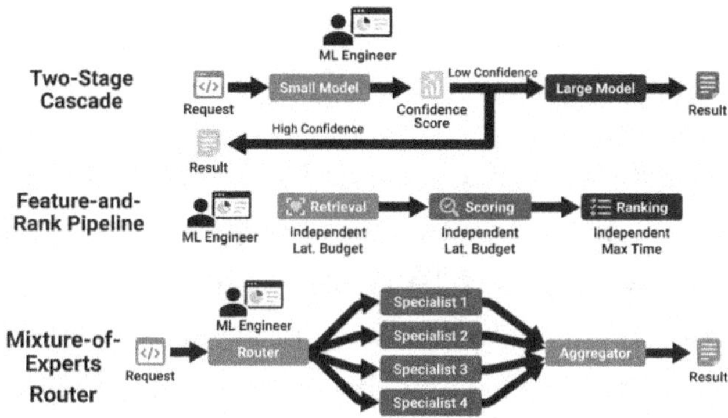

Mixture-of-experts routing at inference time differs from sparse MoE in training (covered in Chapter 10). At the serving level, expert routing means a learned or rule-based router selecting one of several independently deployed models. The router becomes a critical dependency: a failure or latency spike blocks all downstream inference. It must be sized for the aggregate request rate and must define a fallback when the selected expert is unavailable.

Ensemble aggregation, where multiple models score the same input, and their scores are combined, introduces a consistency problem when aggregation logic runs on the client. Each client may implement it slightly differently, producing inconsistent predictions. The canonical fix is to implement aggregation as an explicit serving stage with versioned aggregation logic, deployed and rolled back as a first-class model version.

Latency vs. Throughput Optimization

Latency and throughput are in fundamental tension at the serving layer. A system optimized for throughput batches requests to maximize GPU utilization, adding queuing latency. A latency-optimized system processes each request immediately, minimizing latency at the cost of low utilization. Batching and scheduling patterns navigate this trade-off deliberately rather than picking an extreme.

Static batching collects requests until a fixed batch size B is reached, then dispatches. It is easy to implement but poorly matched to variable traffic. At low rates, each batch takes longer to fill, adding queuing latency proportional to the inter-arrival gap. At high rates, batch sizes may drift away from the model's optimal GPU utilization point. Static batching suits predictable, nearly constant traffic, which is rare in production.

Dynamic batching collects requests up to a maximum batch size B or until a maximum wait time W elapses. This bounds queuing latency, but reduces batch efficiency when traffic is sparse. Dynamic batching outperforms static batching in most conditions but still suffers from long-request blocking: a request generating 1,000 tokens holds a batch slot for many times longer than a request generating 10 tokens, stalling shorter requests that arrive later.

Static vs. Dynamic vs. Continuous Batching Latency Profiles

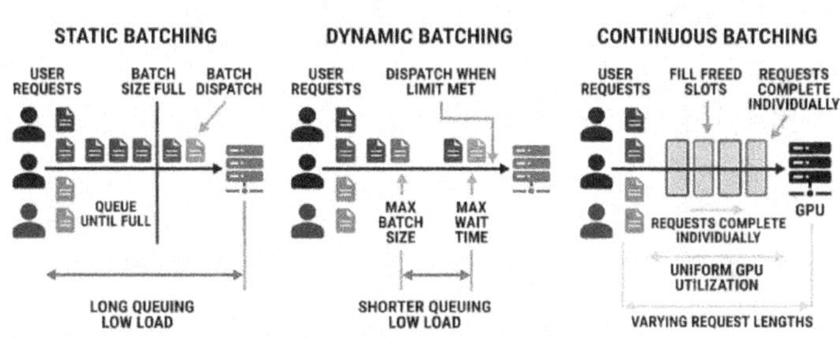

Continuous batching (also called iteration-level scheduling) solves the long-request blocking problem by decoupling the batch from the completion boundary. Instead of treating the batch as a unit that enters and exits the GPU together, continuous batching treats the GPU as a slot pool, where each slot corresponds to a single request in flight. When a request completes its current generation step, its slot is immediately made available to the next request waiting in the queue. The GPU is never idle, waiting for stragglers to finish; throughput approaches the hardware's theoretical maximum. The cost is more complex scheduling logic and additional overhead for managing variable-length sequences in a single kernel dispatch.

Request coalescing applies when many requests share a common prefix. A question-answering system in which all requests begin with the same system prompt can coalesce prefix computation across concurrent requests, rather than recomputing it for each request, thereby requiring only the computation of the diverging suffix for each request.

Admission control prevents the serving system from accepting more work than it can complete within latency budgets. Without it, queued requests grow without bound under overload, producing unbounded latency. An admission control policy rejects requests when estimated queue time exceeds a threshold, trading a predictable rejection rate for a bounded latency guarantee. Rejection responses should carry a retry-after signal to prevent clients from amplifying the overload.

Model Parallelism for Inference

Chapter 2 introduced the canonical parallelism patterns for training: data parallelism, tensor parallelism, pipeline parallelism, and sequence parallelism. These patterns carry over to inference, but with different trade-off profiles because the constraints differ. Training optimizes for throughput on a fixed dataset; inference must minimize per-request latency while maintaining throughput over a continuous stream. This asymmetry changes which parallelism patterns are useful and how they must be configured.

Data parallelism for inference runs multiple independent replicas of the full model, each handling a separate set of requests. Scaling is linear, routing is stateless, and failure isolation is clean. It is the dominant pattern when the model fits in the memory of a single accelerator and does not apply when the model is too large for a single device.

Tensor parallelism for inference shards the weight matrices of each layer across multiple devices in the same node, reducing the per-device memory footprint at the cost of all-reduce communication between devices on every forward pass. The all-reduce becomes the bottleneck when the batch size is small, which is common at low traffic rates. Tensor parallelism is most effective when the batch size is large enough to amortize communication overhead, and is often paired with continuous batching to maintain high batch occupancy. The rule of thumb is to use tensor parallelism within a node (where

NVLink or equivalent high-bandwidth interconnects are available) and avoid crossing node boundaries with tensor-parallel shards.

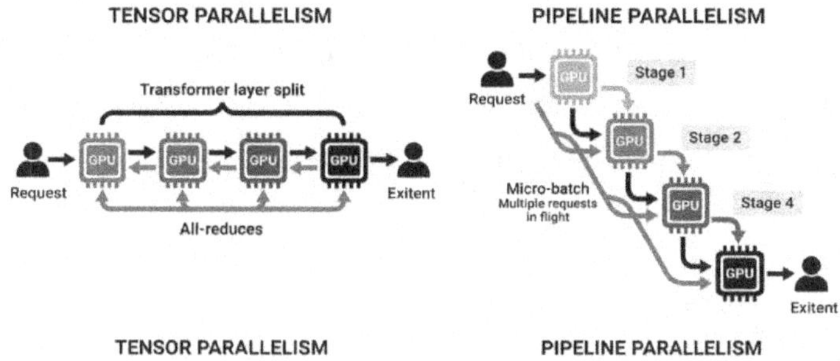

Pipeline parallelism for inference assigns consecutive layers of the model to different devices or nodes, so that different stages of the forward pass run simultaneously on different hardware. Pipeline parallelism reduces per-device memory pressure and can span multiple nodes, making it applicable to models that are too large for tensor parallelism alone. The fundamental cost of pipeline parallelism in inference is pipeline bubble latency: the first stage cannot start processing the next request until the last stage has received the current request's activations, creating idle time at stage boundaries. Micro-batching mitigates the bubble but adds queuing complexity.

Expert routing parallelism is specific to mixture-of-experts architectures. In a sparse MoE model, each token is routed to a subset of expert layers, and those experts may reside on different devices. During inference, the routing decision triggers device-to-device communication to gather the selected expert weights or activations. The serving system must balance expert load in real time; if a popular expert is overloaded, requests queue behind it while other experts are idle. Expert-aware load balancing and capacity reservation for high-demand experts are the canonical mitigations.

Sequence parallelism distributes the attention sequence dimension across devices and is less commonly applied in inference than in training. It becomes relevant when serving models with very long context windows, such as those exceeding 100,000 tokens, where attention computation can account for a significant fraction of total inference time.

The interaction between parallelism strategy and tail latency is the most important consideration for serving. Tensor parallelism reduces per-device memory but adds synchronization latency on every forward pass, which inflates P99. Pipeline parallelism reduces memory pressure and increases throughput, but it adds pipeline-bubble latency, which is visible at P95 and above. Data parallelism with independent replicas yields the cleanest latency profile but is ineffective when the model does not fit on a single device. The canonical recommendation is to prefer data parallelism for models that fit on a single device, apply tensor parallelism within a node when the model does not fit, and use pipeline parallelism across nodes only when the model exceeds the aggregate memory capacity of a single node.

Load Balancing and Request Routing

Generic load balancers distribute requests across replicas without knowledge of the replicas' state. For ML serving, where replicas have varying memory pressure, varying queue depths, and varying warm-up states, a generic load balancer is a latency liability. An accelerator-aware load balancer uses metrics from the serving replicas, such as GPU memory utilization, current queue depth, and active request count, to route each incoming request to the replica best positioned to serve it quickly.

The power-of-two choices algorithm is a simple accelerator-aware routing heuristic. For each incoming request, the load balancer selects two replicas at random and routes the request to the replica with the lower queue depth. This produces a near-optimal load distribution with only two health-check reads per request and no global coordination. It significantly outperforms round-robin under heterogeneous load distributions, where some replicas are temporarily slower than others due to a large batch or a recent model swap.

Accelerator-Aware Load Balancing with Power-of-Two Choices

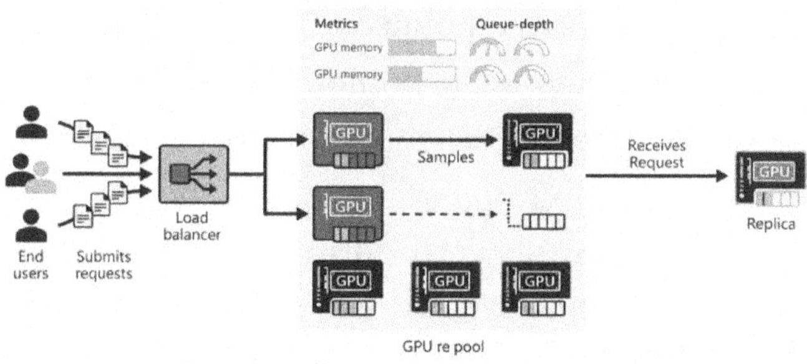

Sticky-session routing is necessary when KV-cache locality matters. A language model serving system that maintains a prefix cache, where previously computed key-value pairs for a shared prompt prefix are stored in GPU memory, benefits from routing requests with the same prefix to the same replica. A round-robin or power-of-two choices policy that ignores prefix affinity defeats the cache. The canonical pattern is to hash the request prefix and use the hash to select a consistent replica, falling back to load-balanced routing when the preferred replica is overloaded.

Weighted routing supports progressive traffic shifts during canary and blue-green deployments. The load balancer maintains a weight vector across replicas or replica groups and routes requests in proportion to those weights. During a canary deployment, the new model version receives a small initial weight (typically one to five percent of traffic); the weight is increased as latency and quality metrics pass guardrails, and the weight reaches 100% when the deployment is declared complete. A rollback reduces the new version's weight to zero and increases the old version's weight correspondingly.

Queue and priority handling become important when the serving system must serve multiple request classes with different latency requirements. A common pattern is a three-tier priority queue: real-time requests from user-facing endpoints have the highest priority, batch inference requests from asynchronous pipelines have the middle priority, and background recomputation or cache warming has the lowest priority. The serving system drains higher-priority queues before dispatching lower-priority requests. Priority queuing without admission control can starve low-priority requests indefinitely under sustained high-priority load; the canonical mitigation is a maximum wait time for low-priority requests after which they are either escalated to a higher priority or rejected with an appropriate response.

Token streaming changes the routing contract for language model serving. A streaming request holds a connection to a specific replica for the duration of a generation, which may be several seconds. A replica with ten streaming connections in progress can be at full capacity even if its instantaneous request rate is low. Load balancers that count only requests per second, without accounting for connection duration over the route to streaming replicas, leading to queue buildup and latency violations.

Caching Patterns for Large Models

Caches for ML serving fall into four categories: result caches, KV caches, embedding caches, and prefix caches. Each operates at a different layer of the serving stack, has different consistency requirements, and produces different efficiency gains. A production serving platform for large models will typically deploy all four, tuned to the workload's specific traffic characteristics.

Result caches store the complete output of a model call, keyed on the input. They are most effective when the input distribution is highly repetitive, such as a question-answering system where a large fraction of users ask the same small set of questions. The cache hit rate for a result cache depends critically on how inputs are canonicalized before the cache key is computed; a query with trailing whitespace should produce the same cache key as the same query without trailing whitespace. Result caches are the cheapest cache to implement and the easiest to reason about, but they offer no benefit for workloads with high input diversity.

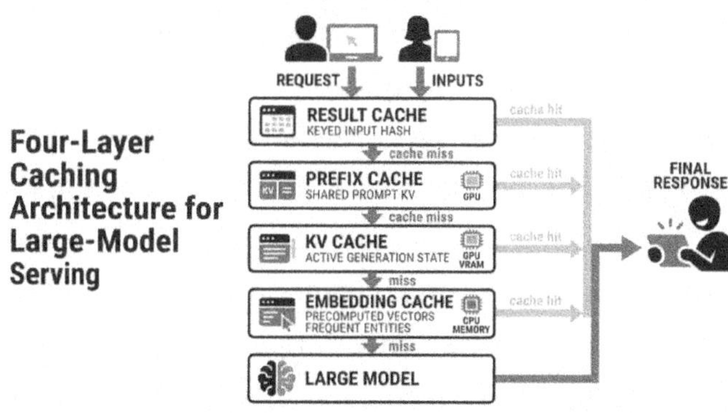

KV caches store the key and value tensors computed for each token during the attention forward pass. Without a KV cache, each new token generated during autoregressive decoding would require recomputing the attention over all previous tokens, making decoding cost quadratic in sequence length. With a KV cache, the key and value tensors for each previously generated token are stored in GPU memory and reused in the next decoding step, reducing per-step cost to linear in the number of new tokens. KV cache management is therefore not a performance optimization for large-model serving; it is a requirement for correctness and feasibility.

KV cache memory pressure drives serving capacity planning. A thirty-two-layer transformer with an eight-thousand-token context window in float16 precision requires approximately three gigabytes of KV cache per concurrent request at full context. A 40-gigabyte replica allocating 24 gigabytes to model weights leaves 16 gigabytes available for the KV cache pool, limiting concurrent requests to roughly 5. Understanding this arithmetic is a prerequisite for fleet sizing.

Paged attention addresses KV cache fragmentation. In a naive KV cache implementation, memory is pre-allocated for the maximum possible sequence length for each request. When actual sequence lengths are shorter, the pre-allocated memory is wasted. When sequence lengths vary widely across concurrent requests, the fixed allocation causes severe fragmentation and limits the maximum number of concurrent requests. Paged attention manages the KV cache as a pool of fixed-size pages, allocating pages to requests on demand and reclaiming them when requests complete. This increases effective KV cache utilization from thirty to fifty percent in typical workloads to seventy to ninety percent, directly increasing the number of concurrent requests a given replica can serve.

Prefix caches extend the KV cache concept to shared prompt prefixes. When many requests share a long system prompt, the KV pairs for that prefix are computed once and reused for all matching requests. Prefix caches are most effective when the system prompt spans hundreds to thousands of tokens and traffic is high enough that the prefix is recomputed multiple times per second without caching. The operational requirement is to invalidate the prefix cache when the system prompt changes, which requires treating the system prompt as a versioned artifact.

Embedding caches store precomputed vectors for frequent entities such as user IDs, item IDs, or common query terms. At scale, embedding table lookups for high-cardinality entities become a bottleneck for retrieval. Covering the top one percent of entities by frequency can reduce embedding table access volume by thirty to fifty percent. Embedding caches require a staleness policy because embeddings of frequently updated entities may change faster than the cache TTL allows.

Real-Time, Batch, and Streaming Inference

Not every prediction needs to be made in real time. Choosing the right serving mode for a workload is a first-order architectural decision that determines infrastructure costs, latency guarantees, and the system's failure surface. The three canonical serving modes are real-time inference, batch inference, and streaming inference, and each is appropriate for a different class of workloads.

Real-time inference delivers predictions in response to individual requests, with latency measured in milliseconds to low seconds. It is the appropriate mode for user-facing features where the result is consumed immediately: search ranking, content recommendation, fraud detection, and question answering. Real-time inference requires a continuously running serving fleet, a load balancer, and a tight latency SLO. It is the most expensive per prediction and the most demanding in operational requirements.

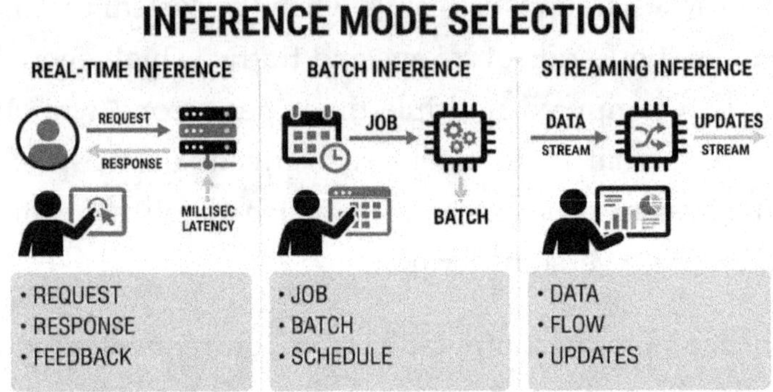

Batch inference collects predictions into scheduled jobs that process large volumes of data offline. It is appropriate to consume predictions asynchronously, such as during nightly recommendation refreshes, daily fraud score updates, or weekly churn propensity score updates. Batch inference decouples serving throughput from serving latency, allowing the system to process far more predictions per GPU-hour than real-time serving because there is no per-request latency overhead, and batch sizes can be maximized. The failure mode of batch inference is data staleness: a model trained on today's distribution runs over yesterday's data and produces predictions that are stale by the time they are consumed. Batch inference requires explicit freshness SLOs in addition to throughput SLOs.

Streaming inference processes events from a continuous data stream and emits predictions as events arrive, with latency measured in seconds to minutes. It is appropriate for workloads that require fresher predictions than batch inference can provide, but cannot justify the cost of a real-time serving fleet. Common examples include session-level recommendation updates, streaming fraud detection, and continuous anomaly scoring. Streaming inference is typically implemented as a consumer of a message queue or event stream, with the serving system consuming events, running inference, and publishing results to a downstream topic.

Mode migration is reversible but not free of charge. Moving from real-time to batch reduces infrastructure cost but requires downstream systems to tolerate stale predictions. Moving from batch to streaming requires a real-time data pipeline, stream-compatible serving infrastructure, and lag monitoring. Moving from batch to real-time requires the most investment: a new serving fleet, a load balancer, latency SLO documentation, and often model retraining. Documenting the mode decision and its assumptions at initial deployment reduces the cost of revisiting those decisions when they change.

Edge and Hybrid Serving

Cloud-only serving is not always the optimal architecture. Latency-sensitive workloads, bandwidth-constrained deployments, privacy-restricted scenarios, and offline-capable applications all benefit from some degree of edge inference. Edge inference runs a model on a device or on infrastructure close to the user, rather than in a centralized cloud cluster, and sends only the results back to the cloud. Hybrid serving combines edge and cloud inference in a topology that uses each tier for the predictions it is best suited for.

Model splitting distributes model layers between the edge device and the cloud. The first N layers run on the edge, producing an intermediate representation that is transmitted to the cloud, where the remaining layers complete inference. This reduces bandwidth compared to transmitting raw inputs and cloud computing compared to running the full model there. The cost is the tight coupling between the two halves, which complicates independent deployment and versioning.

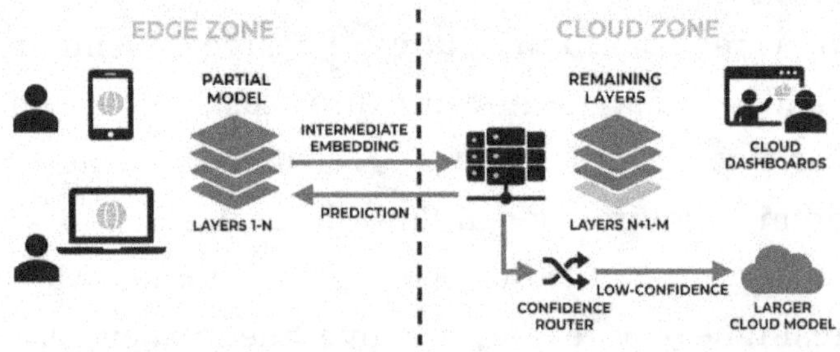

On-device caching handles repeated requests without a round-trip to the cloud. A conversational agent on a mobile device can cache KV pairs for the system prompt and conversation history, reducing each new turn to only the computation for the new user input. Cache invalidation must be consistent with model updates, requiring a reliable over-the-air versioning mechanism.

Confidence-based routing is the canonical hybrid serving pattern. The edge device runs a small, fast model and checks the confidence score. High-confidence results return immediately from the edge; low-confidence requests are forwarded to a larger cloud model. The threshold controls the split between edge and cloud traffic. Setting it too high forces most requests to the cloud; setting it too low accepts low-confidence edge predictions. Calibration requires a labeled dataset representative of edge-device traffic.

Network variability is the dominant operational concern for hybrid serving. An edge device with poor connectivity cannot reach the cloud fallback, causing silent failures or timeouts. The serving system must define explicit degraded-mode behavior: return the edge prediction regardless of confidence, cache the last available cloud result, or surface an appropriate error. Degraded-mode paths must be tested explicitly.

Deployment Patterns: Canary, Shadow, Blue-Green, and Rollback

Safe deployment of model updates is the pattern that determines whether a serving team can ship changes with confidence or whether every deployment is a rehearsal for an incident. The four canonical ML deployment patterns are canary deployment, shadow deployment, blue-green deployment, and rollback. They are not alternatives; they are layers. A mature serving architecture uses all four, with each layer providing a different class of protection.

Shadow deployment runs the new model version in parallel with the current model, receiving a copy of production traffic but not returning predictions to users. Outputs are logged and compared offline. Shadow deployment catches regressions that are invisible in offline evaluation but become apparent under real traffic: latency regressions, shifts in output distribution affecting downstream components, and edge-case error modes absent from test data.

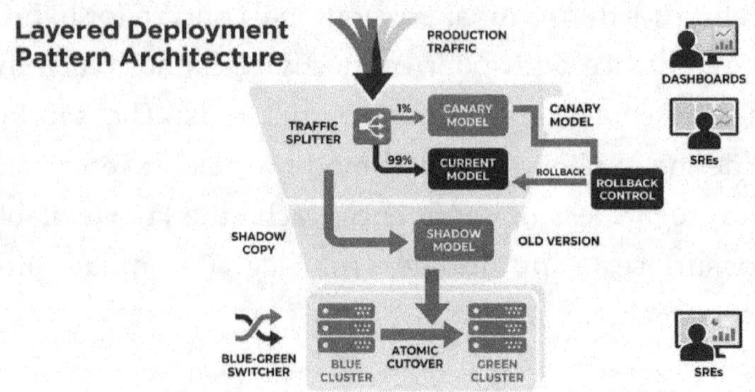

Canary deployment exposes a small fraction of live traffic to the new model version before the full rollout. The canary fraction starts small, typically one to five percent, and is increased progressively as quality and latency metrics pass automated guardrails. The guardrails must be specified before deployment begins, not adjusted during the rollout; post hoc adjustment of guardrails defeats the purpose of the canary stage. The canonical guardrails are a P99 latency bound, an error rate bound, and at least one quality metric that can be evaluated quickly, such as click-through rate or session engagement, rather than a metric that requires days of data collection.

Blue-green deployment maintains two complete serving environments. Traffic cuts over atomically from the current version (blue) to the new version (green) when canary results satisfy the guardrails. The cutover eliminates version mixing, simplifying debugging: every request is handled by one version or the other, never both. The operational cost is the cost of running two full environments simultaneously during the transition. Blue-green is most appropriate when version mixing would cause correctness problems in downstream systems.

Rollback must be a first-class procedure. The canonical rollback pattern uses the same traffic-weight mechanism as canary, in reverse: the new version's weight is set to zero, and the old version's to 100% in a single automated operation. This requires the old model version to remain running during the canary and blue-green stages. Teams that shut down the old version immediately after a successful cutover sacrifice fast rollback capability and must accept the full cold-start cost if a regression emerges later.

Automated guardrails transform rollback from a reactive gesture into a structural requirement. The serving system continuously monitors canary metrics and automatically triggers a rollback when a threshold is breached. The on-call engineer receives a notification that a rollback occurred, not a notification that a metric is degraded. Mean time to recovery drops from the time required to notice, investigate, and act, to under one minute for an automated detection and execution cycle.

Autoscaling and Capacity Planning for Inference

Generic autoscalers designed for CPU-backed stateless services perform poorly on accelerator-backed ML serving for three reasons. First, spinning up a GPU-backed replica is not instantaneous: it requires provisioning an accelerator, pulling a container image, loading model weights into device memory, and running warm-up inference to JIT-compile graphs and fill caches. Cold-start latency for a large model can be on the order of minutes. Second, accelerator capacity in cloud environments is more constrained than CPU capacity; an autoscaler that assumes immediate availability will fail during peak demand. Third, the latency-throughput relationship is nonlinear: doubling replica count does not halve latency when the bottleneck is GPU kernel execution.

Predictive autoscaling addresses the cold-start problem by scaling ahead of demand rather than in response to it. A time-series model trained on historical traffic patterns predicts the traffic rate for the next N minutes and provisions replicas before the demand arrives. The provisioning lead time must exceed the cold-start latency for predictive scaling to be effective; a model with a three-minute cold-start requires a prediction horizon of at least five minutes to account for provisioning time and a safety margin. Predictive scaling is most effective for workloads with regular diurnal or weekly traffic patterns. It is less effective for workloads with irregular spikes, such as social media virality events, where a reactive layer must handle the residual demand.

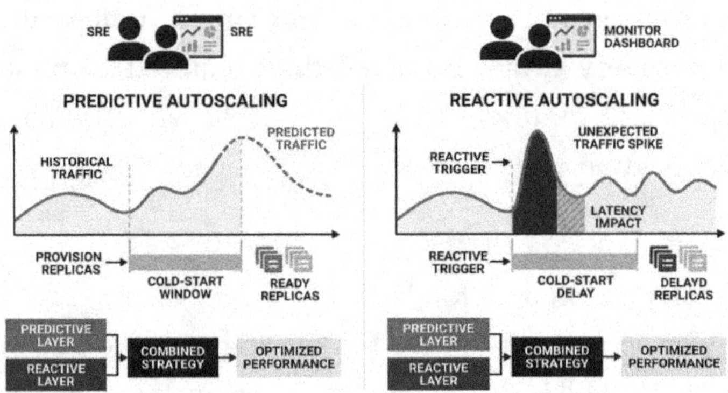

Reactive autoscaling responds to current traffic signals. For ML serving, queue depth is more reliable than raw request rate: it directly measures whether the current replica count is keeping pace with demand, whereas request rate requires a capacity model to translate into replica count. A queue-depth autoscaler that targets a maximum of N queued requests scales up when the queue exceeds N and scales down after the queue has been empty for a stabilization window.

Warmup is a structural requirement for correct autoscaling. A replica that receives live traffic before loading weights and completing warm-up inference will fail requests or produce extremely high latency. The autoscaler must not mark a replica as ready until warm-up completes, and the serving framework must explicitly implement warm-up rather than relying on the first live request to trigger compilation and cache initialization.

Capacity planning requires reasoning about three resources simultaneously: accelerator compute, accelerator memory (weights, KV cache, and activations), and network bandwidth. A fleet is at capacity when any one resource is exhausted, regardless of the others. A common error is sizing for compute utilization while ignoring KV cache memory pressure, producing a fleet that appears underutilized at 50% compute but is at capacity because all KV cache pages are allocated.

Scale-to-zero suits endpoints with sparse or unpredictable traffic, such as internal tools, staging environments, and low-traffic models, across a large catalog. The first request after scale-to-zero incurs full cold-start latency, which is unacceptable for user-facing real-time serving. Critical endpoints that require cost-efficient scaling should maintain a floor of 1 warm replica and scale down to 1, not 0, during low-traffic periods.

Anti-Patterns: Serving Choices That Look Right and Are Not

Serving is where distributed machine learning meets the end user, and it is therefore where a wrong architectural choice is felt most directly: not as a slower training run that finishes a day late, but as a latency regression that a customer experiences on every request. The anti-patterns below share a property that makes them dangerous — each one is locally reasonable. Every one of them was, at some point, a defensible decision under a narrow assumption that later stopped holding. The cost of getting it wrong is not paid once; it is paid per request, continuously, until the underlying choice is reversed.

The most expensive serving anti-pattern is tensor parallelism across nodes for inference. A team that shards a single model's tensors across multiple machines to fit a large parameter count will discover that every token generated now requires a cross-node all-reduce on the critical path, and that cross-node bandwidth, not compute, dominates token latency. The diagnostic signal is the token-generation time, which does not improve when faster accelerators are substituted because the bottleneck was never the accelerator. The fix is to keep tensor parallelism within a single high-bandwidth node and use pipeline or replica parallelism across nodes, accepting a smaller per-node model rather than a larger one whose tokens crawl across the network.

The second anti-pattern is static batching at large batch sizes on a low-latency endpoint. A team optimizing for throughput sets a large fixed batch and a generous batch-collection window, and aggregate throughput does indeed rise — while tail latency collapses, because the first request in a batch waits for the batch to fill before any computation begins. The diagnostic signal is a P99 latency several times the median, while the GPU reports high utilization. The fix is continuous or in-flight batching, which admits requests into a running batch rather than forming a batch and then computing it, decoupling throughput from the latency any single request endures.

The third anti-pattern is a cold KV cache on every request, which treats the cache as per-request scratch memory rather than as a shared resource with a lifetime. Each request recomputes attention over a prefix that an earlier request already computed, and the redundant work scales with prompt length. The diagnostic signal is a prefill cost that grows with traffic that shares common prefixes — such as system prompts and few-shot exemplars — rather than remaining flat. The fix is prefix-aware cache management that retains and reuses shared KV blocks across requests, sizing the cache as a managed pool with an eviction policy rather than allocating it fresh per call.

The fourth anti-pattern is single-model autoscaling driven by aggregate request volume rather than per-model queue depth. When several models share an autoscaling group keyed to total requests, a surge in one model triggers scaling for all of them. At the same time, the aggregate obscures the truly saturated model's own queue. The result is head-of-line blocking: requests to a hot model stall behind a scaling decision that provisioned capacity for the wrong models. The diagnostic signal is rising latency on one model, while cluster-wide utilization looks comfortable. The fix is per-model queue-depth signals that drive per-model scaling, so the saturated model scales while the idle ones do not.

The fifth anti-pattern is canary on volume without canary on cost. A team that promotes a new model version after it passes a volume-based canary — error rate acceptable, latency acceptable — can ship a silent regression in the cost-per-request curve, because a model that is correct and fast can still be markedly more expensive per token. The diagnostic signal is a serving bill that increases after a deployment, for which all quality metrics are approved. The fix is to make cost-per-request a first-class canary metric alongside latency and error rate, and to gate promotion on it, so an economically regressive version is caught before it reaches full traffic.

Anti-Pattern Symptom → Root Cause

A Serving Pattern Catalog

The serving patterns introduced in this chapter can be organized

Symptom (observable)	Likely Root Cause	First Mitigation
Token latency does not improve with faster accelerators	Tensor parallelism spans nodes; cross-node bandwidth dominates	Confine tensor parallelism to one node; use pipeline or replica parallelism across nodes
P99 latency is many times the median at high GPU utilization	Static batching with a large fixed batch and fill window	Switch to continuous/in-flight batching
Prefill cost grows with shared-prefix traffic.	KV cache treated as per-request scratch, recomputed each call	Prefix-aware cache reuse with a managed, evicting pool
One model's latency rises while cluster utilization looks fine	Autoscaling on aggregate volume, not per-model queue depth	Drive per-model scaling from per-model queue-depth signals
Serving bill climbs after a quality-approved deployment	Canary gated on volume and latency, but not cost	Add cost-per-request as a gating canary metric

into a catalog that maps workload characteristics to pattern choices. The catalog is not a rigid decision tree; it is a vocabulary and a starting point for design discussions. Every workload has specific constraints that modify the generic recommendation, and the catalog's value lies in explicitly naming those constraints so the team can debate them.

The catalog begins with the decision on the serving mode. A workload in which predictions are consumed synchronously by a user or a real-time system falls into the real-time serving category. A workload in which predictions are consumed asynchronously, and freshness is measured in minutes to hours, falls into the batch-serving category. A workload in which predictions are consumed as a continuous stream with sub-minute freshness requirements falls into the streaming serving category. When a workload is initially placed in the wrong category, the migration cost is proportional to the extent to which the wrong assumption is embedded in downstream systems.

Serving Pattern Catalog: Workload-to-Pattern Decision Flow

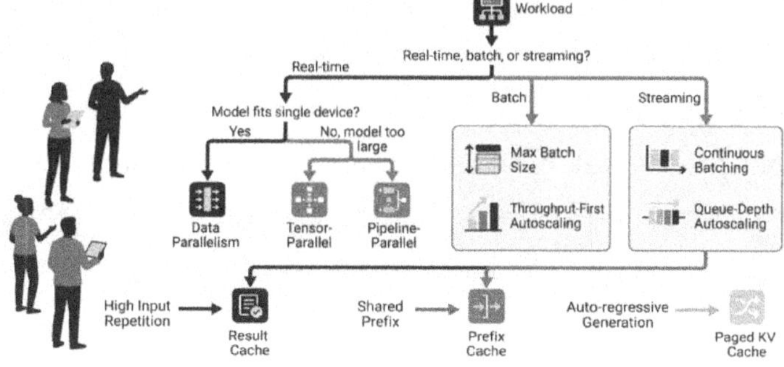

The topology selection follows the mode decision. For real-time serving, the model-per-replica topology is appropriate when the number of models is small (fewer than twenty), and each model has sustained traffic. Model multiplexing is appropriate when the number of models is large, and traffic is unevenly distributed, with some models receiving near-zero traffic for extended periods. Dynamic loading with result caching is appropriate for very large model catalogs where the majority of models are dormant at any given time.

The parallelism pattern follows the model size. A model that fits comfortably in the device memory of a single accelerator should be served with data parallelism across replicas. A model that exceeds a single device's memory but fits within the aggregate memory of a single node should be served using tensor parallelism within the node and data parallelism across nodes. A model that exceeds single-node memory requires pipeline parallelism across nodes, with tensor parallelism within each node where interconnect bandwidth permits.

The batching pattern follows the request rate and latency budget. When the request rate is very low (fewer than one request per second), static or dynamic batching with a generous timeout maximizes GPU utilization, but at the cost of additional queuing latency. When the request rate is high and requests have variable completion times, particularly for auto-regressive language model generation, continuous batching is the appropriate choice. When requests share a common prefix, coalescing the prefix computation and prefix-caching the KV pairs reduces cost without adding latency.

The deployment pattern is not optional. Every model deployment must include a shadow stage for online-offline comparison, a canary stage with automated guardrails, a blue-green or atomic cutover mechanism, and a documented rollback procedure with an automation target of under sixty seconds. Teams that skip any of these layers accept the risk of the opening scenario: a quality regression that remains invisible until users notice it, and a rollback that takes longer than it takes for the incident to escalate.

The autoscaling pattern follows the traffic shape. Regular diurnal traffic benefits from predictive autoscaling when the provisioning lead time exceeds the model cold-start time. Irregular or bursty traffic requires a reactive layer on top of the predictive layer. Sparse or low-priority endpoints use scale-to-zero or scale-to-one economics. In all cases, the autoscaler must be load-tested before it is relied upon for production capacity; an autoscaler that has never been exercised at the scale it needs to reach is not an autoscaler; it is a configuration that has not been validated.

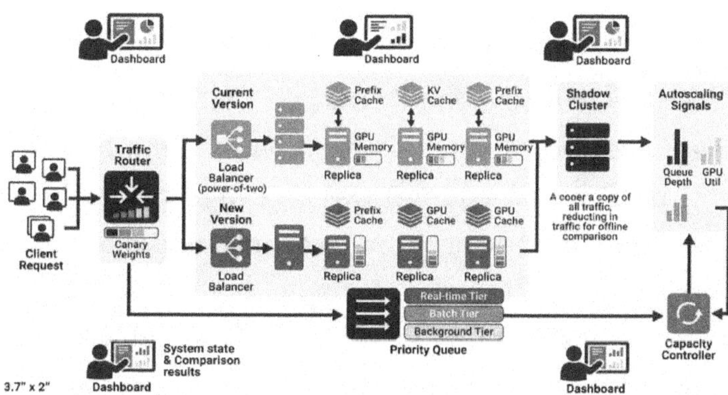

The serving pattern catalog is a living document. Workloads migrate between modes as traffic patterns and quality requirements evolve. New infrastructure primitives change the trade-off surface. The catalog's value comes from the team practice of consulting it when new workloads are designed and updating it when patterns are adopted or retired. Teams that maintain a catalog report faster engineer onboarding, more consistent architecture reviews, and lower frequency of the class of incident that opened this chapter.

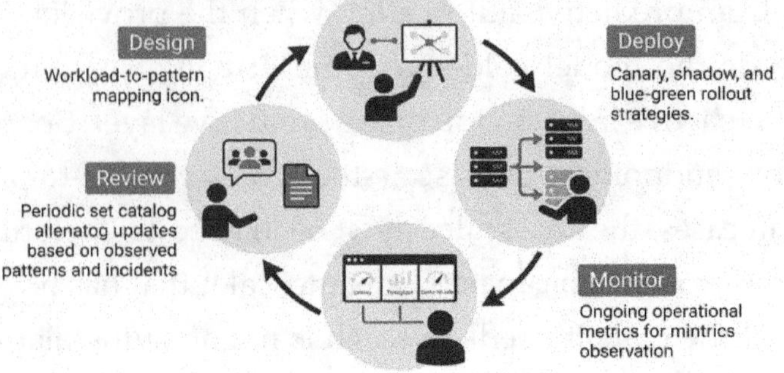

Technical Checklist

- Latency budget is documented at every percentile that matters (P50, P90, P95, P99, P99.9), not only the average.
- The serving topology (model-per-replica, multi-tenant, multiplexing, or dynamic loading) was chosen based on model count, traffic shape, and latency budget, not inherited from a tutorial or a previous project.
- Dynamic or continuous batching is in place where the workload shape supports it; static batching has been explicitly evaluated and accepted or replaced.
- The batching policy has been load-tested at peak traffic to verify that batch fill time does not add unacceptable queuing latency.
- Load balancing is accelerator-aware: the policy uses queue depth, GPU memory pressure, or active request count rather than round-robin or random selection.
- Sticky-session routing is implemented where KV-cache locality matters; the fallback behavior when the preferred replica is overloaded is documented and tested.
- Result, KV, embedding, and prefix caches have each been evaluated against the workload; each has been either adopted with a documented eviction and consistency policy or declined with a documented reason.

- Page attention or equivalent KV cache memory management is in use for auto-regressive models with variable sequence lengths.
- Canary deployment is a structurally required stage in the deployment pipeline, with pre-defined guardrails for latency and quality metrics that cannot be adjusted after the deployment begins.
- Shadow deployment has been used to validate the current model version against at least one production traffic sample before canary promotion.
- Blue-green or equivalent atomic cutover is implemented for deployments where version mixing in downstream systems would cause correctness problems.
- Rollback is automated: a single operator action or automated guardrail breach reduces the new version's traffic weight to zero within sixty seconds, without requiring a manual image swap or rolling restart.
- Autoscaling policies have been load-tested to verify that the scaling trigger fires with sufficient lead time for the model cold-start latency; the cold-start latency is documented.
- Scale-to-zero or scale-to-one policies are applied to non-critical endpoints; critical endpoints maintain at least one warm replica at all times.
- The serving inference mode (real-time, batch, or streaming) is documented with the assumptions behind the choice; a migration plan exists for the case where those assumptions change.
- Edge or hybrid serving deployments define explicit degraded-mode behavior for the case where the cloud tier is unreachable.
- A serving pattern catalog exists, is versioned, and is referenced in design reviews when new serving workloads are introduced.

- Priority queues are in place where the serving system must handle multiple request classes with different latency requirements; the queue starvation policy for low-priority requests is documented.

Team Conversation

11. Which of our serving topologies were chosen deliberately based on model count, traffic shape, and latency budget, and which ones did we drift into because that was how the first model was deployed?
12. When was the last time a model deployment was rolled back? How long did the rollback take from decision to restored traffic, and was that time consistent with our documented rollback SLO?
13. Where in our serving stack would a single hot replica still be able to take down a launch, and what would it take to remove that single point of failure?
14. Are we caching everything we could be caching at the application layer (result cache), the model layer (prefix and KV cache), and the request layer (embedding cache)? What is preventing us from adopting the caches we skipped?
15. Do our autoscalers respect accelerator constraints, including cold-start latency and device memory capacity, or do they thrash when traffic shifts unexpectedly?
16. Have we ever moved a workload from real-time to batch or streaming inference? If yes, what did we learn about the migration cost and the assumptions we had to change? If no, which current real-time endpoints are candidates for that move?
17. Which deployment pattern, shadow, canary, or automated rollback, would have caught our most recent serving incident

before the customer saw it, and is that pattern now structurally in place?

18. How confident are we that our latency SLOs reflect what users actually experience, including queuing time, cold-start time, and streaming token latency, rather than just the model's execution time in isolation?
19. If our largest model doubled in parameter count overnight, which parts of our serving architecture would fail first, and how would we know before the failure became a customer incident?
20. Does our team have a shared vocabulary of serving patterns that we use consistently in design reviews, or do we reinvent the same topology decisions repeatedly for each new workload?

Key Takeaway

Serving is the layer where ML meets the rest of the engineering organization, and it is the layer where the largest gap exists between how teams think about model quality and how users experience it. A model that performs well in offline evaluation but is slow, unreliable, or unsafe to deploy in production is not a production model. The patterns in this chapter, from batching and caching to deployment safety and autoscaling, are the infrastructure that closes that gap. They are not advanced topics for specialist serving teams; they are the baseline for any team shipping ML to users at scale.

The latency-throughput curve is bent by patterns, not by hardware. Together, continuous batching, paged attention, prefix caching, and accelerator-aware load balancing can increase serving throughput by an order of magnitude and reduce per-prediction cost by a comparable factor on the same hardware used by a naive serving implementation. The same is true for deployment safety: canary, shadow, and automated rollback do not require additional infrastructure investment than an ad hoc deployment pipeline, but they transform the risk profile of every model update from an uncontrolled experiment into a mission-aligned, responsible-by-design procedure.

The serving pattern catalog is the most leveraged artifact a serving team can maintain. It encodes the team's collective knowledge of which patterns apply to which workloads, documents the trade-offs behind each decision, and provides new engineers with a map of the serving architecture before they touch it. A catalog that is maintained, consulted, and updated as the workload evolves turns the tribal knowledge of the most experienced engineers into shared operational clarity for the whole team. That is the goal of this chapter and the goal of the book: patterns that travel with the reader to any framework, any cloud, and any team, because the patterns are the spine, and the frameworks are just the current illustration.

8 Resource Management and Cost Optimization

Opening Scenario

The quarterly business review started normally enough. The platform engineering lead pulled up the cloud cost dashboard, expecting the familiar upward slope the team had been explaining away for six months. What she found instead was a near-vertical line. The ML organization now owned 63% of the company's infrastructure bill, up from 41% two quarters earlier. The CFO's first question was simple: What are we getting for it? The second question was harder: Can you show me the number by team and by workload? Neither question had a clean answer.

A spot check of the primary training cluster told the story. Average GPU utilization over the past thirty days was 31%. Two research teams had standing reservations for 40 nodes each; combined, they had consumed fewer than 8 node-days of actual compute in the quarter. The autoscaler configuration, which no one could clearly point to as their responsibility, had quietly doubled the serving fleet's minimum replica count during a high-traffic incident three months earlier and had never been rolled back. Storage costs for intermediate checkpoints were growing at twice the rate of model count because the retention policy was aspirational rather than enforced. Egress from cross-region dataset replication was invisible in the bill because it rolled up under a generic networking line item that no ML engineer reviewed.

The platform team spent the next two weeks building a cost-attribution model from scratch. They discovered that roughly 35% of the current bill could be recovered using patterns the team already knew: right-sizing, spot-instance migration for fault-tolerant training jobs, enforced quotas, and a minimal showback dashboard that made idle reservations visible to the teams holding them. None of the required changes were exotic. All of them had been postponed in favor of feature work.

The scenario is composite, but its shape repeats across organizations of every size. The pattern is always the same: cost grows faster than value because resource management is treated as an operational detail rather than an engineering discipline. This chapter is the canonical reference for the patterns that close that gap, covering the full stack from accelerator utilization to FinOps governance without anchoring to any single tool or cloud provider.

The goal is operational clarity at every layer of the resource stack: clear ownership, measurable utilization, predictable spend, and a quarterly tactics catalog that converts cost awareness into shipped improvements. Cost optimization is not a one-time audit; it is a recurring engineering practice with the same rigor as reliability engineering or security posture.

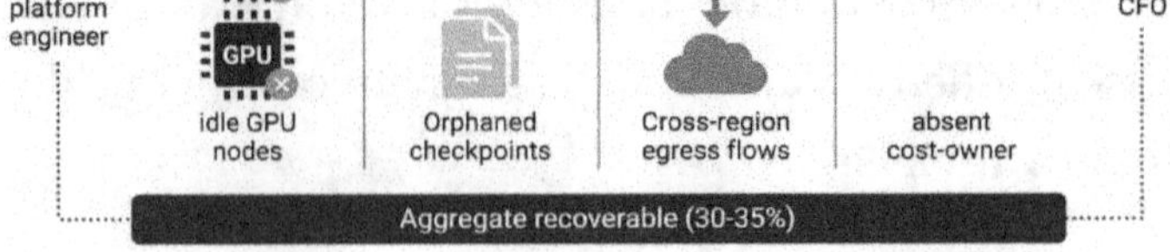

The Cost Leak Anatomy of a Distributed ML Platform

Why It Matters

Distributed ML compute is one of the few line items in software engineering where the marginal cost of a bad decision is measured in thousands of dollars per day. A poorly configured training job that wastes forty percent of its allocated GPUs does not produce a bug report; it produces a cloud invoice. Because the feedback loop between engineering decisions and financial outcomes is long and indirect, resource management is systematically underinvested until the bill becomes politically visible. By that point, the patterns that would have prevented the problem are retrofitted under pressure, and the recovery cost is far higher than the prevention cost would have been.

The workflow-level impact is not limited to finance. Idle accelerators are accelerators that other teams cannot use, which means poor resource management directly degrades the pilot-to-production pathway for every downstream team. A researcher who cannot get a GPU allocation for three days is a researcher whose experiment runs three days later, whose insights arrive three days later, and whose model ships three days later—the latency compounds across teams and quarters. Resource management is not a FinOps concern bolted onto the platform; it is a product concern that determines how quickly the organization can move.

This chapter is mission-aligned in a specific sense: it equips the platform team to answer the questions that every engineering organization eventually faces. How much does it cost to train this model? Who is responsible for that cost? What would it take to cut the bill by twenty percent without disrupting the research roadmap? These are engineering questions with engineering answers, and the patterns in this chapter make those answers auditable, repeatable, and defensible.

GPU and Accelerator Utilization: MFU, Occupancy, and Idle Time

Every cost-optimization conversation in distributed ML starts with utilization, because it is the ratio that converts hardware cost into value delivered. Three distinct metrics together describe how well a workload uses its allocated accelerators, and conflating them is a common source of misdiagnosis.

GPU utilization, as reported by most monitoring tools, is a binary occupancy metric: it reports the fraction of time the GPU's compute units are executing any kernel. A utilization reading of ninety percent sounds healthy. Still, it can coexist with catastrophically low throughput if the kernels being executed are memory-bound, poorly shaped, or waiting on synchronization barriers. Utilization is necessary but not sufficient as a health signal.

Model FLOP Utilization (MFU) is the correct primary metric for training workloads. MFU is the ratio of the actual floating-point operations performed per second to the theoretical peak FLOP/s of the hardware, expressed as a percentage. A well-tuned large model training job on modern accelerator hardware should achieve MFU in the range of forty to sixty percent. MFU below twenty-five percent almost always indicates a structural problem: data starvation, poor kernel selection, excessive host-device synchronization, or batch sizes that do not fill the hardware's computational tiles. MFU is the metric that connects hardware procurement decisions to engineering outcomes.

Idle time is the third axis. Unlike occupancy and MFU, which measure compute quality. At the same time, the job runs, idle time measures the intervals when the GPU is allocated but doing nothing: waiting for data from a checkpoint restore, blocked on a distributed barrier, or sitting between job submissions in a multi-step pipeline. Idle time is the simplest form of waste to measure and often the simplest to recover. A cluster where GPUs are idle for 15% of the wall-clock time effectively pays for 15% more hardware than it uses.

The pattern for addressing utilization problems follows a diagnostic sequence. First, measure idle time at the cluster level using utilization histograms with one-minute granularity or finer. Idle intervals exceeding 5% of wall time are actionable. Second, measure MFU for each job class (training, fine-tuning, distillation) and set a floor target for it. Third, investigate occupancy only when MFU is below target despite low idle time, which points to a compute-efficiency problem rather than a scheduling problem. This sequence ensures that engineering effort is applied to the highest-leverage problem first.

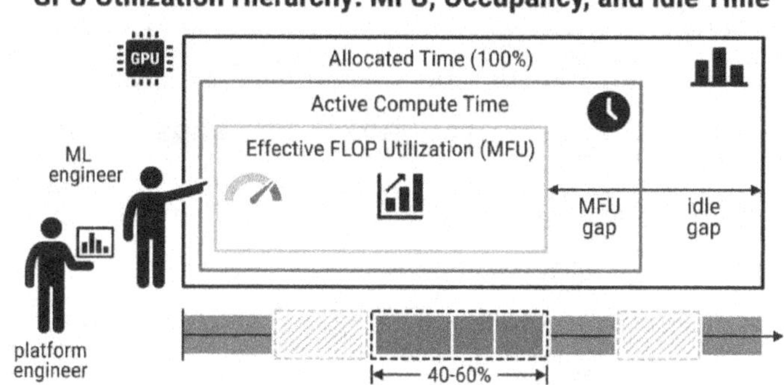

Cluster Sizing Methodology: Right-Sizing from First Principles

Cluster sizing in most organizations is not a methodology; it is a habit. Teams request what they requested last quarter, scaled up by a gut-feel growth factor. The result is a cluster that is simultaneously too large (in steady state) and too small (at peak), because the requests are not grounded in measurements of actual workload shape. Right-sizing means replacing habit with a repeatable analytical process.

The sizing methodology has four steps. The first step is workload profiling: for each workload class on the cluster (pre-training, fine-tuning, batch inference, online serving, evaluation), measure the GPU-hours consumed, the distributions of wall-clock durations and inter-arrival times, and the peak parallelism required. These measurements should be taken over at least four weeks of representative traffic to capture weekly cycles. Profiling is the input to every subsequent decision.

The second step is peak-versus-steady-state analysis. Distributed ML workloads are highly bursty: a team submits a large training run, saturating the cluster for hours or days, then the cluster sits idle while the team analyzes results. Sizing to the peak demand of the busiest concurrent scenario produces a cluster that is idle for most of its life. The correct approach is to size to the eighty-fifth or ninety-fifth percentile of demand (depending on the cost of queuing delay) and then handle the remaining peak with burst capacity, spot instances, or a reservation held by a separate pool.

The third step is headroom modeling. Every cluster needs headroom: unallocated capacity that absorbs job startup spikes, prevents fragmentation from making large jobs unschedulable, and provides a safety margin for unexpected demand. A headroom rule of fifteen to twenty percent is a reasonable starting point for most training clusters. Serving clusters, which have latency-sensitive workloads, typically require twenty to thirty percent headroom to absorb traffic spikes before autoscaling can respond. The headroom target should be documented and justified rather than assumed.

Right-sizing is not a one-time exercise. Workload mix evolves as the ML organization matures, and a cluster sized for a predominantly training workload will be systematically wrong once inference and fine-tuning workloads dominate. The sizing methodology should be treated as a living artifact reviewed each quarter alongside the cost attribution report. Teams that do this consistently find that the savings compound: a correctly sized cluster is also easier to fill with spot instances, schedule efficiently, and attribute costs to.

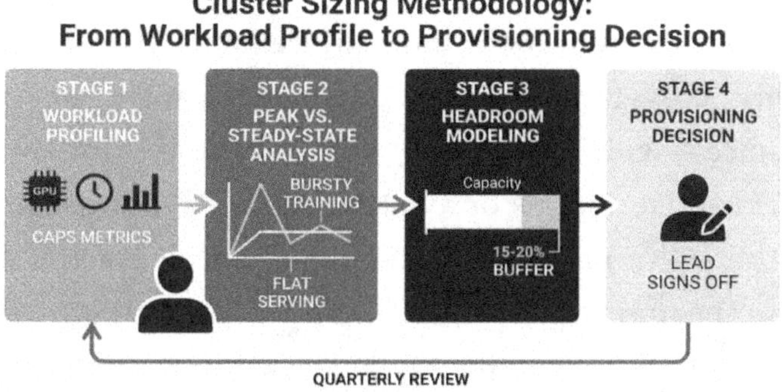

Spot and Preemptible Economics with Checkpoint Amortization

Spot instances (called preemptible VMs on some clouds) are compute capacity offered at a significant discount, typically sixty to ninety percent below on-demand pricing, in exchange for the provider's right to reclaim the instance with short notice, usually two minutes or less. For distributed ML training workloads, spot instances represent the largest single cost reduction available without changing the model architecture or training algorithm. The pattern for exploiting them is checkpoint amortization.

The core economics work as follows. A training job that runs for twenty hours on on-demand instances has a fixed cost. The same job on spot instances, assuming an average spot discount of 70%

and an average preemption rate of 1 interruption every 6 hours, will experience 3 or 4 preemptions over its total wall-clock duration. Each preemption triggers a restart from the most recent checkpoint. If checkpointing is performed every 30 minutes, the maximum rework per preemption is 30 minutes of compute time. At a seventy percent discount, the spot job can absorb several preemptions and still be cheaper than the on-demand alternative. The break-even analysis is explicit and auditable.

Mixed-instance pools extend the spot pattern to reduce the fragility of any single instance type becoming unavailable. A pool that requests capacity across three or four compatible instance types (for example, instances with equivalent GPU memory but from different hardware generations or availability zones) dramatically reduces the probability of a simultaneous preemption across all pool members. The training job must be able to reconfigure its topology on restart to match the available instance mix, which requires a coordinator pattern that separates the job topology from the hardware it runs on.

Multi-region and multi-zone arbitrage is an extension of the mixed-instance pool pattern. Spot pricing varies by region and zone because it reflects local supply and demand for compute capacity. A platform that monitors spot pricing across regions and submits training jobs to the lowest-cost region with sufficient capacity can achieve additional savings beyond the base spot discount. The operational cost of this pattern is higher, primarily because cross-region data transfer fees and latency must be accounted for. Still, for long-running training jobs where the dataset is already in object storage, the arbitrage is frequently worthwhile.

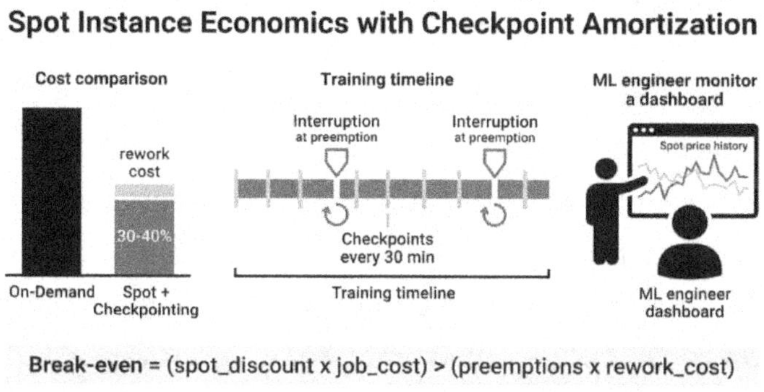

Bin Packing, MIG Partitioning, and Resource Fragmentation

Resource fragmentation is the distributed ML equivalent of heap fragmentation in memory management: the cluster has enough total capacity to run a job, but that capacity is scattered across nodes in shapes that cannot be assembled into the contiguous allocation the job needs. Fragmentation is invisible in aggregate utilization metrics, which is why it is chronically underestimated. A cluster reporting 70% utilization can simultaneously be unable to schedule a job that requires 16 GPUs because no single node has more than 4 GPUs available.

Topology-aware bin packing is the primary pattern for controlling fragmentation. The scheduler, rather than placing jobs on any available nodes, preferentially places jobs on nodes that minimize the fragmentation they leave behind. A job requiring 12 of a node's 16 GPUs is scheduled before a job requiring 4 of 16, leaving the node fully occupied rather than partially occupied. Combined with gang scheduling (where all workers in a distributed job start simultaneously or none start), topology-aware packing dramatically reduces the probability that a cluster reaches a state in which capacity exists but cannot be assembled.

GPU partitioning via Multi-Instance GPU (MIG) technology, available on modern accelerator generations, allows a single physical GPU to be divided into up to seven isolated instances, each with its own memory partition and compute slice. MIG is the correct pattern for serving workloads, inference-time evaluation jobs, and lightweight fine-tuning runs where the model is small enough to fit in a partition. A cluster that runs eight small serving replicas per physical GPU, rather than one replica per GPU, achieves dramatically better hardware utilization at the cost of some operational complexity in partition management.

The right-sizing of resource requests is the upstream complement to bin packing. Jobs that request more CPU, memory, or GPU memory than they actually consume create phantom reservations that prevent other jobs from being scheduled into the unused space. The pattern is to measure actual resource consumption per job class over four or more representative runs, set requests to the ninety-fifth percentile of observed consumption, and set limits to a generous ceiling above requests. Requests that are set to limits without an empirical basis are a common source of fragmentation.

Quantifying the cost of fragmentation converts it from an abstract concern to a trackable metric. The fragmentation rate of a cluster is defined as the ratio of unschedulable capacity (the capacity on nodes that cannot be assembled into a schedulable allocation) to the total capacity. A fragmentation rate above 10% signals that bin-packing and right-sizing improvements are warranted. The platform team should publish this metric alongside utilization in the efficiency dashboard, because utilization without fragmentation rate gives a misleadingly optimistic picture of cluster health.

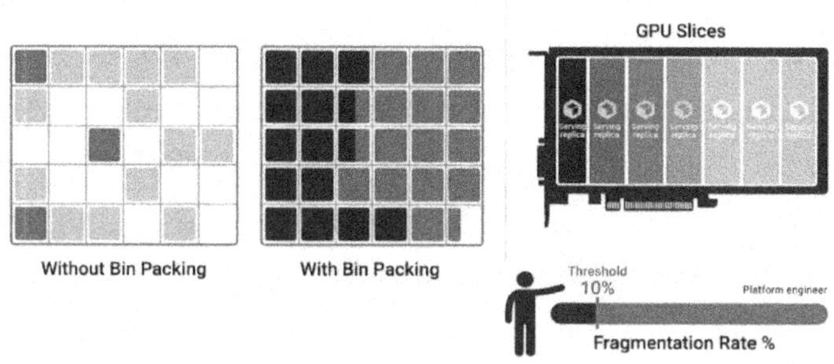

Capacity Planning, Reservations, and Mixed-Capacity Pools

Capacity planning is the practice of aligning the organization's committed infrastructure spending with its expected workload over a planning horizon, typically 1 to 3 years. Done well, it converts a reactive cloud bill into a predictable cost structure with significant discounts. When done poorly, it locks the organization into a capacity that does not match the workload it eventually faces.

The capacity mix pattern divides the fleet into four tiers based on cost and commitment levels. Reserved instances (or committed-use discounts on cloud providers that prefer that term) are purchased one to three years in advance at discounts of thirty to sixty percent over on-demand. They are appropriate for the steady-state serving fleet, for any training workload with a predictable weekly cadence, and for infrastructure components that are always-on regardless of workload (control planes, monitoring, storage). On-demand instances cover the unpredictable component of the workload and provide headroom. Spot instances cover fault-tolerant training and batch inference. A fourth tier, capacity reservations without usage commitment, is available on some cloud providers and guarantees availability without a cost commitment, which is valuable for predictable-but-infrequent large training runs.

The training-versus-serving fleet-sharing pattern reduces the total fleet size by allowing training jobs to use serving capacity during off-peak serving hours, and vice versa. This requires the scheduler to understand both job classes and to preempt training jobs when serving latency thresholds are at risk. The pattern works well when training jobs are checkpoint-safe (so preemption is cheap) and when serving traffic has a clear diurnal cycle. It is less appropriate when serving SLOs are tight or when training jobs have hard deadlines.

Multi-region and multi-zone capacity planning adds a geographic dimension. Cloud providers price compute differently by region, and capacity availability varies significantly for new accelerator generations in the months after launch. Organizations that are willing to run training workloads in secondary regions often find significant price arbitrage during high-demand periods. The constraint is data locality: moving a large dataset across regions to exploit a price difference is worthwhile only if the egress and transfer costs are lower than the price difference. Egress costs are a hidden component of cloud bills that often undermine naive arbitrage strategies.

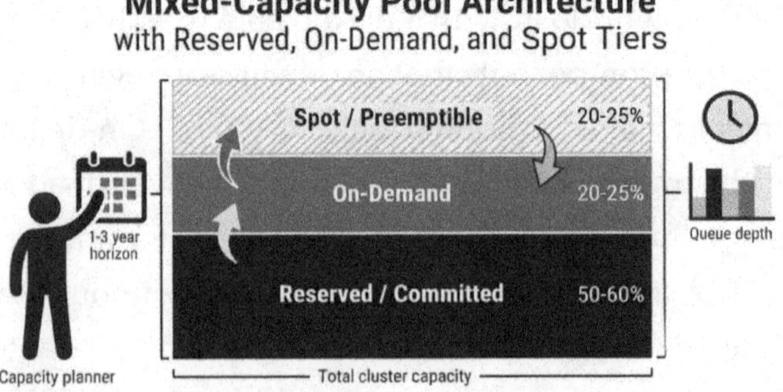

Multi-Tenant Governance: Quotas, Priorities, and Fair Scheduling

A shared ML cluster is not a neutral resource. Without explicit governance, the team with the most experience submitting jobs, the most aggressive retry logic, or the most political capital will consume a disproportionate share of the cluster. Fair scheduling is the engineering discipline that converts "fair" from a social negotiation into a definition that the platform team can defend with numbers.

Namespace-level quotas are the first line of defense. Each tenant (team, project, or cost center, depending on the attribution model) is assigned a quota for GPU count, CPU, memory, and storage. Quotas have two components: a guaranteed minimum that the scheduler will always honor, and a burst ceiling above the guaranteed minimum that the tenant can use when capacity is available. The gap between the guaranteed minimum and the burst ceiling is the slack that the scheduler uses to improve overall utilization by lending unused capacity to other tenants.

Hierarchical quotas extend the flat namespace model to organizations with multiple teams under a single organizational unit. A business unit quota sets the aggregate limit; team-level quotas partition that limit among teams in the unit. This structure allows cost attribution at multiple organizational levels without requiring every chargeback conversation to reach the cluster level. Hierarchical quotas are harder to implement but dramatically reduce the political cost of quota enforcement, because each level of the hierarchy enforces only within its own span of control.

Priority classes partition workloads by urgency and consequence of delay. A typical four-tier model uses the following structure:

production inference serving at the highest priority (never preempted), production training at the next level (preempted only by serving), research training at the third level (preempted by either production tier), and batch evaluation or exploratory runs at the lowest priority. The priority model should be documented, communicated to all tenants, and enforced by the scheduler rather than by convention.

Dominant Resource Fairness (DRF) is the theoretical foundation for multi-resource fair scheduling. DRF extends max-min fairness from a single resource to multiple resources by allocating shares based on each tenant's dominant resource (the resource for which their share is largest). A team whose jobs are GPU-heavy is allocated based on their GPU share; a team whose jobs are memory-heavy is allocated based on their memory share. DRF prevents the pathological case in which one tenant monopolizes a scarce resource by requesting large amounts of a cheap resource in parallel.

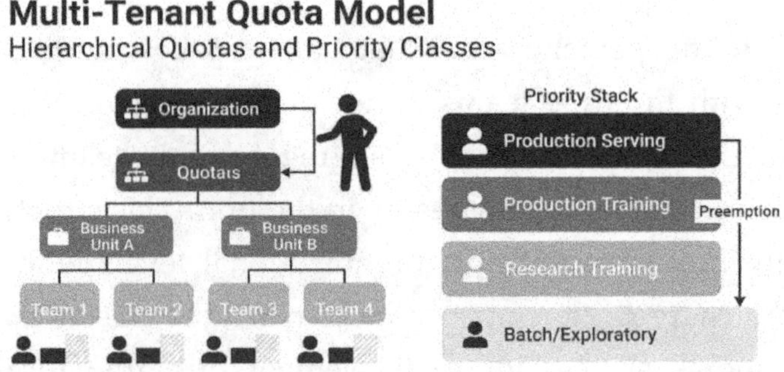

Cost-Aware Autoscaling Patterns

Generic autoscalers are designed for stateless web services with fast startup times and linear scaling relationships. Accelerator-backed ML workloads violate every one of those assumptions. GPU nodes take minutes to start, not seconds. Training jobs are stateful and cannot be trivially scaled out during a run. Serving workloads have non-linear latency profiles because batching dynamics vary with replica count. A generic autoscaler applied to these workloads produces the exact failure mode described in the opening scenario: a minimum replica count is doubled during an incident and never rolled back.

Queue-driven autoscaling is the correct pattern for training workloads. Rather than scaling based on CPU or memory utilization, the autoscaler monitors the depth and age of the job queue. When jobs are waiting for more than a configurable threshold, new nodes are provisioned. When the queue is empty, and existing nodes have been idle for a configurable minimum period, nodes are deprovisioned. The key parameters are the queue-depth trigger, the idle-before-scale-down interval, and the maximum scale-out rate. These parameters should be tuned empirically using production traffic data rather than default values.

Latency-driven autoscaling is the correct pattern for serving workloads. The autoscaler monitors p95 or p99 inference latency and scales out when latency exceeds the SLO threshold, scales in when latency is well below the threshold and utilization is low, and scales in when latency exceeds the SLO threshold and utilization is low. Because GPU serving has batching dynamics (throughput increases up to the optimal batch size, then degrades), the latency signal must be measured at the inference server level rather than at the load balancer. Request-rate autoscaling is simpler to implement but less precise because it does not account for changes in request complexity or model batch size.

Predictive autoscaling extends reactive autoscaling with a time-series forecast of expected demand. For serving workloads with strong diurnal patterns (most consumer-facing ML services), a simple regression or seasonal decomposition model trained on historical traffic can predict demand one to four hours ahead, allowing the cluster to scale out before demand arrives rather than after latency degrades. Predictive autoscaling requires at least four to eight weeks of historical data to be reliable and must include a reactive fallback for unexpected demand spikes.

Deadline-driven autoscaling applies specifically to training jobs with business deadlines: a model that must be ready for a Monday deployment must finish training by Friday. The autoscaler provisions enough parallel workers to meet the deadline, given the current progress, adjusting the worker count as the job runs. This pattern is more complex than queue-driven autoscaling because it requires estimating training throughput per worker count, which must be measured per job class. The operational benefit is that deadline-driven autoscaling converts a binary question (will this finish in time?) into a continuously managed parameter and naturally integrates with spot instances by increasing the worker count when spot prices are low.

Scale-to-zero is the aggressive end of the autoscaling spectrum. For clusters that support it, a scale-to-zero policy removes all worker nodes during extended idle periods (nights, weekends, long public holidays). It recreates them when the next job arrives. The cost savings are proportional to the idle fraction of the week; for a cluster that is idle 40% of the time, scale-to-zero delivers a 40% reduction in compute cost before any other optimization. The trade-off is a cold-start latency on the first job after a scale-to-zero event, which must be acceptable to the teams using the cluster. For research clusters with flexible time horizons, it usually is.

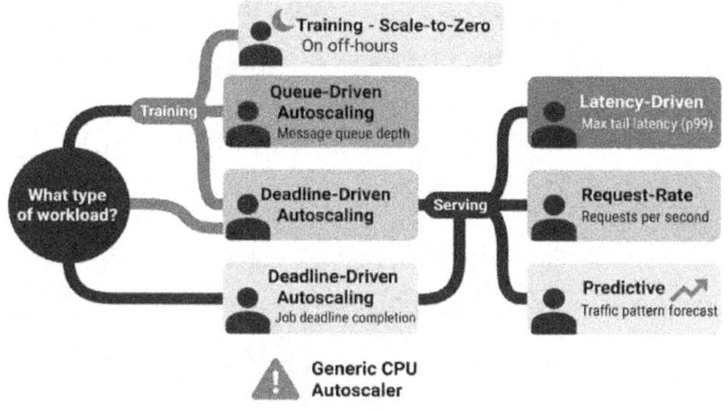

FinOps for ML: Showback, Chargeback, and Cost Attribution

FinOps is the practice of bringing financial accountability to cloud infrastructure decisions. In a distributed ML context, it means converting a single cloud bill into per-team and per-workload cost signals that engineers and managers can act on. FinOps is not an external discipline imposed on the engineering team; it is an engineering responsibility, and the patterns that make it work are engineering patterns. The platform team is the natural owner of FinOps for ML because it controls the attribution infrastructure.

Showback and chargeback are the two primary attribution patterns, and they are not interchangeable. Showback is visibility without financial consequence: each team sees its cost allocation, but the money does not move. Showback is the correct starting pattern for organizations that lack cost-attribution infrastructure, because it builds the measurement foundation and the behavioral feedback loop (teams reduce waste when they can see it) without requiring the organizational agreement needed for actual financial transfers. Chargebacks are visible and carry financial consequences: each team's actual cloud spend is transferred to their budget. Chargeback is more powerful but requires a higher level of measurement accuracy, organizational alignment, and conflict-resolution process. Showback is the floor; chargeback is the ceiling.

Cost attribution requires tagging infrastructure at a granularity that matches the organization's accountability structure. Every job submitted to the cluster must carry tags for team, project, workload type, and priority tier. Every storage bucket and volume must carry the same tags. These tags propagate through the billing system and become the dimensions of the cost attribution report. Organizations that implement tagging after the cluster is built spend weeks untangling mixed-tenant resources; organizations that enforce tagging as a scheduling prerequisite never face this problem.

Unit economics are the metric layer above raw cost attribution. A cost attribution report that says 'Team A spent $43,000 last month' does not tell a team lead whether that number is good or bad. Unit economics connect spending to value produced: cost per training step, cost per thousand predictions, cost per model trained. These metrics are the language of a productive FinOps conversation because they make cost comparable across teams with different model sizes and training durations. A team whose cost per prediction is falling while its prediction volume is rising is doing the right things, even if its absolute cost is growing.

The cost owner per workload is the organizational pattern that makes FinOps durable. Every production workload should have a named cost owner: a specific engineer or team lead who receives the cost attribution report for that workload and is responsible for managing it against the unit-economics target. Without a named owner, cost visibility produces awareness without accountability. With a named owner, visibility produces action. The cost owner should be the same person who owns the workload's production

SLOs, because cost and reliability are not separable concerns in a distributed ML system.

The forecast-and-actuals cadence is the operational heartbeat of FinOps for ML. On a monthly or quarterly basis, each workload's cost owner forecasts expected spending for the next period, compares actuals from the previous period to the prior forecast, and documents the delta. Forecasts that consistently miss by more than twenty percent in either direction indicate that the workload's scaling behavior is not well understood. The cadence creates the social accountability structure that makes showback and chargeback sustainable by converting cost management from an annual audit into a recurring engineering practice.

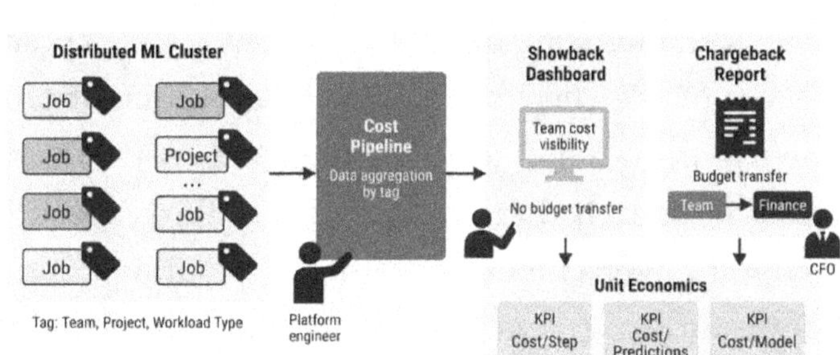

Hidden Costs: Egress, Storage, and Idle Clusters

The visible components of the cloud ML bill are compute and accelerator costs. The invisible components are often 10 to 20 percent of the total and receive almost no engineering attention: network egress, storage growth from unmanaged checkpoints and datasets, and the cost of idle but allocated clusters. These hidden costs are not inherently difficult to control; they are difficult to see. Once instrumented, most of them yield to simple policies.

Network egress fees are charged when data crosses cloud provider boundaries, such as between regions or between cloud providers, or when data flows from the cloud to the internet. In a distributed ML context, the primary egress sources are cross-region dataset replication, model artifact export for on-premises serving, and experiment-tracking data sent to external services. Egress fees often roll up under a networking line item that ML engineers do not review, which is why they are missed. The first step is to break out egress by origin workload in the cost attribution report. The second step is to evaluate whether the egress is necessary: many cross-region dataset copies were created for a specific experiment and never deleted.

Storage cost grows monotonically unless actively managed. The two primary drivers in a distributed ML platform are model checkpoints and intermediate dataset versions. Checkpoints accumulate at a rate of the checkpoint interval multiplied by the number of concurrent training jobs. A platform with twenty concurrent training jobs, each checkpointing every thirty minutes, generates a checkpoint every ninety seconds. Without a retention policy, checkpoint storage grows without bound. The standard pattern is a tiered retention policy: keep the last N checkpoints in hot storage, move older checkpoints to cold storage after a configurable period, and delete checkpoints from finished jobs that a production model has superseded. The same tiering logic applies to dataset versions.

Idle cluster cost is the most recoverable hidden cost. An idle cluster still incurs compute cost for its always-on components: control plane nodes, monitoring agents, persistent volume controllers, and any always-on services the platform provides. For a moderate-sized cluster, this overhead can be 10 to 15 percent of the full utilization cost. The mitigation is a cluster-level idle policy: components that can be suspended are suspended after the cluster has been idle for a threshold period (typically one to four hours, depending on the workloads' cold-start tolerance). This pattern requires the autoscaler to distinguish between true idle (no jobs queued) and transient idle (a brief gap between job submissions), which it does by setting a minimum idle duration before taking any scale-down action.

The efficiency dashboard is the instrument panel that makes hidden costs visible. An effective ML efficiency dashboard tracks, at minimum, a GPU utilization histogram (not just the average), fragmentation rate, idle time fraction, checkpoint storage growth rate, egress by workload tag, and cost per unit of work by workload class. The dashboard should be reviewed weekly by the platform team and monthly by cost owners. A dashboard that is built but not reviewed is not an efficiency instrument; it is decoration. The operational clarity provided by the dashboard is a prerequisite for the quarterly cost-optimization tactic catalog.

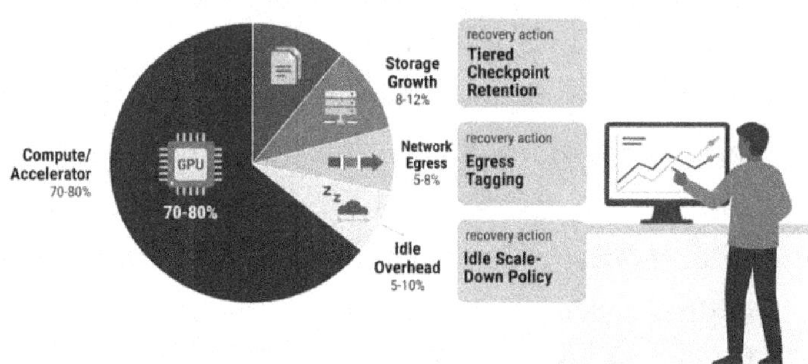

Right-Sizing Accelerators: The Smaller-Faster-Cheaper Curve

The instinct to match the largest available accelerator to every workload is expensive and often wrong. Accelerator right-sizing is the pattern of matching the accelerator class (memory capacity, compute throughput, interconnect bandwidth) to the workload's actual requirements rather than its theoretical aspirations. A serving workload that fits comfortably in eight gigabytes of GPU memory does not benefit from a GPU with eighty gigabytes of memory; the additional memory is idle, and the organization is paying for it at the full rate.

The smaller-faster-cheaper curve describes the relationship between model size, serving latency, and cost per prediction as a function of accelerator class. For small-to-medium models (up to a few billion parameters), smaller, lower-cost GPUs with fast memory bandwidth frequently outperform larger, more expensive GPUs on a cost-per-prediction basis because the workload is memory-bandwidth-bound rather than compute-bound. The correct metric for selecting a serving accelerator is not raw FLOP/s but memory bandwidth relative to the model's weight size. This calculation is explicit and auditable: model weights in bytes divided by memory bandwidth in bytes per second gives the minimum memory-bound latency for a single forward pass.

Mixed-instance pools for serving extend right-sizing to a fleet level. A serving fleet that hosts multiple model sizes allocates smaller-class GPUs to smaller models and reserves larger-class GPUs for models that genuinely require them. This requires the serving scheduler to be model-size-aware and to route inference requests to the appropriate replica class. The operational complexity is manageable when the serving infrastructure uses a unified model registry that tracks memory requirements alongside model artifacts.

Evaluating newer accelerator generations is a recurring cost-optimization opportunity that most teams execute too infrequently.

New hardware generations typically offer better performance per dollar than the previous generation, but the benefit is not uniform across workload types. A training workload that is compute-bound benefits strongly from a generation with higher FLOP/s; a serving workload that is memory-bandwidth-bound benefits more from a generation with higher memory bandwidth. The evaluation methodology is a small-scale benchmark of representative workloads that compares cost per unit of work across generations, rather than raw throughput. Teams that run this benchmark annually consistently find meaningful cost reductions available from generation transitions.

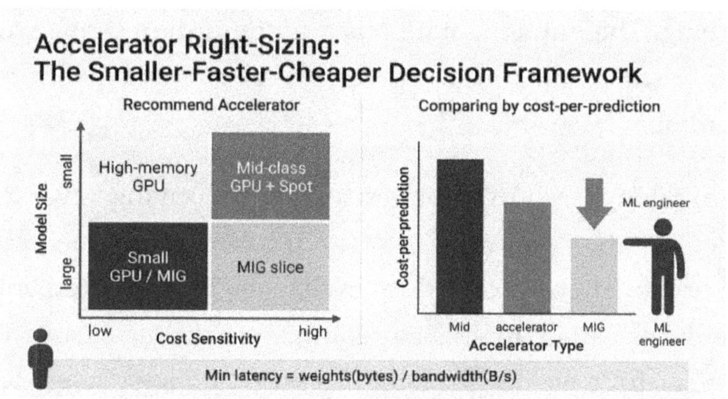

Cost-Optimization Tactic Catalog

A tactic catalog turns cost optimization from a fire drill into a backlog. The following tactics are ordered by expected savings impact relative to implementation effort, from highest to lowest leverage. Each tactic includes a savings range, an operational cost rating, and a risk level to help the team prioritize. The catalog is not exhaustive; it is the minimum viable backlog for a platform team starting a cost-optimization program. The team should add and remove tactics as measurements change the relative rankings.

Tactic 1: Spot Migration for Fault-Tolerant Training. Expected savings: fifty to seventy percent on training costs. Operational cost: medium (requires checkpoint hygiene and spot-aware job submission). Risk: low if checkpointing is solid, medium if it is not. This is almost always the highest-leverage tactic because training is typically the largest component of the bill, and the workloads are naturally suited to interruption.

Tactic 2: Idle Cluster Scale-Down and Scale-to-Zero Policies. Expected savings: 10-40% of compute costs, depending on the idle fraction. Operational cost: low (autoscaler configuration change). Risk: low if cold-start latency is acceptable, medium for latency-sensitive serving. The implementation is a configuration change to an existing autoscaler, making it the lowest-effort tactic given its savings potential.

Tactic 3: Right-Sizing Resource Requests. Expected savings: five to twenty percent of the computing cost by recovering fragmented capacity. Operational cost: low (due to changes in measurement and quota policies). Risk: low. This requires measuring resource consumption for four weeks per job class and a policy change to base requests on measurements rather than estimates.

Tactic 4: Reserved Instance or Committed-Use Conversion. Expected savings: thirty to sixty percent on the reserved fraction of the fleet. Operational cost: medium (requires capacity planning work to size the commitment). Risk: medium (over-commitment locks spend if workloads shrink). The prerequisite is a stable forecast of steady-state demand, which the sizing methodology provides.

Tactic 5: Checkpoint Storage Tiering and Retention Policy. Expected savings: five to fifteen percent of storage cost. Operational cost: low (retention policy configuration). Risk: low if the retention policy is tested against recovery scenarios before enforcement. The easiest implementation is a lifecycle policy on the checkpoint storage bucket that moves objects to cold storage after a configurable age.

Tactic 6: MIG Partitioning for Serving Workloads. Expected savings: up to 70% of the cost of the serving GPU for small models. Operational cost: medium (requires changes to partition management and the serving scheduler). Risk: low for stateless serving. This tactic delivers the highest per-GPU savings of any serving optimization and is particularly effective for organizations running many small-to-medium models on expensive large GPUs.

Tactic 7: Egress Audit and Co-location. Expected savings: variable, typically 2-8% of the total bill. Operational cost: low (audit) to high (data migration). Risk: low. The first step is always the audit: tag egress by workload, identify the top three sources, and evaluate whether co-location with the data eliminates the egress.

Tactic 8: Accelerator Generation Evaluation and Migration. Expected savings: ten to thirty percent on compute cost for workloads that benefit from the new generation. Operational cost: high (due to benchmarking and migration work). Risk: medium (driver and software stack compatibility). This tactic has the longest lead time of any in the catalog, which is why it should be scheduled annually regardless of whether there is an immediate cost pressure.

Tactic 9: Training-Serving Fleet Sharing with Priority Preemption. Expected savings: five to fifteen percent by reducing the total fleet size. Operational cost: high (requires scheduler integration and SLO monitoring for serving). Risk: medium (serving latency degradation if preemption triggers are misconfigured). This tactic is appropriate for platforms with a strong diurnal serving pattern and checkpoint-safe training jobs.

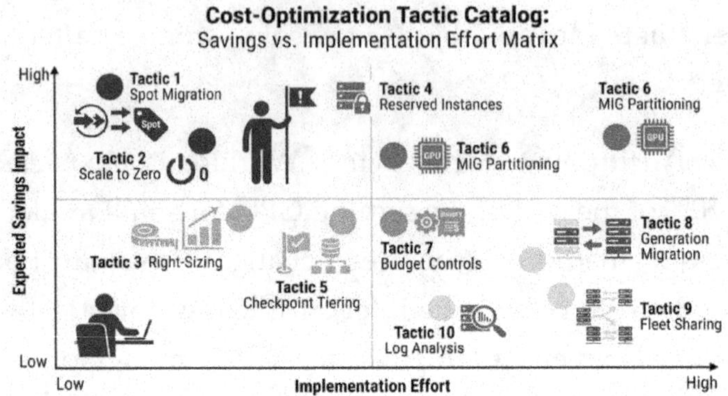

Efficiency Dashboards and Operational Reporting

An efficiency dashboard is not a monitoring dashboard. A monitoring dashboard reports the current state of the system: node health, job status, and queue depth. An efficiency dashboard reports the current quality of resource utilization relative to targets, and it is the primary instrument for the recurring cost-optimization practice. Without a well-designed efficiency dashboard, the tactic catalog lacks a feedback mechanism or a measure of progress.

The core metrics for an ML efficiency dashboard fall into four categories. Utilization metrics capture how well allocated resources are used: GPU utilization histogram (not just average), MFU per job class, and idle time fraction per node pool. Allocation metrics capture how well the cluster is filled: fragmentation rate, queue depth per priority tier, and the ratio of headroom consumed to target. Cost metrics capture financial efficiency: cost per unit of work by workload class, spot-to-on-demand ratio by job type, and storage growth rate by bucket. Attribution metrics capture organizational accountability: cost by team and project, cost-owner coverage rate (fraction of workloads with a named cost owner), and forecast accuracy.

The dashboard review cadence determines whether the dashboard drives action or becomes decoration. The recommended cadence is a weekly fifteen-minute review by the platform team focused on utilization and fragmentation anomalies, and a monthly thirty-minute review by cost owners focused on unit economics and tactic progress. The monthly review should produce one concrete action item per cost owner: a quota adjustment, a checkpoint retention policy update, an autoscaler parameter change, or a job right-sizing task. Small, regular improvements compound into significant annual savings.

Anomaly detection on efficiency metrics converts passive monitoring into active alerting. An autoscaler that doubles a minimum replica count and never reduces it is a detectable anomaly: the minimum replica count parameter should not increase without a corresponding increase in traffic baseline. A checkpoint storage bucket growing faster than the number of active training jobs is a detectable anomaly; it indicates that the retention policy is not being enforced. These alerts are responsible-by-design cost controls that prevent the accumulation of hidden costs described in the opening scenario.

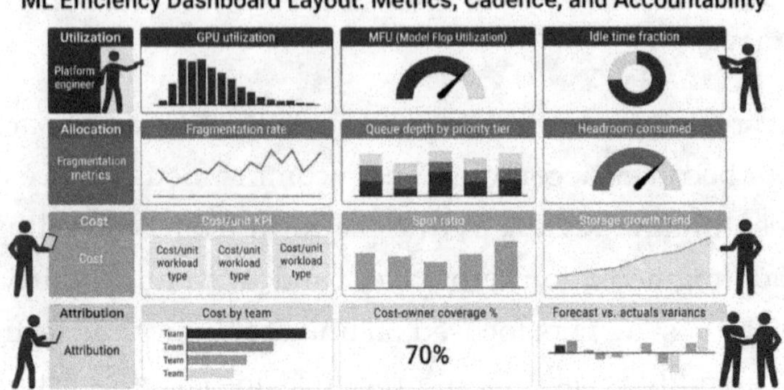

Technical Checklist

Use this checklist as a release gate before declaring a resource management posture production-ready.

- Every production workload has a documented cost owner by name, and that name is current and reachable.
- Every production workload has a unit-economics target (cost per training step, cost per thousand predictions) and a measurement mechanism to track it.
- Cluster sizing is the output of a four-step methodology (profile, peak/steady analysis, headroom modeling, provisioning decision), and the methodology artifact is version-controlled.
- The sizing methodology is reviewed quarterly or whenever the workload mix changes by more than twenty percent.
- GPU utilization is measured with histogram granularity, not just average. The utilization histogram is reviewed weekly.
- MFU is measured per job class, and a floor target is documented. Any job class with MFU below twenty-five percent has an open investigation.
- Idle time fraction is tracked per node pool. Any pool with idle time above 5% of wall-clock time over a rolling 7-day window has an open remediation task.

- Autoscalers are configured with accelerator-aware parameters: queue-depth trigger for training, latency-driven trigger for serving. Generic CPU-utilization autoscaling is not used for accelerator-backed workloads.
- Autoscaler minimum replica counts are reviewed quarterly and after every incident that modifies them. Changes are tracked in version control.
- Fragmentation rate is tracked, and the platform team has a documented target to keep it below ten percent.
- All jobs carry required tags (team, project, workload type, priority tier) enforced at admission time. Untagged jobs are rejected, not silently admitted.
- Quotas reflect actual measured demand, not requests. Quotas are reviewed quarterly and adjusted if the gap between the quota and actual consumption exceeds 40%.
- Spot instance migration is implemented for all fault-tolerant training workloads. Checkpoint hygiene (clean restart, deterministic data loading, optimizer state serialization) is verified for each migrated job class.
- The checkpoint retention policy is documented, enforced by automation (not convention), and tested against recovery scenarios quarterly.
- Egress costs are broken out by workload tag in the cost attribution report and reviewed monthly.
- Storage growth rate is tracked by bucket. Any bucket growing faster than the model count growth rate has an active retention policy review.
- Showback is in place at team granularity and is distributed to team leads on a monthly cadence.

- The capacity mix (reserved, on-demand, spot) is documented with a justification for each tier's sizing. The documentation is reviewed when commitments come up for renewal.
- The accelerator generation evaluation benchmark has been run within the last twelve months, and the results are documented.
- The cost-optimization tactic catalog is reviewed quarterly, each tactic has an owner and an estimated completion date, and at least one tactic is shipped each quarter.
- The efficiency dashboard is reviewed on the documented cadence by the documented reviewers. The last review date is visible in the dashboard header.
- Anomaly detection alerts exist for: minimum replica count increases without traffic baseline increase, checkpoint storage growth anomalies, and cost spikes above twenty percent week-over-week.

Team Conversation

Use these questions in team meetings, architecture reviews, or one-on-ones to pressure-test alignment on resource management and cost optimization.

21. What are our unit economics for training a model and for serving a thousand predictions, and when did we last measure them? If we cannot answer within five minutes, what does that tell us about our cost-attribution infrastructure?
22. Which of our current quotas are enforced by the scheduler and which are enforced by convention? What would happen if a team decided to ignore the conventional ones?
23. How much of last quarter's bill could we have avoided with patterns we already knew but had not implemented? What specifically prevented us from implementing them?

24. Who is the cost owner for our most expensive workload? Would our finance team recognize that person's name, and would that person recognize the number attributed to their workload?
25. When did we last evaluate a newer accelerator generation against our current hardware on actual workloads? Did we benchmark cost per unit of work, or just raw throughput?
26. Are we treating FinOps as a partner discipline with a regular review cadence, or as a recurring inconvenience we address only when the CFO asks a question we cannot answer?
27. What is the next tactic in our cost catalog, who owns it, and what is blocking it from being shipped this quarter?
28. If our training cluster were suddenly subject to a hard cap of half its current GPU count, which workloads would we prioritize and why? Does our priority model reflect those choices, or does it need to be updated?
29. Are our autoscalers configured for the workloads they actually manage, or are they still set to the defaults that were configured when the cluster was first provisioned?
30. If a researcher on a competing team had access to our cost attribution data, what inefficiencies would they find that we have normalized and stopped seeing?

Key Takeaway

Resource management is an engineering discipline with the same structure as reliability engineering: clear ownership, measurable targets, a recurring review cadence, and an explicit backlog of improvements. The patterns in this chapter are not exotic. Spot migration, right-sizing, bin packing, fair quotas, showback, and an accelerator right-sizing methodology are all known techniques. The reason they deliver large savings when finally implemented is not that they are difficult to understand, but that they require the organizational discipline to prioritize them alongside feature work.

Most cost overruns in distributed ML stem from patterns the team already knew but had not applied under the pressure of delivery.

The compound effect of these patterns is the most important practical point. A right-sized cluster is also easier to fill with spot instances. A cluster with well-configured bin packing has lower fragmentation, which means more of its capacity is schedulable. Lower fragmentation, combined with spot migration, produces a cluster in which a significantly larger fraction of total spend goes to actual computation rather than overhead, idle time, or fragmented waste. Showback and chargeback provide the organizational feedback loop that keeps the improvements from eroding over time. These patterns are not independent; they reinforce each other, and the savings compound quarter over quarter when the team maintains the practice.

The tactic catalog at the center of this chapter is designed to be a living engineering artifact, not a document read once and filed away. The team that reviews it quarterly, ships one tactic per quarter, and updates it when new measurement data changes the relative rankings is the team that maintains operational clarity on cost without requiring a crisis to motivate action. The pilot-to-production pathway for cost optimization is the same as for any engineering discipline: measure, prioritize, ship, measure again. Cost is not a finance problem with an engineering consequence; it is an engineering problem with a finance consequence, and the engineering team is the right owner.

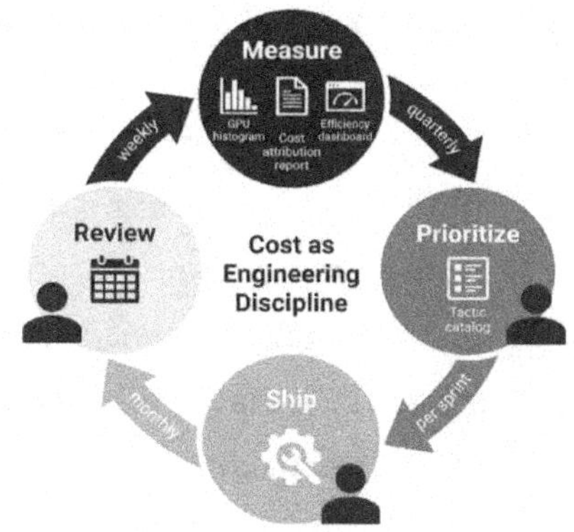
Measure
GPU histogram
Cost attribution report
Efficiency dashboard
quarterly
Prioritize
Tactic catalog
per sprint
Ship
monthly
Review
weekly
Cost as Engineering Discipline

9 Observability and Debugging Distributed Jobs

Opening Scenario

The alert arrives at 02:14 on a Tuesday. A training run that has been progressing normally for 31 hours on a 512-accelerator cluster has dropped to roughly 28% of its expected per-step throughput. The on-call engineer opens the monitoring dashboard and sees accelerator utilization at 94%, validation loss still trending downward, and host CPU well within headroom. Nothing looks broken. Everything looks slow.

The first hypothesis is a data pipeline stall. The engineer checks the data-loader metrics and confirms that prefetch queues are full and I/O wait times are normal. The second hypothesis is a memory pressure event. Memory headroom on every host is adequate. The third hypothesis is a gradient explosion, but gradient norms are tracking in the expected band. After ninety minutes of working through single-node explanations for a fundamentally distributed problem, the engineer escalates.

The senior engineer who joins asks one question: "What does per-rank all-reduce latency look like?" Nobody can answer it, because nobody tracks it. A targeted trace of the all-reduce barrier reveals that four ranks co-located on a single leaf switch are taking six to nine times longer than the median to complete each round. The leaf switch underwent a firmware update forty hours earlier and has silently downgraded its negotiated link speed. Every global synchronization point in the training loop now waits for the four slow ranks.

The fix is a switch configuration correction and a firmware rollback. It takes eleven minutes once the root cause is identified. The investigation took four hours. The four hours were not wasted on the wrong tools. They were wasted on dashboards designed for single-node systems that lacked the network-aware, rank-level observability that distributed ML jobs require. The loss was real: thirty-one accelerator-hours times 512 accelerators, plus four hours of senior engineering time, plus the compounding schedule risk of a delayed training run.

This chapter is the playbook the team will build next. It covers the observability vocabulary, the metric taxonomy, the tracing and profiling disciplines, the SLO and error-budget patterns, the drift and quality monitors, and the debugging runbooks that turn a vague "something is wrong" into a precise, actionable diagnosis.

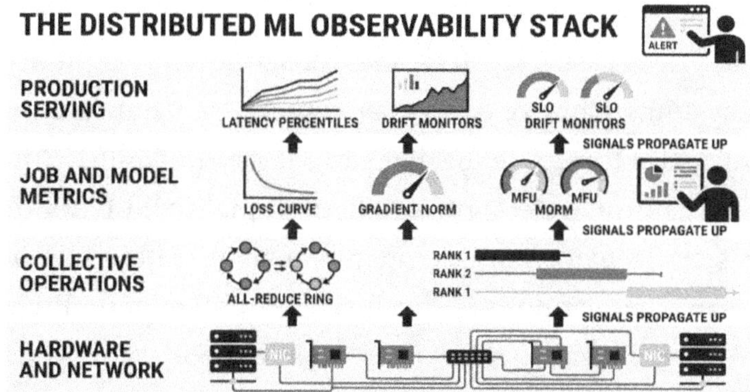

Why It Matters

Distributed ML systems compose more failure surfaces than single-node systems. A compute node can fail, but so can the network fabric, the scheduler, the storage system, the collective library, and the shared parameter server. Each surface produces distinct failure signatures, and most are invisible to dashboards that serve traditional engineering workloads. An engineer who reaches for a generic CPU and memory dashboard to debug a training throughput regression will spend hours examining non-diagnostic signals. The operational clarity that distributed ML requires is not a refinement of generic infrastructure observability. It is a distinct discipline.

The workflow-level impact of poor observability compounds at every phase of the ML lifecycle. During training, a degraded-but-running job may consume accelerator-hours for days before anyone notices the throughput regression. During serving, a latency regression that appears only at a specific batch size may never be captured by coarse-grained average latency metrics. Still, it will affect the tail of the user population. During data processing, a schema drift in a feature pipeline may produce training-serving skew that degrades model accuracy for weeks before any downstream signal connects cause to effect.

The mission-aligned case for first-class observability is that it converts ML systems from opaque cost centers into transparent, improvable infrastructure. A team that can see per-rank all-reduce latency, gradient norm trajectory, prediction-quality SLI trends, and cost per step is a team that can make deliberate trade-off decisions. A team that cannot see those signals will eventually be surprised by them. The patterns in this chapter make the responsible-by-design choice the default, and give every team member the shared vocabulary to discuss distributed system health with precision.

The Four Pillars of Distributed ML Observability

Observability for distributed ML rests on four complementary pillars: metrics, logs, traces, and profiles. Each captures a different slice of system behavior, and each has a different cost-signal trade-off.

Metrics are aggregated time-series measurements sampled at fixed intervals. They are cheap to collect, cheap to store, and well-supported by existing alerting infrastructure. Their weakness is that aggregation destroys the context needed to diagnose root causes. A cluster-level all-reduce bandwidth metric tells the team bandwidth is low; it does not tell them which rank is the bottleneck, which operation within the step is slowest, or whether the bottleneck is the NIC, the switch fabric, or the collective algorithm. Metrics are best used for alerting and for high-level health dashboards.

Logs are append-only, structured event records. They carry rich context per event and suit infrequent, discrete state transitions: job launch, checkpoint completion, optimizer step, worker failure, and gradient clipping event. Their weakness is volume. A thousand-rank job emitting one log line per step produces millions of entries per hour. Structured, sampled, and severity-gated logging is the discipline. Every log event should include the job ID, rank, step, and timestamp as mandatory fields to enable post-hoc correlation.

Traces are causally connected records of operations across process and machine boundaries. A distributed trace of a training step captures the full timeline from the moment a batch enters the data loader to the moment the optimizer writes updated parameters, spanning every collective operation, every host-to-device transfer, and every synchronization barrier in between. Traces are the primitive that makes a distributed system navigable as a single timeline. Their cost is non-trivial: storing a full trace of every step is infeasible, so trace sampling strategies are required.

Profiles are fine-grained, time-windowed recordings of hardware and kernel activity. A GPU profile captures kernel launch latency, memory bandwidth utilization, tensor core occupancy, and PCIe transfer time at microsecond granularity. Profiles are expensive to collect and analyze, but they are the only pillar that can distinguish a compute-bound operation from a memory-bandwidth-bound one, or identify a kernel that is fragmenting its memory accesses. The pattern is to collect a baseline profile at job launch, a comparison profile at each major checkpoint, and a targeted profile on demand when metrics indicate a regression.

The operational discipline is to route each investigative step to the cheapest pillar that can answer the current question: alert on metrics, triage with logs, diagnose with traces, and confirm hardware hypotheses with profiles.

Distributed Tracing for ML Workloads

A distributed trace is a directed acyclic graph of spans, where each span represents a named unit of work with a start timestamp, an end timestamp, and a set of attributes. In a distributed ML training job, spans may represent data-loader prefetching, host-to-device memory copies, forward passes, backward passes, all-reduce for gradients, optimizer steps, or checkpoint writes. The spans from all ranks of a single training step, stitched together by a shared trace context, form a single timeline that makes the full cost of a step visible.

Span propagation across workers is the foundational pattern for distributed tracing. Each worker receives a trace context (a trace ID and a parent span ID) when assigned a unit of work. It creates a child span, annotates it with local measurements, and propagates the context to any downstream workers it invokes. In a data-parallel training job, the propagation path runs from the orchestrator through the data loader, each rank's forward and backward passes, the all-reduce collective, and finally back to the optimizer. In a pipeline-parallel job, propagation also traverses pipeline stage boundaries, making inter-stage communication latency visible.

GPU-aware traces extend the standard span model to capture GPU kernel activity. A GPU-aware span includes device-side kernel execution time, PCIe transfer time for memory copies, and tensor core utilization during the span, in addition to the standard host-side timestamps. The pattern is to use GPU-aware spans for critical path operations (forward pass, backward pass, all-reduce) and lightweight host-only spans for supporting operations (data loading, checkpoint writing).

Trace sampling strategies must account for volume and diagnostic value. Three patterns are common. Rate-limited sampling collects one complete trace every N steps, providing the team with a statistical sample of normal behavior. Error-triggered sampling collects a full trace whenever a monitored metric exceeds a threshold, providing a high-fidelity record of abnormal behavior as it occurs. Adaptive sampling raises the sampling rate when anomaly detectors signal elevated uncertainty, then reduces it when behavior returns to baseline. The combination of rate-limited baseline sampling and error-triggered diagnostic sampling covers most production needs without the storage cost of full trace capture.

Distributed Trace Anatomy for a Training Step

Context propagation in heterogeneous pipelines requires a trace context format that all components understand. The W3C Trace Context standard (traceparent and tracestate headers) works across HTTP, gRPC, and message-queue boundaries. Using a standard format means a single trace can span from the feature store query that assembles a training batch, through the distributed training job, to the serving endpoint that deploys the resulting model, giving end-to-end causal visibility across the full ML lifecycle.

Cluster-Level Metrics and the Communication Fabric

The cluster-level metric taxonomy for distributed ML is distinct from that for general-purpose compute clusters. Generic dashboards track CPU utilization, memory, network bytes transferred, and disk I/O. Those metrics are necessary but not sufficient. The metrics unique to distributed ML workloads fall into four categories: collective-operation metrics, NIC-saturation metrics, topology-aware bandwidth metrics, and collective-library health metrics.

Collective operation metrics measure the performance of the synchronization primitives that enable distributed training. The primary metrics are all-reduce latency per rank, all-reduce latency variance across ranks, all-gather latency, reduce-scatter latency, and barrier wait time. Per-rank breakdown is essential: a cluster-level average all-reduce latency of 45 milliseconds may conceal a single rank at 300 milliseconds and 31 ranks at 42 milliseconds each. The average is not an anomaly. The rank-level distribution is. Alert thresholds should be set on the 99th percentile of the per-rank distribution, not on the mean.

NIC saturation metrics measure the state of the network interface cards connecting accelerator hosts to the fabric. Relevant signals are transmitted bytes per second, received bytes per second, transmit queue depth, receive queue depth, retransmit count, and link error count. NIC saturation is a common root cause of all-reduce latency regressions because the all-reduce collective is fundamentally a network operation. A rank whose NIC transmit queue is consistently at maximum depth is network-bottlenecked regardless of what GPU utilization reports.

Topology-aware bandwidth metrics quantify the available bandwidth between each pair of topology levels: within a host (PCIe or NVLink), between hosts on the same leaf switch (intra-rack), between leaf switches connected to the same spine switch (inter-rack), and across spine switches (inter-pod). A topology-aware bandwidth metric makes it immediately apparent when a job's communication pattern drives traffic across a boundary that its algorithm was not designed to cross.

NCCL timeout and error metrics capture the health of the collective library layer. NCCL timeouts occur when a rank fails to complete a collective operation within the configured timeout window and are almost always a symptom of a slower underlying cause: a failed host, a degraded network link, a memory error that stalls a rank, or a persistent straggler. NCCL timeout counts, retry counts, and associated rank identifiers should all appear on the cluster-level dashboard. A single timeout is a signal to investigate. Ten on the same rank in an hour is a signal to evacuate that rank's workload.

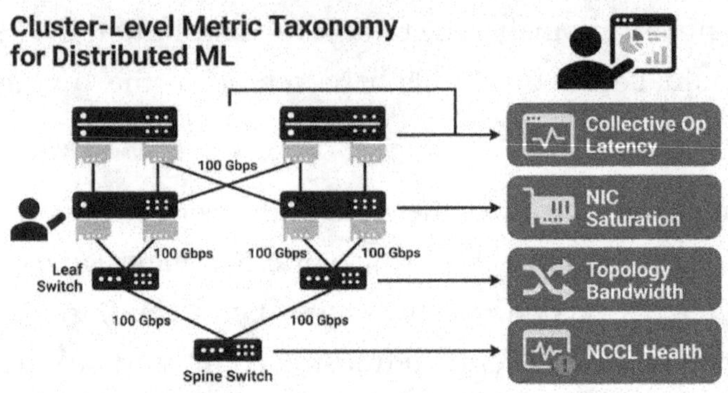

Per-Rank Metrics and Straggler Detection

A straggler is a worker who is consistently slower than its peers in completing each synchronization barrier. Because distributed data-parallel training synchronizes gradients across all ranks after each backward pass, a single straggler extends the wall-clock time of every step for every rank in the job. A straggler that is twice as slow as its peers effectively halves the throughput of the entire job.

The minimal per-rank metric set for straggler detection includes step time per rank, backward-pass time per rank, all-reduce wait time per rank (the time each rank spends blocked while waiting for the slowest rank to arrive), optimizer step time per rank, and memory utilization per rank. Step time and all-reduce wait time are the two most diagnostic signals: a rank with a high all-reduce wait time is not the straggler, but its peers are waiting for one; a rank with a high step time and a low all-reduce wait time is likely the straggler.

Straggler detection algorithms operate on the per-rank time distributions. The simplest effective algorithm is a rolling z-score on step time: a rank whose step-time deviation exceeds 2.5 standard deviations from the mean over a window of N steps is flagged as a candidate. A more robust algorithm uses the interquartile range rather than the standard deviation because the standard deviation is sensitive to outliers that the algorithm is trying to detect. Either should emit a structured event with a rank identifier, deviation magnitude, and a per-rank metric snapshot when a straggler is detected.

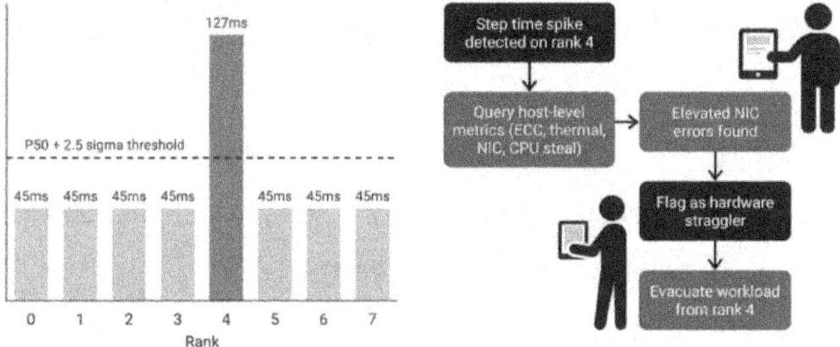

The root causes of stragglers fall into three broad categories. Hardware-related stragglers arise from NIC degradation, memory errors that increase ECC correction overhead, thermal throttling, or PCIe link width reduction due to cable or slot degradation. Software-related stragglers arise from uneven data distribution (a shard with more samples), load-imbalanced pipeline stages, or GC pauses in the data loader. Cluster-related stragglers arise from noisy-neighbor effects in multi-tenant clusters, topology mismatches between the job's communication pattern and the switching hierarchy, or fair-share scheduling decisions that transiently reduce one host's CPU or network allocation.

Host-level observability distinguishes hardware-related stragglers from software-related ones. When a straggler is detected, the investigation should immediately query host-level metrics for the affected rank: CPU steal time (noisy-neighbor contention), thermal sensors, ECC error counters, PCIe link speed, and NIC error counters. A host with elevated CPU steal is a noisy-neighbor victim. A host with elevated ECC errors is approaching hardware failure. A host with a degraded PCIe link is memory-bandwidth-constrained at the host-device transfer layer.

GPU Utilization, Model FLOPS Utilization, and Efficiency Diagnostics

GPU utilization, as reported by hardware performance counters, measures the fraction of clock cycles during which at least one warp is executing. A GPU reporting 92% utilization may be performing useful tensor computation, or it may be spinning on a memory-bandwidth-bound kernel waiting for data. High GPU utilization is necessary but not sufficient for efficient computation.

Model FLOPS utilization (MFU) fills the gap. MFU measures the fraction of the device's theoretical peak floating-point throughput achieved by the model's forward and backward passes. For a dense transformer on a modern accelerator, MFU above 50% is generally good; above 70% is excellent. MFU below 30% indicates a significant efficiency gap worth investigating before scaling the job to more accelerators, because scaling an inefficient job multiplies the waste rather than eliminating it.

The efficiency diagnostic ladder begins at MFU and descends to progressively finer-grained signals. If MFU is low, the next question is whether the bottleneck is compute-bound or memory-bandwidth-bound, requiring a kernel-level profile. If the kernel profile shows high memory bandwidth utilization and low tensor core utilization, the bottleneck is data movement; the remediation is kernel fusion or a change in data layout. If MFU is acceptable but overall throughput is below target, the bottleneck is likely the communication fabric, routing the investigation back to collective operation metrics.

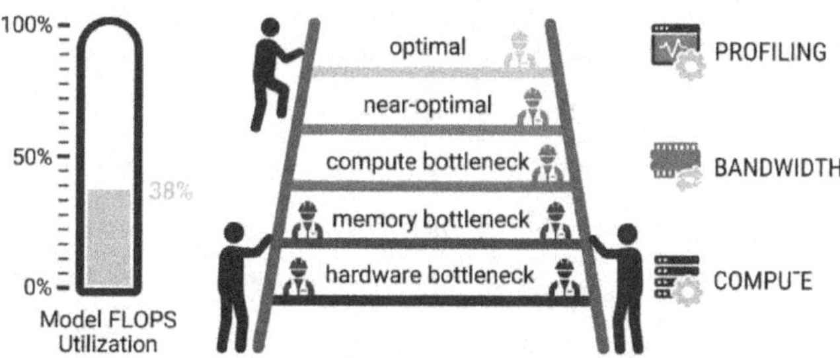

Arithmetic intensity is the ratio of compute operations to memory operations in a kernel, and it determines whether a kernel is compute-bound or memory-bandwidth-bound on a given hardware platform. Dense matrix multiplications are high-arithmetic-intensity and typically compute-bound on modern accelerators. Element-wise activation functions are low-arithmetic-intensity and typically memory-bandwidth-bound. The Roofline model, which plots kernel performance against its arithmetic intensity relative to the hardware's compute and memory bandwidth ceilings, is the canonical tool for classifying kernels and identifying the remediation class for each efficiency gap.

The practice of measuring MFU at the start of every training job, before scaling to the full cluster size, is among the highest-leverage activities in distributed ML. A job achieving 35% MFU on 8 accelerators, scaled to 512 without an efficiency investigation, will achieve roughly 35% MFU on 512 accelerators, plus the additional overhead of inter-node communication at scale.

NaN and Inf Detection, Gradient Norm Tracking, and Divergence Diagnostics

Training divergence is one of the most consequential failure modes in large-scale distributed ML. A training run that diverges after days of computation represents a significant loss of accelerator budget, and a divergence detected late (after the model has been checkpointed in a diverged state) may require rolling back to an earlier checkpoint and repeating the affected training work. Early detection requires instrumentation of signals that precede divergence: NaN and Inf values in activations and gradients, as well as the gradient-norm trajectory.

NaN and Inf propagate predictably through floating-point computation. A NaN in a single activation produces NaN outputs for all subsequent operations consuming that activation, eventually producing a NaN loss. The instrumentation pattern registers a hook on the backward pass of each layer (or a representative sample for large models) to check the gradient tensor for NaN and Inf values after each backward computation. When detected, the instrumentation should log the layer name, step number, rank identifier, and fraction of elements affected, and emit a structured alert. The alert policy should distinguish between transient events (a NaN that clears in the next step) and progressive events (growing across steps).

Gradient norm tracking is the early warning system for divergence events that have not yet produced NaN values. The gradient norm is the L2 norm of the flattened gradient tensor across all model parameters. For a well-configured training run, the gradient norm should be roughly stable within a band characteristic of the learning rate, batch size, and model architecture. A monotonically increasing gradient norm is a signal of impending divergence. A sudden spike that returns is a transient numerical instability. A sudden drop to near-zero is a vanishing gradient.

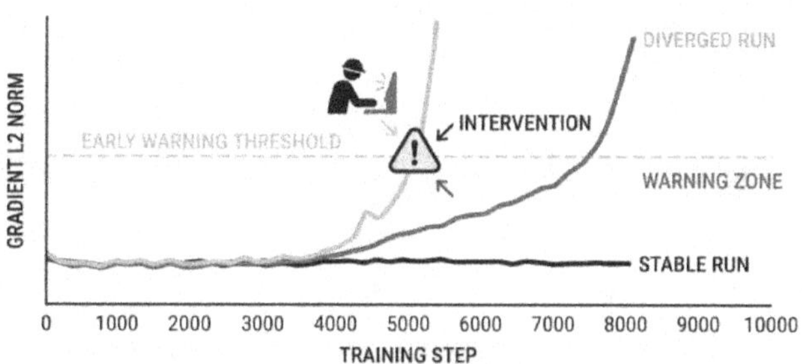

Gradient clipping is the standard intervention for preventing gradient-norm explosions, and the clip event itself is worth instrumenting. A training run that clips gradients at every step uses a learning rate that is too high for the current level of gradient noise. A run that clips fewer than 1% of steps is well-configured. The clip rate (clips per N steps) is a simple, derived metric that gives the team a continuous view of training regime stability. The root-cause analysis for divergence follows a structured decision tree: growing gradient norms before NaN values suggest an excessive learning rate or a precision issue; NaN values without preceding growth in gradient norms suggest a bad data batch or a hardware memory error.

Profiler Integration and Cross-Rank Performance Analysis

Profiling is the observability discipline that connects system-level metrics to hardware-level behavior. A profiler captures a time-ordered record of every kernel launch, memory transfer, and synchronization event on a device at microsecond granularity. The resulting profile, visualized as a timeline, makes the actual execution schedule of a training step visible: which kernels ran in parallel, which kernels were serialized due to dependencies, which synchronization events blocked device execution, and where device idle time was hidden by host-side overhead.

Framework-level profilers group kernel activity under the named operation that scheduled it (forward pass, attention, all-reduce), exposing an annotated timeline that corresponds to the programmer's model of the computation. Kernel-level profilers go deeper, capturing raw hardware performance counters for each kernel: FLOP throughput, memory bandwidth utilization, L1 and L2 cache hit rates, shared memory transactions, and warp efficiency. Framework-level profiles are the right starting point; kernel-level profiles confirm specific hardware hypotheses.

Cross-rank profiling collects profiles from all ranks simultaneously and aligns them on a common clock. The aligned multi-rank profile makes the communication pattern of the distributed job visible as a whole: which ranks are consistently ahead, which behind, where the all-reduce introduces the longest barrier, and how communication overlaps (or fails to overlap) with computation. The pattern of computation-communication overlap is one of the most important efficiency levers in distributed training, and it is only visible in a cross-rank profile.

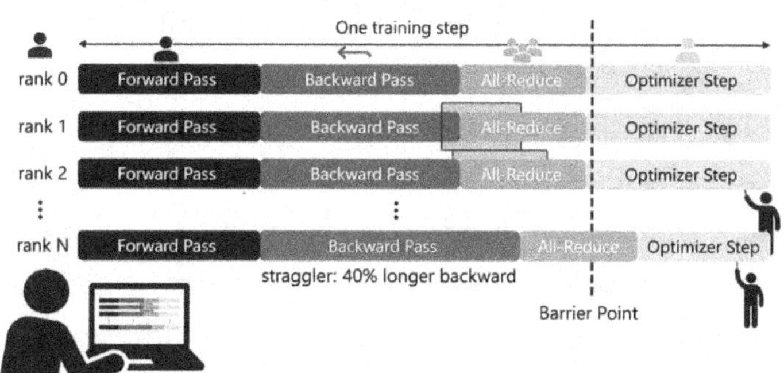

Profiling overhead is a practical constraint. A full framework-level profile of a large training job may add five to fifteen percent overhead to step time, and a kernel-level profile significantly more. The pattern that minimizes overhead while preserving diagnostic value is: enable profiling for a fixed number of steps (typically 5 to 20) at a regular interval (every N thousand steps or at each checkpoint), collect profiles on a representative subset of ranks (at minimum one per node type, plus any identified straggler), and store profiles in compressed format. Profile collection should be triggered automatically by the training harness because a profile requiring manual setup will not be collected when it is most needed.

Continuous profiling is the practice of collecting a lightweight profile of every training step, storing only summary statistics rather than the full kernel-level trace, and using those statistics to detect performance regressions across steps and runs. Continuous profiler metrics (step time, all-reduce time, forward pass time, backward pass time, peak device memory, key hardware counters) can be collected with negligible overhead and stored in a time-series database alongside training metrics. A regression in step time that is unexplained by changes in model size, batch size, or learning rate is a signal to trigger a full diagnostic profile.

SLOs, Error Budgets, and ML-Aware Service Health

Service-level objectives for ML serving workloads require a different vocabulary from that of SLOs for stateless web services. A web service SLO typically covers request latency at multiple percentiles and availability. An ML serving SLO must also cover freshness (how recently the model was deployed), prediction quality (whether the model's output meets a minimum quality bar), and throughput. Each dimension has a service-level indicator (SLI) that must be measured and reported continuously.

Latency SLIs should be defined at the percentiles meaningful for the use case. P50 describes the typical request experience. P99 describes the worst one percent. For interactive use cases (search, recommendation, conversational AI), P99 latency governs the long-tail user population. For batch inference, throughput SLIs (requests per second or tokens per second) may be more appropriate than latency percentiles.

Freshness SLIs measure the age of the deployed model relative to the training data cutoff. A stale model has not been retrained since the data distribution shifted enough to degrade prediction quality. The freshness SLI is typically expressed as the maximum acceptable staleness in hours or days, and the alert fires when the deployed model's training data cutoff falls outside that window. Freshness SLOs matter most in domains where the target distribution evolves rapidly, such as news recommendation, financial fraud detection, and real-time auction systems.

Prediction-quality SLIs measure the model's performance on a live or held-out evaluation set. Operationalizing a prediction-quality SLI requires an online evaluation pipeline that samples predictions, collects ground-truth labels, and reports the resulting metric (accuracy, AUC, NDCG, or a business-specific KPI) as a time series. The prediction-quality SLI is the most directly mission-aligned observability signal in the ML stack: it measures whether the model is doing its job, not just whether the serving infrastructure is alive.

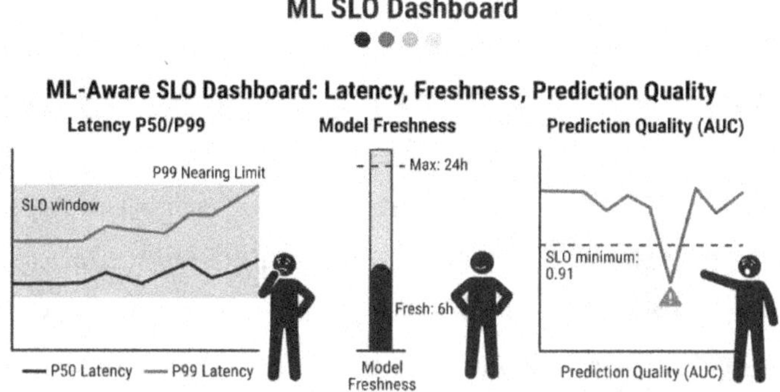

Error budgets translate SLOs into an operational decision framework. An error budget is the complement of the SLO availability target: a system with a 99.9% latency SLO has a budget of 0.1% of requests that may breach the latency target before the SLO is violated. When the error budget is full, the team can take on risk through aggressive releases. When nearly depleted, the team should freeze risky changes and focus on reliability improvements. Error budget policies codify this logic, making release and change management data-driven. ML error budgets must account for delayed ground-truth signals: a deployment that degrades prediction quality may not trigger a latency alert, and the budget accounting must include prediction-quality SLI breaches even when detected with a delay.

Drift, Skew, and Quality-of-Prediction Observability

Training-serving skew is the discrepancy between the data distribution on which a model was trained and the data distribution it encounters during serving. It is one of the most common causes of post-deployment model degradation and one of the most frequently undetected, because it produces no serving errors, latency alerts, or hardware anomalies. A model experiencing training-serving skew continues to serve predictions at normal latency; those predictions are less accurate.

The canonical sources of training-serving skew are feature computation mismatches (a feature computed differently in the training pipeline than in the serving pipeline), temporal distribution shift (the serving population has changed since the training data was collected), and label leakage remediation (a feature improperly included in training is excluded from serving). Each source produces a different signal. Feature-computation mismatches lead to distributional differences across individual features. Temporal shift produces a gradual drift in population-level feature statistics. Label leakage remediation produces a sudden quality degradation when the leaking feature is removed.

Feature drift monitoring tracks the distribution of each input feature in serving traffic and compares it to the distribution in the training data. The comparison typically uses statistical distance metrics: KL divergence for continuous features, population stability index (PSI) for binned distributions, or chi-squared test for categorical features. The practical challenge is alert volume: a model with hundreds of features generates many false positives if every feature is monitored at the same sensitivity. Alert design should prioritize features with high predictive importance and use adaptive thresholds that account for natural seasonal variation.

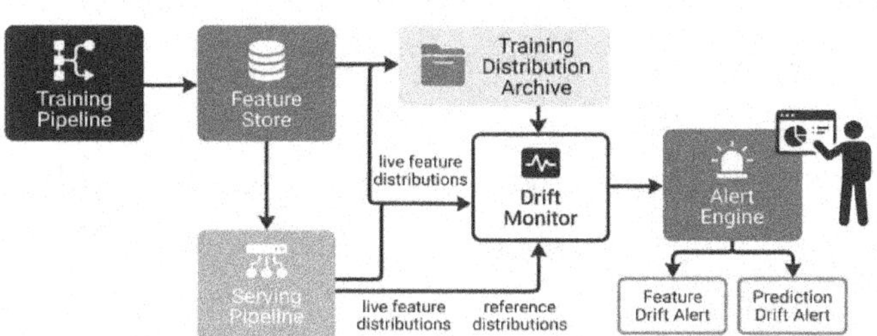

Prediction drift monitoring tracks the distribution of the model's output scores or labels in serving traffic and compares it to the historical baseline. A shift in the output distribution without a corresponding shift in any single input feature signals a compositional distributional shift that no individual feature monitor catches. Prediction drift is a more robust early warning signal for compositional shifts and a useful sanity check on model behavior after deployment.

Ground-truth delayed-label pipelines address the latency between when a prediction is made and when the ground-truth label becomes available. A credit fraud model knows whether a transaction was fraudulent only after a chargeback is processed. A delayed label pipeline collects predictions at serving time (with a unique identifier and a timestamp), joins them with ground-truth labels as they arrive, and computes running quality metrics from the joined dataset. The resulting metrics feed the prediction-quality SLI and close the feedback loop between serving behavior and model performance measurement.

Training Divergence Diagnostics and the RCA Decision Tree

Root-cause analysis (RCA) for distributed training failures is a structured discipline, not an ad-hoc investigation. The structure matters because the state space of possible root causes is large, the available signals are incomplete, and the cost of following the wrong branch is measured in senior engineering hours. A team with a written RCA decision tree for common training failure shapes will diagnose faster, more consistently, and with less dependence on individual expertise than a team that reconstructs the procedure each time.

The primary training failure shapes fall into five categories. Throughput regressions are failures where the job runs but more slowly than expected. Loss stalls are failures where the loss curve has flattened. Divergences are failures in which the loss increases or NaN values appear. Hangs are failures where the job has stopped making progress and is no longer emitting metrics. Crashes are failures in which one or more ranks exit with a non-zero return code.

For throughput regressions, the decision tree begins with communication fabric metrics: is per-rank all-reduce latency variance elevated? If yes, the root cause is likely a straggler or network degradation; proceed to per-rank straggler detection. If no, is MFU below baseline? If so, the root cause is a compute efficiency issue; proceed with kernel profiling. If MFU is acceptable, check data pipeline throughput. For loss stalls, begin with the gradient norm: a near-zero value suggests a vanishing gradient due to an excessively small learning rate or a saturating activation. A normal gradient norm with a stalled loss suggests that the data pipeline is delivering a non-representative subset of the training distribution.

For hangs, the first diagnostic signal is whether all ranks are still running (as reported by the orchestrator) or whether some have exited silently. A hang where all ranks are alive but producing no output is typically a collective operation deadlock: a rank has entered an all-reduce call with a different communicator state than its peers. A hang where some ranks have exited while others wait at a barrier is a stale rendezvous: the surviving ranks are waiting for a rank that has already exited and will never complete the collective.

Root Cause Analysis (RCA) Decision Tree for Distributed Training Failures

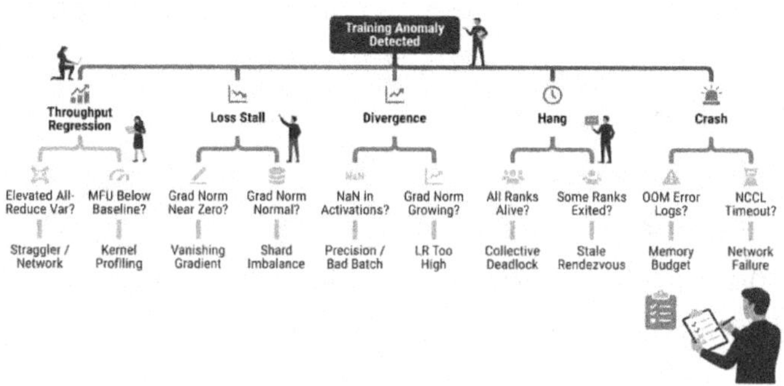

Runbooks translate the RCA decision tree into actionable steps for on-call engineers. A runbook for a specific failure shape includes: the symptoms that trigger the runbook, the first three diagnostic commands to run (with expected output for both healthy and failure cases), the escalation criteria (when to escalate, when to evacuate, when to restart), and the data to collect before the post-mortem (which logs to preserve, which profiles to save, which configuration snapshots to capture). Runbooks should be stored in the incident management system, linked directly from the alert definition, and reviewed after each incident. A runbook that lives only in a wiki and is not linked from the alert it serves will not be used when it is most needed.

Dashboards for Cluster Health and Job Monitoring

Dashboard design for distributed ML is an exercise in information hierarchy. The on-call engineer who opens a dashboard during an incident needs to answer "Is the cluster healthy and is my job making normal progress?" in under 30 seconds. The answer should be visible without interaction: green means proceed, amber means investigate, red means act now. Dashboards that require the operator to calculate or compare values across panels to determine cluster health have failed the thirty-second test.

The canonical dashboard set for a distributed ML platform comprises four dashboards arranged in a hierarchical scope. The cluster health dashboard covers the full fleet: total accelerator count, accelerators in use, accelerators that have failed or are quarantined, NIC error rates across the fabric, switch health indicators, and storage I/O rates. This is a single pane of glass for the SRE team and on-call rotation; it should never require scrolling to confirm cluster health. The job monitoring dashboard covers a single active training run: throughput in tokens or samples per second, MFU, loss curve, gradient norm, all-reduce latency (mean and P99), per-rank step time distribution, and checkpoint status.

The communication dashboard covers the collective operation layer: a per-rank all-reduce latency heatmap (ranks on one axis, time on the other, latency encoded as color intensity), a NIC saturation heatmap, a topology bandwidth-by-rack-pair heatmap, and an NCCL error count. The communication dashboard is the first destination for any investigation into a throughput regression, because the communication fabric is the most common source of such regressions in large, distributed jobs. The cost and utilization dashboard covers accelerator-hours consumed per job, cost per training step, cost per unit of model quality improvement, and utilization by team or tenant.

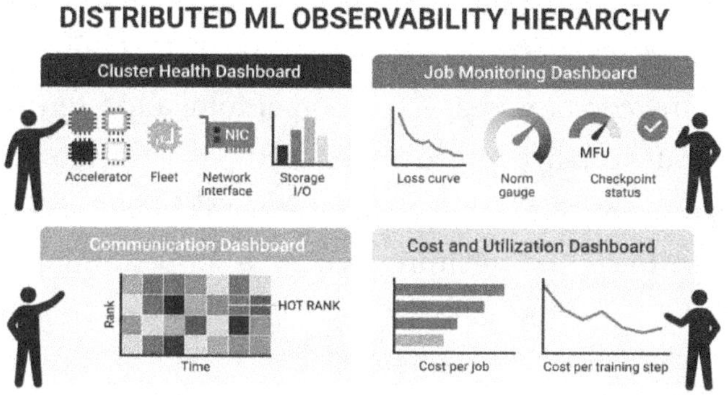

On-call alerts that survive high-stakes production environments share three properties: they are actionable (the receiving engineer knows exactly what to do), they are precise (they fire only when the described condition is genuinely present), and they are linked (each alert links directly to the runbook for the failure shape it describes). An on-call rotation's most valuable asset is the trust that every alert deserves attention, and that trust is destroyed one false positive at a time.

Cost Observability and Unit-Economics Dashboards

Cost observability for distributed ML requires connecting the financial cost of each workload to its operational metrics. An accelerator-hour has a dollar cost. A training run consumes a countable number of accelerator-hours. A training run produces a model of measurable quality. Connecting those three facts in a shared dashboard gives the engineering team and the engineering manager a common view of return on infrastructure investment, which is the foundation of responsible-by-design cost management.

The unit economics metric set for training workloads includes cost per training step (dollar cost divided by steps completed), cost per epoch, cost per unit of model quality improvement (cost per training run divided by the quality improvement from the previous checkpoint), and accelerator utilization efficiency (MFU multiplied by accelerator-hours consumed as a fraction of total allocated). The cost per unit of model quality improvement is the most strategically useful: it makes the cost of diminishing returns visible. It gives the team a principled basis for deciding when to stop training and deploy.

Cost observability for serving workloads includes cost-per-thousand predictions, cost per unit of latency reduction, and cost per quality point (the cost of model updates that improve the prediction-quality SLI). These metrics connect the serving team's engineering decisions to the business outcomes they produce, supporting the pilot-to-production pathway by making the cost of scaling a new model deployment visible before full rollout.

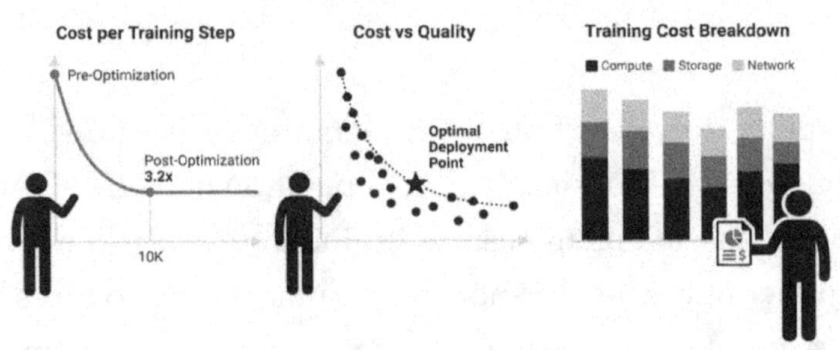

Shared cost dashboards that display cost metrics alongside performance and quality metrics in a single view are the most effective tool for aligning engineering and business decision-making. When the engineering team sees that a training run consumes 40% of the monthly compute budget for a 0.3 AUC improvement. At the same time, the previous run achieved 1.1 AUC improvement for 18% of the budget; the case for an efficiency investigation is self-evident. That alignment, driven by shared data rather than top-down budget mandates, is what operational clarity looks like in practice.

Anti-Patterns: Observability Choices That Look Right and Are Not

Observability for distributed machine learning fails characteristically: not by collecting too little, but by collecting the wrong things and presenting them at the wrong granularity. A cluster that emits millions of metrics can still be unobservable in the only sense that matters — the inability to answer, during an incident, which rank failed and why. The anti-patterns below each look like diligence. Each produces dashboards, alerts, and traces. Each fails precisely when an engineer most needs an answer, because the instrumentation was designed for the steady state rather than for the failure.

The most common observability anti-pattern is per-rank metrics without cross-rank aggregation. A team instruments every worker and emits a metric stream per rank; on a thousand-rank job, the on-call engineer is handed a thousand time series for each quantity of interest. The diagnostic signal is an incident response that begins with scrolling rather than reasoning, because no single panel shows whether the fleet is healthy. The fix is to aggregate across ranks by default — minimum, maximum, and percentile envelopes — and to expose per-rank detail only on drill-down, so the first view answers "is the job healthy" and the second answers "which rank is not."

The second anti-pattern is tracing only the training driver. The driver process is the easiest to instrument and the least informative, because the failures that matter — a hung collective, a corrupted gradient, a straggler — originate in the worker ranks, not in the coordinator that waits on them. The diagnostic signal is a trace showing the driver blocked on a synchronization barrier, with no indication of which worker failed to arrive. The fix is to trace the worker ranks themselves, propagating a common trace context across the collective so that a stalled all-reduce can be attributed to the specific rank that did not participate.

The third anti-pattern is treating raw GPU utilization as the only health signal. A utilization figure near 100% is reassuring and often wrong, because a device can be fully occupied moving data or waiting on a poorly overlapped communication step while doing very little useful arithmetic. The diagnostic signal is high reported utilization alongside throughput well below the hardware's capability. The fix is to track model FLOP utilization as the primary efficiency metric, because MFU measures the fraction of peak arithmetic actually spent on the model and exposes the communication and memory stalls that raw utilization conceals.

The fourth anti-pattern is loss-only divergence alerting. A team alerts when the loss curve misbehaves, which is a lagging indicator: by the time the loss visibly diverges, the run has often been silently degrading for thousands of steps. The diagnostic signal is a post-incident timeline showing gradient norms collapsing or NaN counts rising long before the loss alert fired. The fix is to alert on the leading indicators — gradient-norm trajectory and per-step NaN and Inf counts — at the same priority as a loss spike, so the run can be halted before wasted compute accumulates.

The fifth anti-pattern is building dashboards for the steady state. Most dashboards are designed while the system is healthy and answer the questions a healthy system invites — throughput, utilization, cost per step. The questions that actually matter are asked during incidents: which rank died, when the gradient norm first deviated, and what changed between the last good checkpoint and now. The diagnostic signal is an incident in which the team abandons the dashboards and manually queries raw logs. The fix is to design the primary dashboards around incident questions — recent topology changes, rank liveness, leading-indicator deviations — and to relegate steady-state metrics to a secondary view consulted when nothing is wrong.

Anti-Pattern Symptom → Root Cause

Symptom (observable)	Likely Root Cause	First Mitigation
Incident response begins with scrolling through thousands of series	Per-rank metrics emitted without cross-rank aggregation	Aggregate to min/max/percentile envelopes; drill down to per-rank on demand
Trace shows the driver blocked on a barrier, no failing rank named	Only the training driver is traced, not the workers	Trace worker ranks with propagated context across collectives
High GPU utilization, but throughput far below hardware peak	Raw utilization is used as the sole health signal	Track model FLOP utilization (MFU) as the primary efficiency metric
Loss diverges after thousands of silently degraded steps	Alerting only on the lagging loss curve	Alert on gradient-norm trajectory and NaN/Inf counts at equal priority
Team abandons dashboards and greps raw logs during incidents	Dashboards designed for steady-state questions	Build primary views around incident questions; demote steady-state panels

A Debugging Playbook for Distributed ML

A debugging playbook synthesizes the individual diagnostic tools and decision trees in this chapter into a single, sequential procedure that the on-call engineer follows from the moment an alert fires until the post-mortem is closed. The playbook does not replace expertise; it scaffolds it. A team with a well-maintained playbook and an experienced engineer on call will diagnose faster and more consistently than any individual working alone.

The playbook opens with triage. When an alert fires, the first task is to classify the failure shape from the alert name and the initial metric snapshot: throughput regression, loss stall, divergence, hang, or crash. The classification determines which branch of the RCA decision tree applies and which dashboard to open first. Misclassifying the failure shape is the most expensive mistake in distributed ML debugging because it steers the investigation toward the wrong signals.

The second phase is signal collection. For each failure shape, the playbook specifies the signals to collect in priority order: which metrics to query, which logs to pull, which traces to retrieve, and whether to trigger a targeted profile. Signal collection should be time-boxed to 15 minutes per signal source, with a hard stop at 45 minutes if no root cause has been identified. After 45 minutes without a root cause, the playbook prescribes escalation and a switch to the next signal source in order of priority. The time-box discipline prevents the sunk-cost effect that leads engineers to pursue a single hypothesis for 3 hours rather than reconsider the failure-shape classification.

The third phase is hypothesis testing. For each candidate root cause, the playbook specifies the confirming signal (what to find in the data if the hypothesis is correct) and the ruling-out signal (what to find if the hypothesis is incorrect). A well-written hypothesis test is falsifiable: the engineer either finds the confirming signal and proceeds to remediation, or finds the ruling-out signal and moves to the next hypothesis. Hypothesis tests that require the engineer to "look around and see if anything seems off" are searches, not tests, and searches in the state spaces of distributed systems are expensive.

The Distributed ML Debugging Playbook: Four-Phase Flow

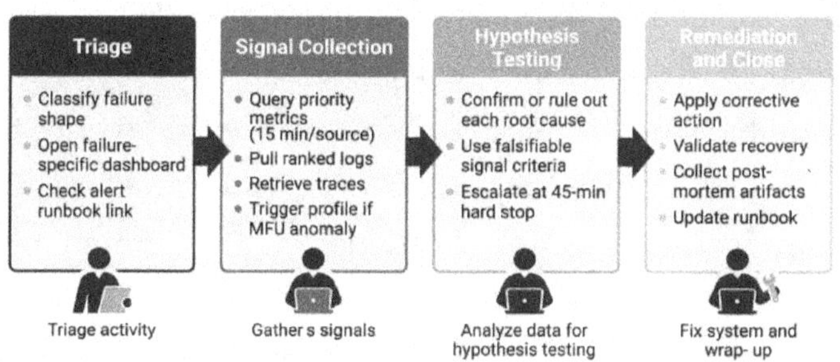

The fourth phase is remediation and closure. The remediation section for each root cause specifies the corrective action, the expected recovery time, the validation step (confirming the remediation worked), and the rollback plan (what to do if the remediation makes the situation worse). The post-mortem closes the loop: every incident requiring escalation or more than thirty minutes of on-call engineering time should produce a post-mortem document covering the incident timeline, a one-sentence root cause statement, contributing factors (instrumentation gaps, configuration decisions, or process gaps that made the incident worse), and tracked action items. A post-mortem without action items is a retrospective. A post-mortem with tracked, completed action items is a learning system.

Technical Checklist

- Every distributed training job emits per-rank metrics (step time, backward pass time, all-reduce wait time, memory utilization) by default, without requiring manual instrumentation by the training author.
- Cluster-level collective operation metrics (all-reduce latency P50 and P99 per rank, all-reduce latency variance, barrier wait time) are collected from every job and visible on the communication dashboard.
- NIC saturation metrics (transmit queue depth, receive queue depth, link error count, retransmit count) are collected from every host and available for per-rack correlation.
- Topology-aware bandwidth metrics instrument each level of the switching hierarchy (intra-host, intra-rack, inter-rack) so that cross-topology communication bottlenecks are immediately visible.
- NCCL (or equivalent collective library) timeout counts, retry counts, and associated rank identifiers are surfaced on the cluster health dashboard and trigger a named alert when the count exceeds a threshold for a single rank.
- MFU is computed and reported for every training job at launch and at each checkpoint; jobs with MFU below the team's configured threshold trigger an investigation ticket before being scaled to full cluster size.
- NaN and Inf detection hooks are registered on the backward pass of representative layers in every training job, and detection events emit a structured alert with layer name, step, rank, and NaN fraction.
- Gradient norm is tracked per step and per rank for every training job; the clipping rate is reported as a derived metric; and a

monotonically increasing gradient norm over a configurable window triggers a divergence early-warning alert.

- Distributed traces are collected for every training job using rate-limited sampling (at minimum one full trace per N steps) and error-triggered sampling (a full trace whenever a monitored metric exceeds its alert threshold).
- Framework-level profiles are collected automatically for a fixed number of steps at each checkpoint and at job launch, stored in compressed format, and linked from the job monitoring dashboard.
- SLOs are defined for every production serving workload and include latency percentiles (at minimum P50 and P99), freshness, and prediction-quality SLIs where ground truth is available within an acceptable delay window.
- An error-budget policy is documented, links release decisions to error-budget consumption, and has been applied to at least one real release decision in the current quarter.
- Feature drift and prediction drift monitors are active for every serving workload, feature importance scores are used to prioritize feature-level alert thresholds, and at least one drift alert has been validated against a known data distribution change.
- A ground-truth delayed label pipeline is in place for any serving workload where labels arrive with a delay, and the pipeline feeds the prediction-quality SLI on the job monitoring dashboard.
- Cost-per-step and cost-per-prediction metrics are visible on a shared dashboard alongside MFU and prediction-quality metrics, and the team reviews them before each major training scale-up or serving expansion.

- The RCA decision tree and associated runbooks are stored in the incident management system, linked from each alert definition, and reviewed after every incident requiring escalation.
- The debugging playbook has been walked through at least once per quarter in a tabletop exercise with the on-call rotation, and findings from the exercise are used to update runbook content.

Team Conversation

1. Walk through the last three incidents your team investigated. At which phase of the debugging playbook (triage, signal collection, hypothesis testing, remediation) did the investigation spend the most time? What would have accelerated that phase?
2. Does your team currently track per-rank all-reduce latency variance? If not, what would need to change in your instrumentation stack to add it, and what would that have been worth in the most recent throughput regression?
3. Review the SLOs defined for your production serving workloads. Do they include freshness and prediction-quality SLIs, or only latency and availability? What would a prediction-quality SLI look like for your primary use case?
4. Which signals in your current observability stack are aggregated at a level that destroys the diagnostic context needed to identify root causes? Identify two and describe the per-rank or per-layer granularity that would replace them.
5. Does your team profile continuously, at each checkpoint, or only when a performance regression is already visible? What is the minimum instrumentation change that would give you a continuous performance regression signal without unacceptable overhead?
6. Describe your current error budget policy. Has it been used to make a release decision? If not, what organizational or process

change would move it from an aspirational document to an operational tool?

7. Where does training-serving skew live in your current stack, and who owns monitoring for it? Is there a feature whose serving distribution you suspect has drifted from its training distribution but that you do not currently monitor?
8. Review your team's last post-mortem. Did it produce action items? Were those action items completed? What process change would make post-mortem action item completion the default outcome rather than the exception?
9. If you were designing the cluster health dashboard for a new distributed ML platform from scratch, what are the five metrics that must be visible without scrolling, and what alert condition linked to each one would give you the highest-confidence signal that human judgment is required?
10. How does your team currently communicate cost-per-step and cost-per-prediction to the engineering manager and to the team? If those metrics are not currently visible on a shared dashboard, what would it take to add them in the next sprint?

Key Takeaway

Distributed ML systems fail in ways that single-node systems do not, and the observability discipline required to diagnose those failures is correspondingly different. The four pillars (metrics, logs, traces, and profiles) each provide a different slice of system behavior, and the operational discipline is to route each investigative step to the cheapest pillar that can answer the current question: alert on metrics, triage with logs, diagnose with traces, and confirm hardware hypotheses with profiles. Collective operation metrics, per-rank straggler detection, and topology-aware bandwidth instrumentation are not refinements of generic infrastructure observability. They are the baseline for any team operating a distributed ML workload at scale.

The SLO and error-budget patterns in this chapter extend the standard reliability engineering vocabulary to the ML-specific concerns of freshness and prediction quality. A serving workload that meets its latency SLO but fails its prediction-quality SLI is not a healthy system; it is failing its users in a way that latency metrics cannot capture. The drift and skew monitors, the delayed-label pipelines, and the prediction-quality SLIs introduced here are the instruments that make the failure visible. They are mission-aligned additions to the observability stack, not optional enhancements.

The debugging playbook and the RCA decision tree are the artifacts that operationalize the patterns in this chapter. Patterns that live in documentation but are not linked to alerts or runbooks will not be used when most needed. The discipline of linking each alert to a named runbook, running tabletop exercises with the on-call rotation before incidents occur, and closing every significant incident with a post-mortem that yields tracked action items is what separates a team that learns from its incidents from one that repeats them. The vocabulary introduced here (from MFU to gradient norm tracking to error budgets to training-serving skew) is the shared language that later chapters rely on to describe observability concerns at the serving, security, and platform levels.

10 Security and Multi-Tenancy

Opening Scenario

It begins as a routine red-team exercise. A security engineer posing as a junior ML practitioner in the analytics tenant provisions a training job using the platform's shared base image. Within twenty minutes, she has read-access to a mounted volume belonging to a different tenant, a regulated healthcare team whose training corpus contains de-identified patient records. The mount was added eighteen months earlier by an engineer who has since left the company. The pull request description reads 'temp fix for data-loader path issue.' The fix shipped and stayed.

The incident response clock starts the moment the finding is escalated. The applicable regulation requires a written report within seventy-two hours if protected health information was exposed to an unauthorized party. The platform team spends the first four hours trying to answer a deceptively simple question: which jobs ran with that image, across which namespaces, over the past eighteen months? The logging system has the data, but it was not instrumented to answer tenancy-boundary questions. The team is writing ad hoc queries under compliance pressure while simultaneously patching the volume configuration, notifying privacy counsel, and fielding questions from business units whose workloads are paused.

The post-mortem surfaces four structural gaps. Namespace boundaries exist on paper but are not enforced by network policy or by admission control. The shared base image is pulled from an internal registry with no signing requirement or bill of materials. RBAC roles were seeded from a template two years ago and have not been reviewed since. Audit logs record API-server events but not volume-mount lineage or cross-namespace traffic. None of these gaps is exotic. All of them are common on platforms that grew fast and were later secured.

The fix is not a single configuration change. It is a set of composable patterns: workload isolation that makes cross-tenant volume access impossible by construction, RBAC that reflects the tenancy boundaries the regulator expects, network policy that defaults to deny, a supply-chain discipline that prevents temporary fixes from becoming permanent attack surfaces, and an audit posture that turns incident response from a fire drill into a structured query. This chapter outlines the design the team will adopt in the future.

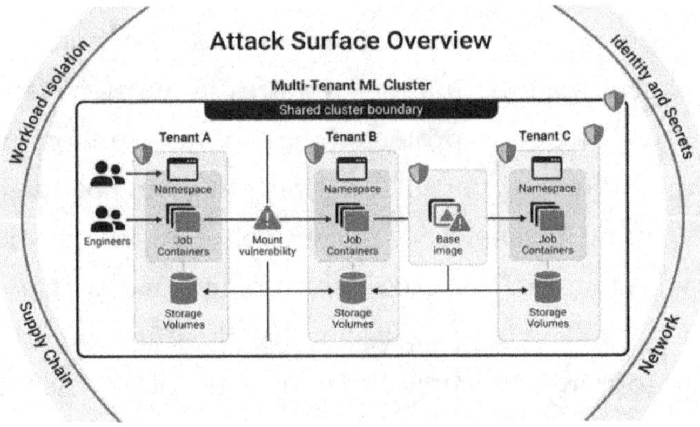

Why It Matters

Distributed ML clusters are attractive targets for three reasons that compound each other. They hold sensitive training data, which is often subject to privacy regulations. They hold trained model weights, which represent months of compute investment and competitive differentiation. And they share expensive GPU capacity across many teams, creating scheduling surfaces where one tenant's misconfiguration can exhaust another tenant's quota or expose another tenant's data. A conventional application cluster concentrates on compute. A distributed ML cluster simultaneously concentrates compute, data, and intellectual property.

The threat landscape for ML platforms has matured from theoretical to operational. Supply chain attacks against ML training pipelines, model weight exfiltration via misconfigured object storage, side-channel attacks on shared GPU memory, and prompt injection at inference endpoints are all documented incident types. Regulatory frameworks are catching up: data-residency requirements, right-to-explanation mandates, and AI-specific risk classifications are all placing new compliance obligations on the platform teams that run these clusters. The workflow-level impact is not abstract. A single misconfigured volume mount can trigger a 72-hour regulatory clock, freeze production workloads, and consume weeks of engineering time for post-mortem analysis and remediation.

The patterns in this chapter are not optional hardening steps to apply after the platform is built. They are structural decisions that determine whether the platform can be operated safely at scale. A team that designs isolation, RBAC, and network policy from the start spends audit time answering questions rather than reconstructing decisions made under pressure. A team that bolts them on later discovers that every remediation requires a change freeze, a compliance review, and a tenant migration. Responsible by design is not a posture; it is a cost-avoidance strategy with a compounding return.

Workload Isolation Patterns

Workload isolation is the foundational pattern. Its job is to make it structurally impossible for one tenant's job to read, write, or interfere with another tenant's resources, regardless of how either tenant's workload is configured. The distinction matters because RBAC and network policy are correctness controls: they fail if they are misconfigured. Isolation is a construction control: it fails only if the isolation mechanism itself is broken. The two work together, and neither is sufficient alone.

Namespace isolation is the first layer. Each tenant gets one or more namespaces, and platform admission control enforces that workloads cannot reference resources outside their assigned namespaces. This is necessary but not sufficient. Namespaces are a logical grouping mechanism, not a security boundary at the kernel level. A container running in namespace A can still make network requests to namespace B unless network policy prevents it, and it can still read host volumes unless the container runtime prevents it. Namespace isolation is the policy layer; it needs enforcement layers below it.

Sandboxed container runtimes provide kernel-level isolation for workloads that require it. Standard container runtimes share the host kernel, so a kernel-level vulnerability in one container affects every container on the same node. Sandboxed runtimes interpose a lightweight hypervisor or a language-based sandbox between the container and the host kernel, at the cost of roughly ten to fifteen percent additional memory overhead and some reduction in I/O throughput. For training jobs handling regulated data, that overhead is a rational trade. For high-throughput inference serving, it may not be the case.

The placement decision is a policy decision made by the platform team on a per-workload-class basis, not a global binary decision. Regulated training jobs run in sandboxed runtimes on dedicated node pools. Batch preprocessing jobs on public data run in standard containers on shared nodes. Inference endpoints run on dedicated nodes for availability reasons, independent of security. The platform enforces this through scheduling constraints and admission webhooks, preventing tenants from overriding the classification assigned to their workload.

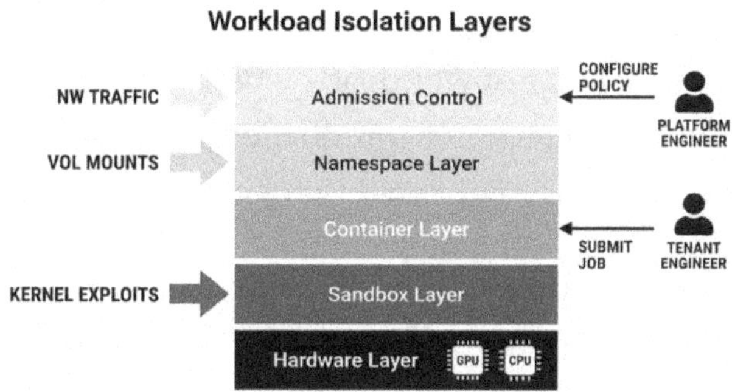

Accelerator-level isolation deserves explicit treatment because GPU memory is not automatically cleared between jobs on the same physical device. A job that terminates abnormally may leave weight fragments or activation buffers in GPU DRAM. A subsequent job on the same device can access that memory through a careless allocation pattern. This is not a theoretical side channel; it is a measurable data leak.

Multi-Instance GPU (MIG) partitioning solves this at the hardware level by dividing a physical GPU into isolated instances, each with its own compute engines, L2 cache partition, and DRAM slice. Memory from one MIG instance cannot be observed from another. MIG incurs a modest throughput penalty relative to running a single large job on the full device, but it provides a hardware guarantee that software isolation cannot match. For multi-tenant inference serving, MIG is the correct default when the GPU supports it.

Time-sliced GPU sharing and Multi-Process Service (MPS) offer higher utilization but weaker isolation. Time-slicing switches context between tenants at a configured interval; memory is not partitioned, and residual data may persist. MPS runs multiple CUDA contexts through a single server process, sharing the GPU more efficiently but sharing memory protection accordingly. Both are acceptable for dev and test workloads where tenants trust each other. Neither is acceptable for production workloads where tenants belong to different compliance domains.

RBAC for ML Platforms

Role-based access control on an ML platform must express a richer role hierarchy than that of a standard application platform. There are at least four distinct principals: the platform team that owns the cluster and its shared services, the tenant admin who manages a team's namespace and quota, the ML engineer who submits and monitors jobs, and the automated service account that runs those jobs. Each principal needs a different permission set, and the failure mode of treating them all as equivalent is quiet privilege creep that remains invisible until an audit or a red-team exercise reveals it.

The least-privilege principle applied to ML platforms means that each role has exactly the permissions required to perform its function and no others. The ML engineer role can create and delete training jobs in assigned namespaces. It cannot list secrets, modify network policies, or access other tenants' namespaces. The service account role that runs the job can read the input data path and write to the output data path for that specific job. It cannot list other secrets, read other tenants' volumes, or call the cluster API for anything beyond its job. These constraints are expressed as binding policies in admission control, not as documentation.

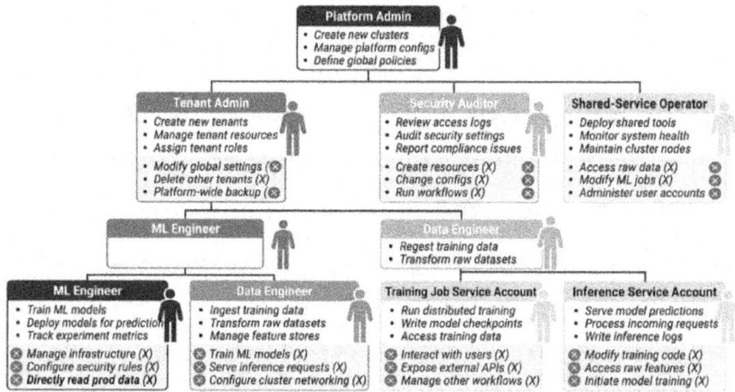

A common anti-pattern is to seed RBAC roles from a template and then layer exceptions on top over time. Each exception is individually defensible, but the aggregate produces roles with permissions that no one on the current team can fully enumerate. The mitigation is a quarterly role review that re-derives each role from a minimal starting point and requires justification for any permission that does not follow directly from the role's stated function. This is not bureaucracy for its own sake; it is the operational pattern that prevents privilege creep from compounding silently.

Workload identity is the pairing pattern for RBAC. Instead of giving a service account a static credential to authenticate to downstream services, the orchestrator issues the service account a short-lived, cryptographically attested identity token tied to the specific workload. The token carries the workload's namespace, name, and scheduling context. Downstream services validate the token against the orchestrator's identity provider rather than against a static password or API key. When the job completes, the token expires. There is no credential to rotate, no secret to leak into an image, and no long-lived surface to phish.

The boundary between RBAC and data governance deserves explicit attention. RBAC controls who can perform cluster-API operations. It does not control who can read a training dataset stored in object storage or a feature table stored in a feature store. Those access controls reside in the data layer, not the orchestration layer, and they must align with the RBAC boundaries at the cluster layer. An ML engineer who cannot list another tenant's secrets via the cluster API should also not be able to read another tenant's training data via the storage API. Maintaining that consistency requires explicit integration between the cluster's identity system and the data platform's access-control system.

Secure Multi-Tenant Cluster Design

Cluster design for multi-tenancy begins with a decision about the tenancy model: hard multi-tenancy, soft multi-tenancy, or dedicated clusters per tenant. Hard multi-tenancy provides strong isolation guarantees using namespace boundaries, network policies, resource quotas, and admission control, all on shared infrastructure. Soft multi-tenancy provides logical separation without enforcing it at the kernel or hardware level. Dedicated clusters provide the strongest isolation, but at the highest operational and cost overhead. Most production platforms land on a hybrid model: a hard, multi-tenant shared cluster for development and batch training, with dedicated clusters or node pools for regulated production workloads.

Per-tenant namespace design is the structural expression of the tenancy model. Each tenant receives a namespace with a resource quota that caps CPU, memory, and accelerator consumption. The platform team sets the quota based on the tenant's subscription tier or internal allocation. Admission webhooks prevent tenants from creating resources that reference objects outside their namespace.

A separate limit range enforces minimum and maximum resource requests per pod, preventing a misconfigured job from consuming unbounded resources and starving neighbors.

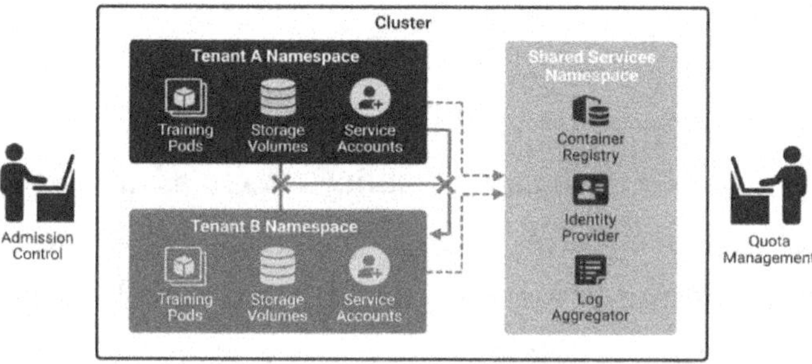

Dedicated node pools serve two purposes in a secure multi-tenant design. They provide physical separation for workloads where namespace isolation is not sufficient (regulated data, sandboxed runtimes, specialized hardware). And they provide scheduling predictability for latency-sensitive workloads, such as inference serving, that cannot tolerate the noise from a training job scheduled on the same node. Node affinity and taints express these constraints to the scheduler. The platform team owns the node-pool taxonomy; tenants request a workload class, and the scheduler places them accordingly.

Shared services can create overlapping blast radii if not carefully designed. A container registry that serves all tenants is a supply-chain dependency for all tenants. An artifact that is corrupted or replaced in that registry propagates to every tenant that pulls it. A logging pipeline that aggregates events from all namespaces into a single index is a data-leakage surface if the tenant does not partition query access. The design principle is that shared services must provide tenant-scoped views of their data and must not create paths that allow one tenant's operations to affect another tenant's data. This is a harder design problem than it appears, and it requires explicit security review of each shared service.

Admission control is the enforcement point where cluster-design policy becomes operational reality. Every resource creation or modification request passes through the admission controller before the API server persists it. Validating webhooks can reject resources that violate the isolation policy, such as a pod that references a volume outside its namespace. This service account requests excessive permissions, a job spec that turns off sandbox requirements. Mutating webhooks can inject required security context fields that tenants are not expected to manage themselves, such as the non-root user requirement or the read-only root filesystem flag. The combination of validating and mutating admission webhooks converts policy into a structural guarantee rather than a guideline.

Data Access Controls and Encryption

Data protection in a distributed ML platform spans three distinct layers of concern: who can access which data (access control), what the data looks like when accessed without authorization (encryption at rest), and what the data looks like in transit between components (encryption in transit). Each layer addresses a different threat model, and all three are required for a defensible posture. Treating encryption as a substitute for access control, or access control as a substitute for encryption, yields a platform that appears secure on paper but fails under real attack conditions.

Access control for training data must be expressed at a granularity that matches the data's sensitivity, not just the tenant boundary. A tenant may own a dataset that combines public benchmark data, proprietary internal data, and third-party licensed data, all of which are subject to contractual restrictions. A data engineer within that tenant needs access to all three categories to build the training pipeline. An ML engineer may need access only to the public benchmark and proprietary segments, not the licensed segment. A row-level or column-level access control policy enforced by the data platform, not by the application code, is the pattern that makes this composable. Application-layer enforcement is fragile because it can be bypassed by accessing the storage layer directly.

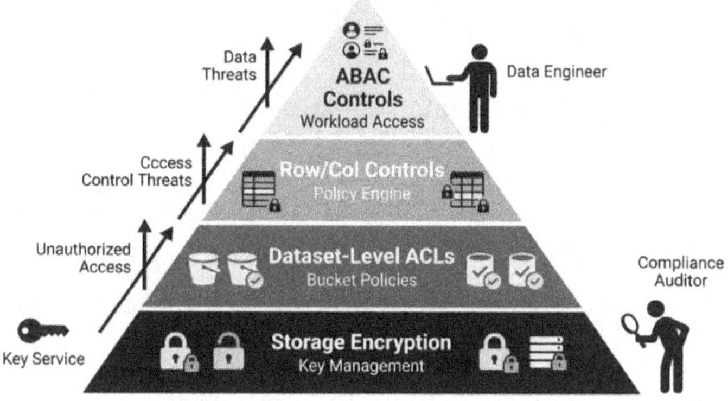

Key management is the load-bearing infrastructure for encryption at rest. Each tenant's data is encrypted with a tenant-specific data encryption key (DEK). The DEK is itself encrypted with a key encryption key (KEK) held in a hardware security module or a managed key management service. The tenant admin can rotate or revoke the KEK for their data without affecting other tenants. The platform team cannot read tenant data because they do not hold the KEK; they can demonstrate to an auditor that the encryption is in place and that the key material is tenant-controlled. This is the envelope encryption pattern applied to ML storage, transforming encryption from a checkbox into a meaningful access-control mechanism.

Encryption in transit must cover both intra-cluster and external traffic. A common gap is that TLS is enforced on ingress traffic from external clients but not on traffic between services inside the cluster. In a distributed training job, gradient communication between worker nodes traverses the cluster network in plaintext unless mutual TLS is explicitly configured. In a multi-tenant cluster, that plaintext traffic is visible to any workload that can access the same network segment. Mutual TLS between all workloads, enforced by the service mesh, closes this gap. The mesh sidecar handles certificate rotation automatically, so operational costs are low once the mesh is in place.

Differential privacy applied at data ingestion is a complementary pattern for workloads where the training data itself contains individual-level sensitive information. Adding calibrated noise to gradient updates during training (DP-SGD) provides a mathematical bound on the contribution of any individual record to the trained model. This makes it provably harder to reconstruct training data from model weights or from inference outputs. Differential privacy has a training cost (more iterations to converge, some accuracy loss) and requires the team to reason about the privacy budget as a resource consumed by training runs. The platform's job is to provide the tooling; the team's job is to decide when the privacy guarantee justifies the cost.

Network Policies and Service Mesh Patterns

The network is the data plane of a multi-tenant ML cluster. Every cross-component communication, every gradient exchange, every model-weight download, and every inference request traverses it. Securing the network layer means defining who can talk to whom, enforcing that definition regardless of how workloads are configured, and detecting when traffic patterns deviate from the expected shape. None of those three requirements can be met without an explicit network policy.

Default-deny is the correct baseline network policy for a multi-tenant cluster. A default-deny policy blocks all ingress and egress traffic for a namespace unless a specific rule permits it. This means that a new workload is isolated by default, and the platform team explicitly opens the paths it needs. The alternative, default-allow, means that new workloads have unrestricted network access unless someone explicitly restricts them. Default-allow clusters regularly accumulate unintended connectivity over time as services are added without corresponding policy updates. Default-deny clusters have exactly the connectivity specified by their policy.

Default-Deny Network Policy Topology

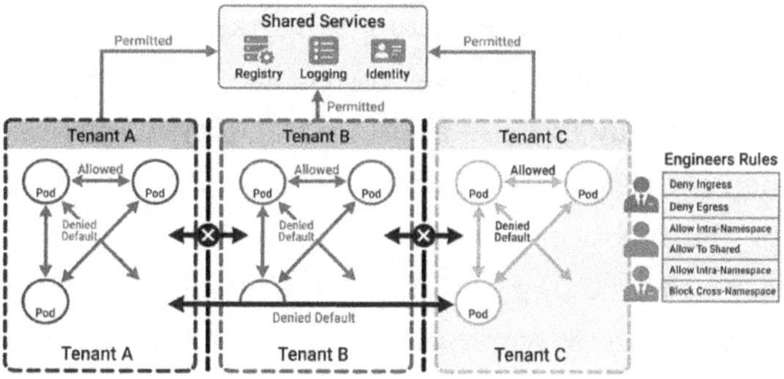

Layer-4 network policies (IP and port level) are necessary but not sufficient. A policy that permits traffic from namespace A to namespace B on port 443 does not validate the identity of the caller. Any process that can reach port 443 from namespace A can send traffic to namespace B. Mutual TLS at the application layer, typically provided by a service mesh, adds identity verification: the caller must present a certificate issued by the cluster's certificate authority that identifies the specific workload, not just the source IP address. The combination of layer-4 network policy and mTLS provides defense-in-depth: the network policy limits which endpoints can communicate, and mTLS verifies that the communicating parties are who they claim to be.

Service mesh authorization policies extend this further by expressing identity-based access control at the application layer. Rather than saying 'traffic from port range X is allowed,' a mesh authorization policy says 'requests from workload identity training-job-a in namespace tenant-a to the storage service are allowed for HTTP GET on path /data/tenant-a/input and HTTP PUT on path /data/tenant-a/output.' This level of specificity is not achievable with network-layer policy alone. It requires the mesh to intercept traffic and evaluate request attributes against the policy.

The operational challenge with service-mesh policy is keeping it in sync with the platform's actual topology as workloads are added, moved, and retired. A policy that was correct six months ago may be over-permissive today because a service it referenced has been decommissioned and its name reused. The mitigation is to generate mesh policy programmatically from the same source of truth that defines the workload topology, rather than maintaining policy as a separate artifact. When a workload is declared, its permitted communication paths are also declared. The mesh policy is derived from those declarations and applied automatically.

Secrets, Identity, and Workload Federation

Secrets management in a distributed ML platform is complicated by the number of systems that workloads need to authenticate to: object storage, feature stores, model registries, experiment trackers, notification systems, and external APIs. The naive pattern is to encode the credentials for each system as environment variables in the job specification or as secrets mounted into the container filesystem. This works and is common. It also creates a credential sprawl problem: dozens of long-lived credentials scattered across job definitions, image build scripts, and configuration management systems, each a potential surface for leaks with an indefinite validity window.

Workload identity federation is the pattern that replaces static credentials with short-lived, attested tokens. The orchestrator issues a service account token to each running workload, cryptographically signed by the cluster's identity provider. The token encodes the workload's namespace, service account name, and job metadata. Downstream services are configured to trust tokens issued by the cluster's identity provider. A training job that needs to read from object storage presents its token to the storage service. The storage service validates the token signature, reads the claims, and applies the access policy for that workload identity. No static credentials are needed. When the job completes and the token expires, the access window closes automatically.

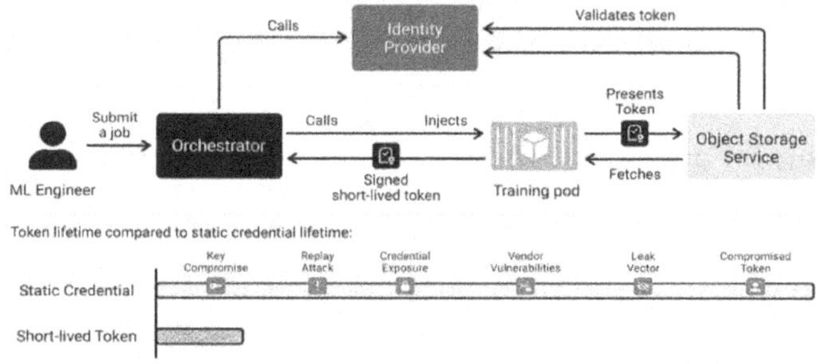

Cloud provider workload identity integration extends this pattern beyond the cluster boundary. A job running inside the cluster can be granted a cloud IAM role that permits it to read from a specific storage bucket and write to a specific metrics namespace, without storing any cloud credentials anywhere in the cluster. The cluster's service account is annotated with the cloud IAM role ARN or equivalent, and the cloud provider's credential helper automatically exchanges the cluster token for a cloud credential scoped to that role. The credential is short-lived, automatically rotated, and scoped to exactly the permissions specified by the annotation. This is the federation pattern at the cloud boundary.

Secrets that cannot be replaced by workload identity (third-party API keys, database passwords, license tokens) must be managed through a secrets store that provides versioning, rotation, and audit logging. The secrets store injects secrets into workloads at runtime via a sidecar or a CSI driver, rather than at image build time. The distinction matters because an image that contains a secret embeds that secret in every layer of its build history. Anyone who can pull the image can extract the secret. A secret injected at runtime is never stored in the image and is not visible in the container specification. Automatic rotation is enabled for all secrets that support it. The rotation event updates the secrets store; the next workload injection picks up the new value without a deployment.

Long-lived credentials are the highest-risk item in most platform inventories because they are often forgotten rather than revoked. A quarterly credential audit that lists every static credential, its age, its last-use timestamp, and the workload it was issued for is the operational hygiene pattern that keeps the inventory manageable. Any credential older than ninety days with no active workload associated with it is revoked. Any credential older than 1 year is flagged for mandatory review, regardless of active workload status. This is not a security-theater exercise; it is the inventory management discipline that makes the workload-identity migration tractable by keeping the static-credential footprint from growing as the migration proceeds.

Supply-Chain Integrity for ML Platforms

An ML platform's supply chain is larger than that of a conventional application platform. It includes the base container images that training and serving workloads run in, the ML framework packages and their transitive dependencies, the pre-trained model weights that fine-tuning jobs start from, the training data and the preprocessing pipelines that produce training-ready datasets, and the model artifacts that result from training runs. Any component in that chain can be tampered with to introduce adversarial behavior, exfiltrate data, or consume unauthorized resources. Supply-chain discipline is the set of practices that makes tampering detectable.

Image signing is the entry point. A trusted key must sign every container image that runs on the platform before it is admitted. The signing step happens in the CI/CD pipeline after the image passes vulnerability scanning and before it is pushed to the production registry. The admission controller verifies the signature at deployment time using the signing authority's public key. An image without a valid signature is rejected before it runs. This provides a strong guarantee that the image running in production is the one produced by the CI/CD pipeline, not a modified version that was substituted after the build.

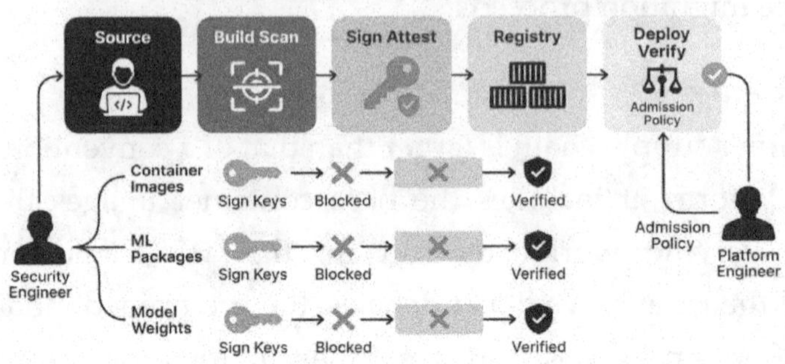

Software Bill of Materials (SBOM) generation closes the visibility gap that image signing does not address. Signing proves that the image has not been tampered with since it was signed. The SBOM documents what the image contains: every package, library, and transitive dependency, along with their versions. When a vulnerability is disclosed in a specific library version, the team can query the SBOM inventory to identify every image in the registry that contains the vulnerable version, rather than scanning the entire image catalog reactively. SBOMs for ML workloads should include not only the container layer but also the Python package environment, because ML framework dependencies are deep and often include packages not managed by the base image maintainer.

Model artifact signing applies the same pattern to trained model weights. A model artifact is signed by the training pipeline at the end of the training run, with the signature covering the weights file, the training configuration, the dataset lineage metadata, and the training run identifier. The serving infrastructure verifies the signature before loading the model. This prevents a scenario in which a model in the model registry is replaced with a modified version that produces different outputs for specific inputs. It also provides the foundation for provenance tracking: a deployed model can be traced back to the specific training run, dataset version, and code commit that produced it.

Provenance tracking is the end-to-end supply-chain pattern that links training data, code, model weights, and deployed models into a single auditable chain. Each step in the chain records its inputs, outputs, and configuration as a signed attestation. The attestation chain can be traversed from any deployed model back to the original training data. For regulated workloads, this chain is the artifact that answers the auditor's question: what data was used to train this model, who authorized its use, and whether anything in the chain has been modified. A team that has provenance tracking in place turns that question into a query. A team that does not turn it into a project.

Privacy-Preserving ML Patterns

Privacy-preserving ML patterns address a class of workloads that conventional access control cannot handle: workloads in which the training data itself must not be accessible to the model trainer, or in which the trained model must not reveal information about specific training records to an adversary who can query it. These are not exotic research scenarios. They arise whenever training data include individual-level health records, financial histories, communication logs, or any other information subject to a privacy protection regime. The pattern the platform selects has significant implications for training infrastructure, cost, and the accuracy of the resulting model.

Differential privacy applied to the training process (DP-SGD) provides a formal privacy guarantee: the trained model is epsilon-delta differentially private if the probability that any given training record can be inferred from the model's outputs is bounded by the privacy budget. The mechanism clips per-sample gradients to bound their sensitivity, then adds calibrated Gaussian noise before aggregating. The privacy budget epsilon is a resource that is consumed by each training step. Running more steps, using more data, or lowering the noise multiplier all increase epsilon, weakening the guarantee. The platform team provides the DP training library and the privacy budget accounting tooling. The ML team decides how to allocate the budget across experiments.

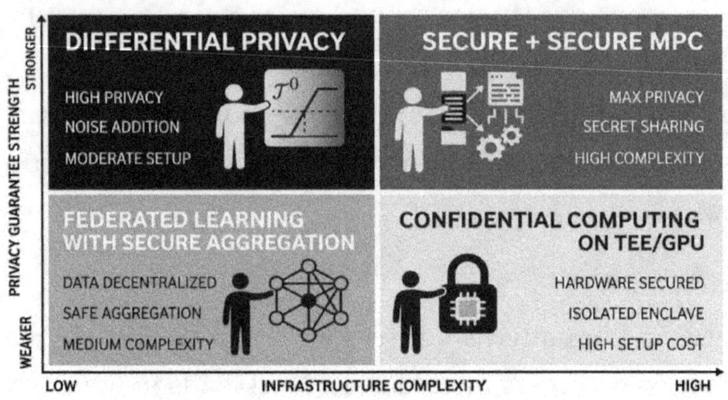

Federated learning is the pattern for scenarios where training data must not leave the node or the organization that holds it. Instead of bringing data to a central training cluster, the training computation goes to the data. Each participant trains a local model update on their local data and sends only the update (not the data) to a central aggregator. The aggregator combines the updates and returns an improved global model. In a multi-hospital clinical training scenario, no patient record ever leaves a hospital's network. The practical challenges are coordination overhead, statistical heterogeneity across participants, and the risk that gradient updates themselves can reveal information about the local data. Secure aggregation, which uses cryptographic techniques to compute the aggregate without revealing individual updates, addresses the last concern at the cost of additional computation.

Confidential computing on accelerators moves the trusted execution environment down to the hardware level. A confidential virtual machine (confidential VM) provides a hardware-attested boundary around the computation: the hypervisor and the cloud provider cannot inspect the memory or computation inside the enclave. GPU confidential computing extends this to the accelerator, so that model weights and activations in GPU DRAM are protected from the host OS and the hypervisor. The attestation report produced by the enclave can be presented to a data owner as proof that their data was processed in a protected environment. This pattern enables a platform team to process third-party data under a data-processing agreement without requiring the data owner to place full trust in the platform operator.

Secure multi-party computation (MPC) enables multiple parties to jointly compute a function of their private inputs without revealing those inputs to one another. In an ML context, this means multiple organizations can jointly train a model on the union of their datasets without any organization seeing the other's data. The computational overhead of MPC is significant: cryptographic protocols operating on large tensors are orders of magnitude slower than plaintext computation. MPC is the correct pattern for high-value, low-throughput scenarios where the privacy guarantee is worth the cost. It is not appropriate for latency-sensitive or high-throughput training workloads.

Compliance Patterns for Regulated Workloads

Compliance for distributed ML platforms is not a property that can be added after the platform is built. It is a set of structural design decisions about data residency, access logging, evidence retention, and operational processes that must be made before workloads are onboarded. A platform not designed for a specific regulatory regime will require significant re-architecting to support that regime. A platform designed for composable compliance can add new regulatory requirements as a configuration rather than as a structural change.

Data residency is the foundational compliance constraint for many workloads. The requirement is that training data, model weights, and inference logs must not leave a specified geographic boundary. In a multi-region cluster architecture, enforcing data residency requires that the scheduler know which data assets have residency constraints and never schedule the workloads that use them outside the permitted region. This is expressed as scheduling constraints on the workload and as data-location metadata on the storage assets. The combination ensures that a training job on a residency-constrained dataset will fail to schedule rather than silently running in a prohibited region.

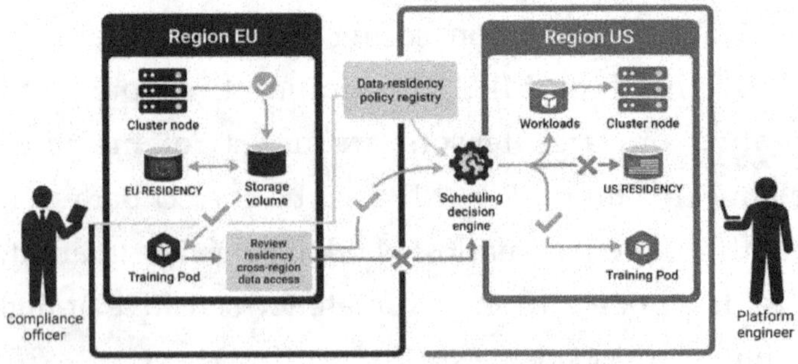

Audit logging is the second foundational compliance requirement. An audit log for a distributed ML platform must answer questions that a generic Kubernetes audit log does not: which job accessed which training dataset, which model was served to which requestor, which user modified which RBAC policy, and which container image was running at the time of any given event.

Achieving this requires that the audit log integrates events from the cluster control plane, the data access layer, the model registry, and the inference serving layer into a unified event stream that can be queried by workload identity, tenant, dataset, and time range.

Evidence collection is the operational pattern that turns audit logging into audit readiness. An audit-ready platform stores signed audit log snapshots according to a retention schedule that matches the applicable regime (typically 1 to 7 years for most regulatory frameworks). The snapshots are stored in an append-only, tamper-evident log store so that their integrity can be verified at audit time. Compliance reports are generated from the same log store automatically on a regular schedule, so the audit process is a review of existing reports rather than an emergency extraction. A team that runs quarterly compliance report reviews before the audit has no surprises at audit time.

PII handling in training data is a specific compliance concern that spans multiple stages of the ML pipeline. PII may enter a training dataset from source data that was not fully de-identified, from data labeling processes that embed annotator identifiers, or from web-scraped data that contains personally identifiable text. The platform's job is to provide scanning tools that detect PII patterns in datasets before they are used for training, and to enforce a gating policy that prevents a training job from launching on a dataset that has not been screened for PII. This is a pipeline-level control, not a model-level control, because model-level de-identification (such as prompting the model not to reveal PII) provides weaker guarantees than preventing PII from entering the training set in the first place.

Inference Security: Prompt Injection and Abuse Prevention

The inference endpoint is the public face of the ML platform, and it introduces a class of security concerns that do not exist at the training layer. Distributed training infrastructure is typically accessible only to authenticated internal users. Inference endpoints, by design, accept requests from external clients. The security model must assume that some fraction of those requests will be crafted to elicit unauthorized behavior, extract information about the model or its training data, consume disproportionate resources, or inject content that affects subsequent users of the same inference session.

Prompt injection is an attack pattern in which an adversarial input causes a language model to override its system prompt or produce outputs outside its intended scope. The attack exploits the model's tendency to follow instructions embedded in its input regardless of whether those instructions come from the authorized system prompt or from user-supplied content. Defense is layered. Structural prompt design separates system and user content with explicit markers that are difficult to spoof in user input. Input validation rejects requests that contain known injection patterns. Output monitoring detects when the model's response deviates from expected content patterns. None of these defenses is sufficient on its own; together, they significantly reduce the attack surface.

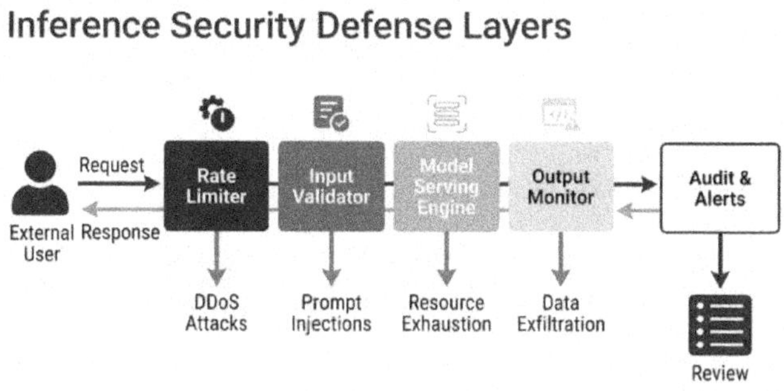

Inference Security Defense Layers

External User
Request
Response
Rate Limiter
Input Validator
Model Serving Engine
Output Monitor
Audit & Alerts
DDoS Attacks
Prompt Injections
Resource Exhaustion
Data Exfiltration
Review

Rate limiting and abuse prevention apply to inference endpoints, which carry higher stakes than conventional API rate limiting because inference computing is expensive and uneven. A single adversarial request to a large language model may consume orders of magnitude more compute than a typical request if it is crafted to elicit a long generation, trigger a long chain-of-thought, or exploit a quadratic attention pattern. Token-based rate limiting, which counts input plus output tokens against a per-client budget rather than counting requests, is a more accurate resource-protection mechanism for this workload class. A client that submits requests designed to maximize token consumption hits their token budget quickly and cannot starve other clients.

Model extraction attacks attempt to reconstruct a proprietary model's behavior by querying it systematically and training a surrogate model on the query-response pairs. The economic damage is the loss of the model's competitive differentiation. Detection relies on monitoring query patterns: a client that submits queries spanning a systematic sample of the input space is behaving differently from a client submitting natural application requests. Flagging and rate-limiting high-volume clients with systematic query patterns is the operational response. Complete prevention is not achievable through inference-layer controls alone;

it requires that the model is valuable enough to steal and that the inference endpoint is not publicly accessible without authentication.

Multi-tenant inference serving raises concerns about context isolation between tenants. In a session-based inference system, one tenant's session state must not be visible to another tenant. In a system that uses a shared KV cache across requests, cache partitioning must ensure that a tenant's request cannot access cached activations from another tenant's request. This requires that the inference serving layer has an explicit model of tenant identity and enforces cache isolation at the session level. Serving frameworks that do not support tenant-aware cache isolation require tenant-per-instance deployment, which reduces the efficiency benefits of a shared KV cache but eliminates cross-tenant information risk.

GPU Side-Channel Exposure and Resource Fairness

Multi-tenant GPU clusters pose two interrelated problems: side-channel exposure due to shared hardware state, and scheduling fairness when tenants compete for scarce accelerator capacity. These are different concerns, but the same layer of the platform addresses them, and the choices made for one directly affect the other.

GPU side-channel attacks in a multi-tenant setting exploit the fact that power consumption, memory bandwidth utilization, and cache behavior are observable properties of a shared physical device. A malicious workload running on the same physical GPU as a target workload can infer information about the target's computation by measuring these side channels. The most practical mitigation is hardware-level isolation through MIG partitioning, which provides separate memory and compute slices that do not share a cache. When MIG is unavailable or impractical, scheduling tenants from different trust domains to different physical nodes eliminates the shared-hardware surface.

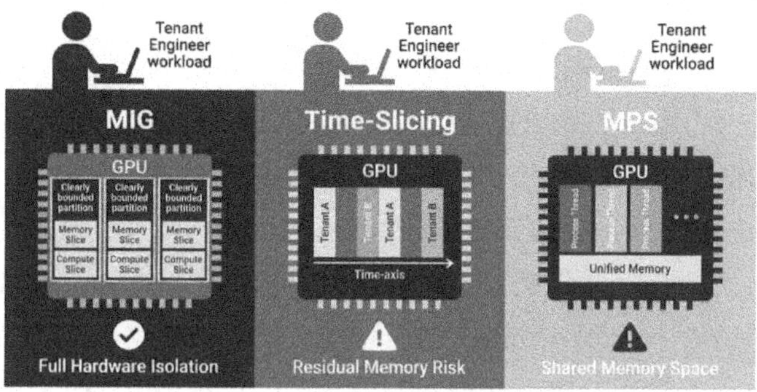

MIG for Regulated Production Time-Slicing for Dev/Test MPS for Trusted Tenants Only

Scheduling fairness in a multi-tenant accelerator cluster requires a quota system that limits how much of the cluster's GPU capacity any single tenant can consume at any given time. Without quota enforcement, a tenant that submits a large training job at the right moment can occupy a disproportionate share of cluster resources while other tenants' jobs queue. The fairness-sharing scheduler pattern addresses this by tracking each tenant's historical consumption and adjusting their scheduling priority accordingly. Tenants who have used less than their fair share get priority over tenants who have used more. Over a rolling window, each tenant receives approximately their allocated share.

Priority classes provide a mechanism for expressing that some workloads are more time-sensitive than others within a tenant's allocation. A production inference endpoint has a higher priority than a batch training job. An urgent fine-tuning run for a breaking product issue takes priority over a routine evaluation sweep. Priority classes allow the scheduler to preempt lower-priority workloads to make room for higher-priority ones when the cluster is under pressure. The platform team defines the priority class taxonomy; tenants classify their workloads accordingly. Misclassification (marking everything as high priority) is prevented by quota enforcement for high-priority slots.

Anti-Patterns: Security Choices That Look Right and Are Not

Security for distributed machine learning fails most often not because of the absence of controls, but because controls are scoped one level too coarsely. A cluster can have authentication, authorization, network policy, and encryption all switched on and still grant a single compromised workload the run of the fleet, because each control was applied at the cluster boundary rather than at the workload boundary. The anti-patterns below are configurations that pass an audit checklist and fail a penetration test. Each looks like security. Each grants more trust than its author intended, because the scope of the grant was never examined, under the assumption that a single workload within the perimeter is already hostile.

The most common security anti-pattern is cluster-level RBAC without namespace isolation. A team configures role-based access control at the cluster scope, satisfying the requirement to govern access. At the same time, every workload still shares a flat namespace in which any pod can enumerate and reach any other. The diagnostic signal is a single misconfigured or compromised workload that can list secrets, services, and pods belonging to unrelated teams. The fix is namespace-per-team or per-tenant isolation with RBAC scoped to the namespace, so that a breach is contained to the blast radius of one tenant rather than the whole cluster.

The second anti-pattern is mounting service account tokens on every pod by default. Most orchestration platforms mount a service-account credential into each pod unless explicitly told not to, and a workload that never calls the control plane therefore carries a credential it does not need, and an attacker would gladly use. The diagnostic signal is lateral movement during an incident review: a compromised inference pod reaching the API server with a token it had no functional reason to hold. The fix is to turn off automatic token mounting by default and grant it only to workloads that demonstrably require control-plane access, thereby removing the free credential that can turn a single-pod compromise into cluster-wide reconnaissance.

The third anti-pattern is PII redaction only at the application layer. A team scrubs personally identifiable information where the application reads it, satisfying the visible requirement. At the same time, the same data persists unredacted in request logs, in error traces, and — most insidiously — in gradients and intermediate activations that can leak training data through inversion. The diagnostic signal is sensitive content discovered in a debug log or a checkpoint that the application layer believed it had sanitized. The fix is defense-in-depth: redact at ingestion, scrub structured logs and traces, and treat gradients and checkpoints as data-bearing artifacts subject to the same handling controls as the raw inputs.

The fourth anti-pattern is network policies that allow by namespace rather than by workload identity. A policy that permits all traffic within a namespace is administratively convenient and grants every pod in that namespace the trust intended for a few of them. Hence, a compromised sidecar inherits the network reach of the application it accompanies. The diagnostic signal is east-west traffic between workloads that have no functional dependency on each other, yet are permitted to communicate. The fix is identity-based policy — mutual TLS or a workload-identity mesh — that authorizes connections between specific services rather than between everything that happens to share a namespace.

The fifth anti-pattern is treating the model artifact registry as trusted. Teams harden the container supply chain by scanning and signing images, and then pull model weights from a registry with none of the same scrutiny, even though a poisoned or backdoored model file is a supply-chain attack that lands as data rather than code. The diagnostic signal is a deployment pipeline that verifies container provenance but loads model artifacts with no signature check. The fix is to sign model artifacts in the build environment and verify those signatures before loading, extending the supply-chain controls already applied to containers to the model files, which now constitute an equal attack surface.

Anti-Pattern Symptom → Root Cause

Symptom (observable)	Likely Root Cause	First Mitigation
One compromised workload can enumerate the whole cluster	Cluster-level RBAC without namespace isolation	Scope RBAC to per-team / per-tenant namespaces
The compromised pod reaches the API server; it never needed	Service-account tokens are auto-mounted into every pod	Disable default token mounting; grant only where required
Sensitive data found in logs, traces, or checkpoints	PII redaction applied at the application layer only	Defense in depth: redact at ingestion, in logs, and in artifacts

East-west traffic permitted between unrelated workloads	Network policy allows by namespace, not identity	Identity-based policy (mTLS / workload-identity mesh)
Pipeline verifies containers but loads unsigned model files	Model artifact registry treated as trusted	Sign model artifacts at build; verify signatures before load

Technical Checklist

- Namespace isolation is enforced by admission control: pods cannot reference volumes, secrets, or service accounts outside their assigned namespace. This has been verified with a red-team exercise or equivalent adversarial test.
- Container runtimes are tiered: regulated workloads run in sandboxed runtimes on dedicated node pools; general workloads run in standard runtimes on shared nodes. The classification is enforced by scheduling policy, not documentation.
- Accelerator isolation is configured per workload class: MIG partitioning for regulated multi-tenant inference, standard time-slicing only for trusted dev/test environments. Time-sliced production environments have been reviewed and approved with documented risk acceptance.
- RBAC roles have been derived from a minimal starting point, reviewed in the past quarter, and validated to contain no permissions not directly required by the role's stated function. No role has been seeded from an older role without review.
- Workload identity federation is in place for all workloads that access cloud resources. No cloud credentials are stored as environment variables, image build-time arguments, or long-lived Kubernetes secrets.

- All secrets that cannot be replaced by workload identity are managed through a secrets store with versioning, rotation, and audit logging. Automatic rotation is enabled for all secrets that support it.
- Encryption at rest is in place for all training data and model artifacts, using envelope encryption with tenant-controlled key encryption keys. The platform team cannot read tenant data.
- Mutual TLS is enforced for all intra-cluster traffic by the service mesh. Layer-4 default-deny network policies are in place and verified. No namespace has a default-allow network policy in production.
- All container images are signed before admission to production. The admission controller rejects unsigned images. A Software Bill of Materials is generated and indexed for every image in the production registry.
- Model artifacts are signed at the end of training runs, covering weights, training configuration, and dataset lineage metadata. Serving infrastructure verifies artifact signatures before loading.
- Provenance tracking links every deployed model to its training run, training data version, and code commit. The provenance chain is auditable and stored in a tamper-evident log.
- Privacy-preserving patterns (DP-SGD, federated learning, confidential computing) have been evaluated against each workload class that handles individual-level sensitive data. The selected pattern and its rationale are documented.
- Data residency constraints are expressed as a scheduling policy, not documentation. Jobs that reference residency-constrained datasets will fail to schedule in prohibited regions.
- Audit logs integrate events from the cluster control plane, data access layer, model registry, and inference serving layer. Logs

are stored in a tamper-evident store with a retention schedule matching the applicable regulatory regime.

- Inference endpoints enforce token-based rate limiting, structural prompt separation, input validation, and output monitoring. All four controls are active in production.
- A security checklist covering isolation, RBAC, encryption, network policy, supply chain, and compliance posture has been signed off on by both platform and security leadership in the current quarter.

Team Conversation

1. Which of our last three security findings would the patterns in this chapter have prevented? For each finding, identify the specific pattern (isolation, RBAC, network policy, supply chain, secrets management) that addresses it.
2. Do our tenancy boundaries at the cluster and data layers match the boundaries our regulators expect? Where do they diverge, and what is the plan to close the gap?
3. Where in our platform is a long-lived credential still in use, and why has it not been replaced by workload identity federation? What would it take to replace it?
4. Is our supply-chain discipline strong enough to detect a compromised base image before it runs in production? Can we trace every currently running workload to a signed image and a verified SBOM?
5. Have we ever rehearsed an incident response for a leaked training dataset? If the regulatory clock started today, how long would it take us to identify which jobs accessed the affected data and which tenants were involved?
6. Which privacy-preserving pattern (differential privacy, federated learning, confidential computing, secure

aggregation) would unlock a workload we currently cannot ship because of data sensitivity constraints?

7. When did our security and platform leaders last sign off on the same security checklist together? Is that checklist the same one we would present to a regulator?
8. If a red team had unrestricted access to our cluster for four hours today, what would they find? Which sections of this chapter's checklist would they expose?
9. How are we managing the GPU isolation configuration for our multi-tenant inference cluster? Do the current settings match the trust relationships among the tenants sharing each physical device?
10. Are our audit logs sufficient to answer a regulator's question about which user accessed which training data over the past year? If not, what is missing, and what is the plan to add it?

Key Takeaway

Distributed ML clusters concentrate three categories of value in a small number of places: sensitive training data, trained model weights that represent months of compute investment, and expensive GPU capacity that many teams share. That concentration is a design feature that enables the efficient use of scarce resources, yet it is also a structural security risk. The patterns in this chapter are the structural responses to that risk: workload isolation that makes cross-tenant access impossible by construction, RBAC that reflects real tenancy boundaries rather than approximating them, network policy that defaults to deny, supply-chain discipline that makes tampering detectable, and an audit posture that turns compliance from a fire drill into a query.

Privacy-preserving patterns deserve special attention because they are often framed as an overhead that privacy-sensitive workloads must bear. The more useful framing is that they are unlocks. A team that cannot federate training across hospital networks due to data-sharing constraints can offload that workload with federated learning. A team that cannot process a third-party dataset without granting the data owner full trust in the platform can unlock that workload with confidential computing. Differential privacy turns a dataset that would fail a privacy review into one that ships. Each pattern converts a blocked workload into a shippable one, and the platform's mission-aligned job is to provide the infrastructure that makes each pattern available without requiring every team to implement it from scratch.

The operational clarity test for this chapter's patterns is a signed-off checklist. A platform team that has worked through the checklist at the end of this chapter, item by item, with both platform and security leadership, has produced a security posture that is both defensible and understood. A team that has not worked through it has a platform whose security properties are only approximately known. That approximation will be tested by a red team, an auditor, or an attacker, in roughly that order. The checklist is not a performance for external parties. It is the mechanism by which the team verifies together that the pilot-to-production pathway for each new workload class passes through the security controls that protect the cluster and the tenants who depend on it.

11 Distributed Patterns for LLMs and Foundation Models

Opening Scenario

The platform team at a mid-sized AI company receives a deceptively simple request: take the fine-tuning recipe the research team has been running on a sixteen-GPU workstation and turn it into a production-grade capability that serves thousands of internal analysts. The model is a sixty-five-billion-parameter instruction-tuned variant of a publicly available foundation model. The production timeline is six weeks.

The first two weeks go smoothly. The team extends the fine-tuning recipe with sharded data parallelism, distributes optimizer state across sixteen accelerator nodes, and cuts per-epoch time from eleven hours to under ninety minutes. Checkpoint quality looks strong. Leadership is confident.

Week three is when the serving problem arrives. The team's existing inference infrastructure assumes stateless, fixed-batch requests. The new model does not work that way. Each active user session requires that the key-value cache from prior tokens remain in GPU memory throughout the conversation. At peak load, 40 concurrent sessions occupy 38 GB of KV cache, leaving almost no room for new requests. The autoscaler spins up new replicas, but each new replica starts cold with no session state. Existing sessions routed to the new replicas lose their context and return incoherent responses.

The fix takes two weeks, not one afternoon. The team must replace stateless serving with an architecture that treats the KV cache as a first-class infrastructure resource: paged attention so cache memory is managed in fixed-size blocks rather than session-sized contiguous allocations, prefix caching so shared prompt prefixes are not recomputed on every request, session-aware routing so users consistently reach the replica that holds their cache, and an admission-control policy that rejects new sessions when memory pressure crosses a defined threshold. The second launch is stable. The team learns that the training patterns and the serving patterns are both LLM-specific compositions of distributed primitives from earlier chapters. Still, they compose in ways that actively break classical serving assumptions. Knowing the pattern names is the difference between a six-week timeline and a six-month one.

LLM Platform Architecture Overview

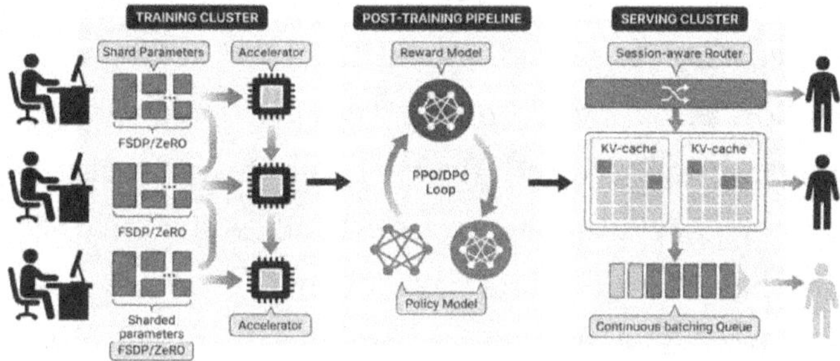

Why It Matters

Large language models are not an exotic edge case in distributed ML. They are the new mainstream. Every team shipping a conversational feature, a document-processing pipeline, or a retrieval-augmented search product is now an LLM operator. The clusters those teams run are the same as those covered in earlier chapters of this book. Still, the workloads have different memory profiles, communication patterns, and failure modes than those of any classical model-serving workload. Treating an LLM deployment as a slightly larger image classifier is the most common and most expensive mistake a platform team can make.

The mission-aligned reason to master these patterns is direct: the gap between a team that knows them and a team that does not is measured in months of instability, GPU costs two to five times higher than necessary, and user-facing regressions are invisible to classic monitoring. A KV cache, treated as a model detail, will crash under peak load. A post-training pipeline without an orchestration plan will produce corrupted reward signals. An agentic workload without a cost ceiling will consume a GPU budget in a weekend. This chapter serves as the reference for building operational clarity into an LLM platform before incidents occur.

Sharded Data Parallelism: FSDP and ZeRO Patterns

Classical data parallelism replicates the full model on every accelerator and synchronizes gradients after each backward pass. A sixty-five-billion-parameter model in BFloat16 occupies approximately one hundred and thirty gigabytes of parameter memory alone, before gradients or optimizer state. No single accelerator holds that. Sharded data parallelism is the answer: each accelerator holds only a shard, and parameters are gathered on demand during the forward and backward pass.

The Zero Redundancy Optimizer (ZeRO) family defines three stages of sharding. ZeRO-1 shards only the optimizer state, leaving parameters and gradients fully replicated. ZeRO-2 adds gradient sharding, reducing gradient memory by a factor of the number of data-parallel ranks. ZeRO-3 shards parameters as well: each accelerator stores only its slice of each parameter tensor, gathering the full tensor before each layer's forward pass with an all-gather, then redistributing gradient shards with a reduce-scatter. Fully Sharded Data Parallelism (FSDP) is the canonical implementation of the ZeRO-3 framework. The critical design decision is the wrapping policy: wrapping at the transformer block boundary, treating each self-attention plus feed-forward sublayer as one FSDP unit, is the standard heuristic.

Activation memory adds a second pressure that parameter sharding alone does not address. Activation checkpointing discards intermediate activations after the forward pass and recomputes them during the backward pass. The trade-off is roughly 30% recomputation overhead per step in exchange for a peak activation memory reduction that can approach the square root of the number of layers. The choice between ZeRO stages depends on model size, cluster topology, and interconnect bandwidth. High-bandwidth interconnects (NVLink, InfiniBand) make ZeRO-3 viable; commodity Ethernet makes it prohibitively slow. Profiling communication overhead before committing to a sharding stage at production scale is the responsible-by-design choice.

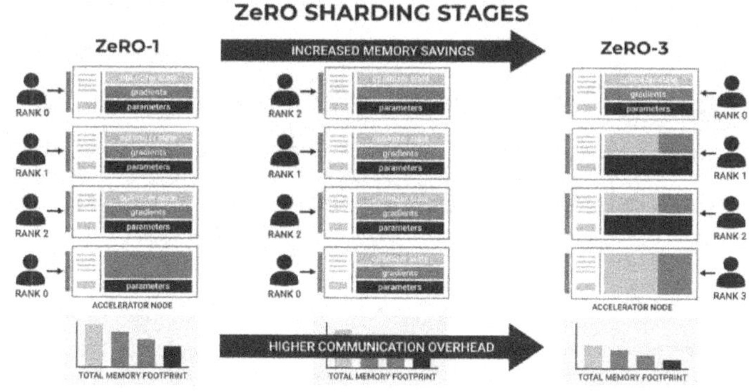

3D and 4D Parallelism for Foundation Models

Sharded data parallelism alone does not scale to the largest training runs. A cluster of ten thousand accelerators cannot run a single ZeRO-3 job across all ranks without saturating the interconnect with all-reduce traffic. The canonical answer is 3D parallelism: a combination of data parallelism, tensor parallelism, and pipeline parallelism that maps different parts of the model and different batch dimensions to different subsets of the cluster.

Tensor parallelism (TP) shards individual weight matrices across accelerators within a node. Column-parallel and row-parallel halves of each matrix multiply on separate accelerators, with an all-reduce at the boundary. Because all-reduce is on the critical path of every forward pass, tensor parallelism is practical only within a node with a high-speed interconnect, limiting the TP degree to 8 or 16 ranks. Pipeline parallelism (PP) assigns different layers to different pipeline stages across nodes. Micro-batches flow through stages like an assembly line under the 1F1B (one forward, one backward) schedule. PP tolerates lower inter-node bandwidth because inter-stage communication is via activation tensors rather than via gradient all-reduces.

Data parallelism operates at the outermost level: independent replicas of the full 3D parallel group each process different samples. The product of TP, PP, and DP must equal the cluster size. The sizing methodology first maximizes TP within a node, then selects PP to match the model depth across nodes, and finally allows DP to fill the remaining parallelism budget. Sequence parallelism (SP) is the fourth dimension for long-context models: it shards the sequence dimension of the attention computation across accelerators using ring attention, in which query, key, and value chunks rotate around a ring, so each accelerator computes its portion of the attention output. A profiling run at one-tenth scale with the target configuration is the only reliable way to validate the memory budget before a full production run.

3D Parallelism Topology: Tensor, Pipeline, and Data Dimensions

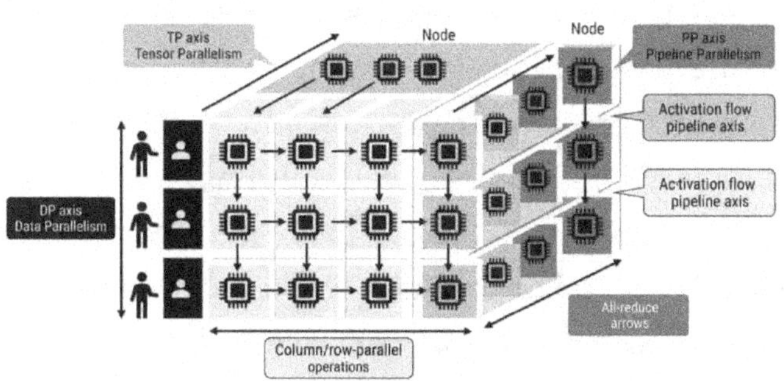

Mixture-of-Experts Training Infrastructure

Mixture-of-Experts (MoE) models replace the dense feed-forward sublayer with a set of expert networks, each processing only a subset of tokens per step. A gating function routes each token to the top-K experts (typically K = 1 or 2) from a pool of eight to thousands of experts. The theoretical compute per token remains comparable to that of a dense model. Still, the total parameter count can be 10 to 100 times larger, enabling a qualitative increase in capacity without a proportional increase in FLOPs per token.

The training infrastructure challenge of MoE is dynamic routing: the gating function assigns tokens to experts at runtime, and when experts live on different accelerators, routing requires all-to-all communication. All-to-all is irregular (different experts receive different numbers of tokens), latency-sensitive (each micro-batch must complete the all-to-all before proceeding), and expensive on long-haul interconnect. Capacity factor is the central design parameter: each expert is pre-allocated a buffer of size the average number of tokens multiplied by the capacity factor. Tokens exceeding the buffer are dropped (token dropping). A capacity factor of 1.25 reduces dropping but wastes buffer memory.

Expert collapse is a failure mode in which the gating function routes most tokens to a small subset of experts. The canonical remedy is an auxiliary load-balancing loss that penalizes the gating function for concentrating routing probability too much. The weight of the auxiliary loss requires careful tuning: too weak and collapse proceeds, too strong and the gating function learns to route uniformly regardless of token content. Expert parallelism (EP) partitions experts across accelerators and composes with TP, PP, and DP in a 4D or 5D configuration. Teams shipping MoE training at scale should plan interconnect topology before model architecture, because the all-to-all volume scales directly with the number of active experts per token and the hidden dimension.

MoE Token Routing and Expert Parallelism

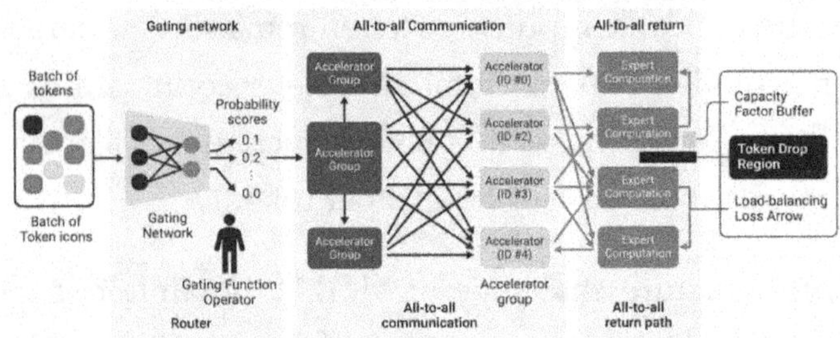

RLHF, DPO, and Post-Training Pipeline Patterns

A pretrained base model is a powerful predictor of next tokens. Post-training closes the gap to a usable product. The two dominant paradigms are reinforcement learning from human feedback (RLHF) and direct preference optimization (DPO), both of which require a distributed infrastructure that differs substantially from that used in pretraining.

RLHF uses a reward model trained on human preference data to provide a scalar signal guiding policy training. The pipeline runs four concurrent model copies: the policy model (being trained), the reference model (a frozen copy of the policy for KL-divergence penalty computation), the reward model (scoring policy outputs), and the value model (a critic estimating expected future reward). Proximal Policy Optimization (PPO) updates the policy using the reward and KL penalty. The rollout phase generates completions using autoregressive decoding and is the most infrastructure-intensive part. The update phase is a supervised training step on the experience buffer using the same sharded data-parallel patterns as pretraining. The separation between rollout and update infrastructure is the central orchestration decision: sharing the same accelerator pool is simpler to operate; separate pools deliver higher throughput.

Direct Preference Optimization (DPO) eliminates the reward model and the PPO loop. The policy model trains directly on preferred and rejected completion pairs using a closed-form objective that implicitly defines a reward function. DPO requires only two concurrent model copies (policy and reference) and removes the rollout phase entirely. The training loop resembles supervised fine-tuning. The trade-off is that DPO cannot incorporate online feedback; all preference data must be collected before training begins. For teams without a large human annotation pipeline, DPO on a curated preference dataset is often the right path from pilot to production. LoRA (Low-Rank Adaptation) and QLoRA reduce memory footprint for supervised fine-tuning by training only low-rank adapter weights adjacent to frozen pretrained weights, enabling fine-tuning of very large models on modest hardware.

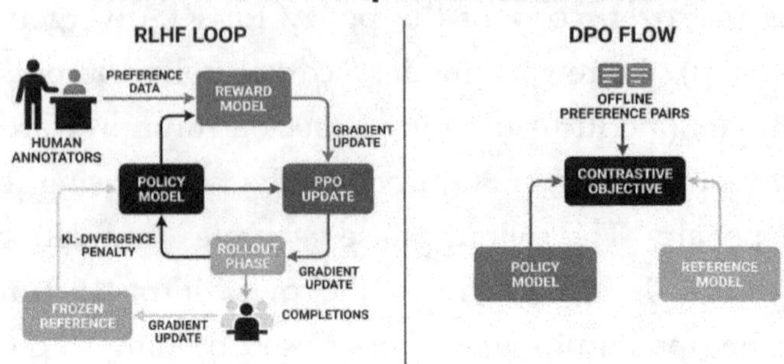

Long-Context Training: Sequence Parallelism and Activation Strategies

Doubling context length quadruples the memory required for the attention computation, because attention scores scale as the sequence length squared. A model with a context length of 128,000 tokens requires fundamentally different training infrastructure than one with a context length of 4,000 tokens, even if all other hyperparameters are identical. The memory budget for long-context training has three dominant components: attention activation memory (quadratic in sequence length), feed-forward activation memory (linear), and backward-pass gradient buffers. Activation checkpointing converts the quadratic memory cost of quadratic attention into a linear one by discarding and recomputing attention activations. Above 32,000 tokens, it is typically mandatory.

Flash attention makes long-context attention computationally tractable by tiling the computation into blocks that fit in SRAM rather than materializing the full sequence-by-sequence score matrix in HBM. This reduces HBM memory footprint for attention from quadratic to linear, proportionally reducing HBM reads and writes. Flash attention is now the default attention implementation for long-context training across all major frameworks. Ring attention extends flash attention for sequence parallelism: the sequence is split across a ring of accelerators, key-value chunks rotate around the ring, and each accelerator computes its portion of the attention output using the local query chunk and the currently available key-value chunk. The ring all-gather can be overlapped with computation for the previously received chunk, hiding most of the communication latency.

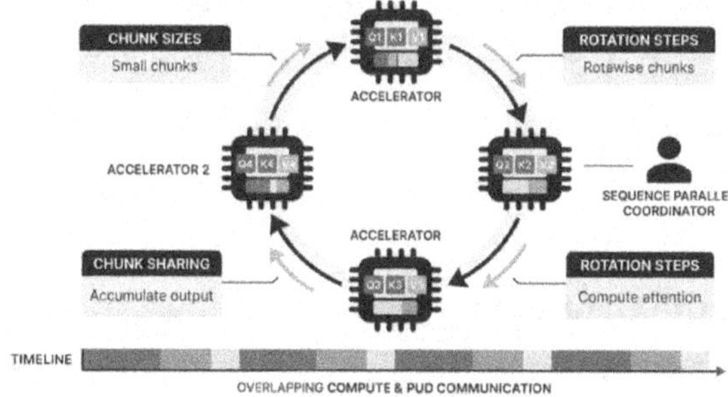

Pretraining Data Curation at Scale

The quality and composition of pretraining data are the highest-leverage variables in foundation model development. A cluster of ten thousand accelerators running a well-engineered training job produces a model bounded by its data. Teams that invest heavily in parallelism configuration and lightly in data curation are making a systematic allocation error.

Deduplication is the most impactful single step in curation. Web-crawled corpora contain massive amounts of near-duplicate content: the same article repeated across news aggregators, boilerplate legal text appearing millions of times, and SEO pages differing in only a handful of words. Near-duplicate content biases the model toward overrepresented material and inflates corpus size without adding information. The canonical approach uses MinHash locality-sensitive hashing to find document pairs with high n-gram overlap, then removes one member of each pair. Decontamination prevents evaluation benchmark data from appearing in training. A thirteen-gram exact-match check against every known benchmark removes any document that could inflate benchmark scores. Quality filtering applies perplexity under a reference language model, n-gram repetition rate, and heuristic rules for known low-quality patterns.

Domain mixing ratios determine the relative proportion of each domain (web text, code, papers, books) in the training corpus. They should be treated as a documented, versioned hyperparameter validated against a held-out benchmark suite before the full run. Streaming data patterns implement domain ratios by adjusting sampling weights at the data loader level, enabling ratio changes between runs without rebuilding the corpus. Provenance tracking records the source URL or origin, the crawl date, the processing steps, and the pipeline version for every document. Provenance enables the reproduction of any prior run, the removal of a specific source in the event of a licensing dispute, and the demonstration of an audit trail to a regulator. Building provenance tracking retroactively is substantially more expensive than designing it in from the start.

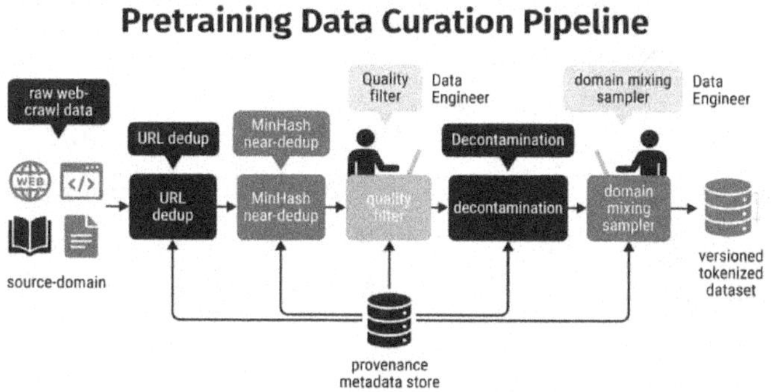

Pretraining Data Curation Pipeline

KV Cache Management and Paged Attention

The key-value cache is the data structure that enables efficient autoregressive generation. Each forward pass computes new query, key, and value tensors for the current token; the serving system caches the keys and values from all previous tokens so they need not be recomputed. The KV cache for a single session grows by one key vector and one value vector per generated token, per layer. For a 65-billion-parameter model with 80 layers and a 16,000-token context window, the KV cache for a single session can exceed 10 gigabytes.

Classical serving systems allocate KV cache memory statically: each session reserves a contiguous block sized to the maximum context length at session creation. This is logically simple but physically wasteful. A session generating 200 tokens wastes the reservation for 15,800 unused tokens. Paged attention solves this by managing the KV cache in fixed-size blocks (pages), analogous to virtual memory. Each page holds the key and value tensors for a fixed number of tokens, typically sixteen or thirty-two. A page table maps logical token positions to physical pages. Pages are allocated on demand and freed when the session ends. Fragmentation is limited to at most one partially filled page per session, and GPU memory utilization approaches the theoretical maximum.

Prefix caching shares pages across sessions that share a common prompt prefix. When all sessions share a fixed system prompt, prefix caching computes the KV cache for that prompt once and references it with copy-on-write semantics. For a system prompt of 1,000 tokens, this eliminates 1,000 tokens of prefill computation per new session. Eviction policies (least-recently-used by default) determine which pages are freed when the total KV cache fills. Capacity planning for KV cache uses the formula: average concurrent sessions × average session length in tokens × 2 × the number of layers × the head dimension × the number of KV heads × the element byte size. This budget must be subtracted from the total GPU memory before any other allocation. Teams that treat KV cache as a residual discover the error at the first traffic spike.

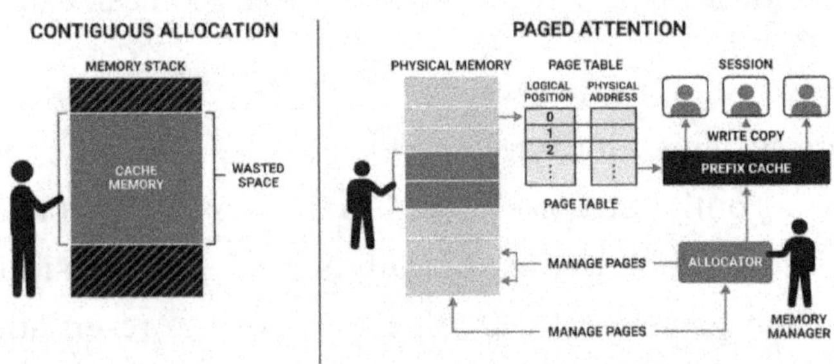

Continuous Batching and Speculative Decoding

Static batching groups requests into fixed-size batches and runs each batch to completion before accepting new requests. When request lengths vary widely, a batch containing one long request and seven short requests leaves seven slots idle while the long request finishes, resulting in poor GPU utilization. Continuous batching treats the serving engine as a token-level scheduler: each forward pass may include tokens from multiple requests at different stages of generation. When a request finishes, its slot is immediately filled by the next request in the forward pass. Throughput improvements of two to ten times over static batching are typical, depending on the variance in request lengths.

Continuous batching requires disaggregating the prefill phase (computing the KV cache for all prompt tokens in parallel, compute-bound) from the decode phase (generating new tokens one at a time, memory-bandwidth-bound). A common scheduling strategy completes a set of decode steps before admitting new prefill requests, preventing prefill from starving decode throughput.

Speculative decoding accelerates the decoding phase by using a small draft model to propose K candidate tokens in K serial steps, which the large target model then verifies in a single parallel forward pass. When the draft model acceptance rate is high, speculative decoding yields a speedup proportional to the average number of accepted tokens per verification step, with no change in output distribution. Lookahead decoding is an alternative that does not require a separate draft model: it runs multiple parallel n-gram hypotheses. It advances the decode position by multiple tokens when a consistent continuation is found. Combining continuous batching with speculative decoding captures both the throughput benefit of dynamic batching and the latency benefit of multi-token verification. The configuration space includes the batch scheduler policy, the draft model choice, the speculation depth K, and the acceptance-rate threshold below which speculation is disabled.

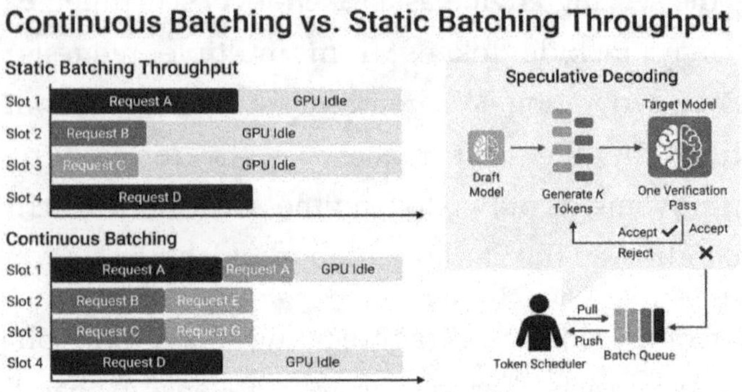

Distributed Inference for Very Large Models

Inference optimizes for a combination of throughput (tokens generated per second across all users) and latency (time-to-first-token and per-token latency for an individual user). The parallelism configuration optimal for training throughput is often suboptimal for inference latency. Tensor-parallel inference distributes computation within a single forward pass: weight matrices are sharded column- and row-parallel, and each forward pass requires an all-reduce after the row-parallel multiplication. Because all-reduce is on the critical path for every generated token, tensor parallelism for inference is strictly limited to accelerators with high-bandwidth intra-node interconnects, typically with TP degrees of 2 to 8.

Pipeline-parallel inference assigns different layers to different stages. For batch workloads with high latency tolerance, pipeline parallelism increases throughput by allowing stages to process different micro-batches concurrently. For online generation with tight latency requirements, pipeline parallelism adds inter-stage activation transfer latency to every forward pass, increasing time-to-first-token. Disaggregated prefill-decode serving routes prefill and decode phases to separate accelerator pools sized for their respective bottlenecks: prefill pools are compute-bound, decode pools are memory-bandwidth-bound. The cost is the KV cache transfer between pools during request transitions, which requires high-bandwidth inter-pool connectivity.

Multi-tenant serving adds isolation requirements: tenant-aware KV cache isolation ensures one tenant's cache pages are not readable by another, even when physical memory is shared. Prefix caching must be scoped per tenant to prevent information leakage. Token streaming delivers the character-by-character output experience users expect. The serving engine streams each generated token to the client via server-sent events or WebSocket, which require stateful connections that persist throughout generation. A session mid-generation cannot be freely rerouted to a different replica without losing the streaming connection, so the load balancer must implement session affinity for streaming requests.

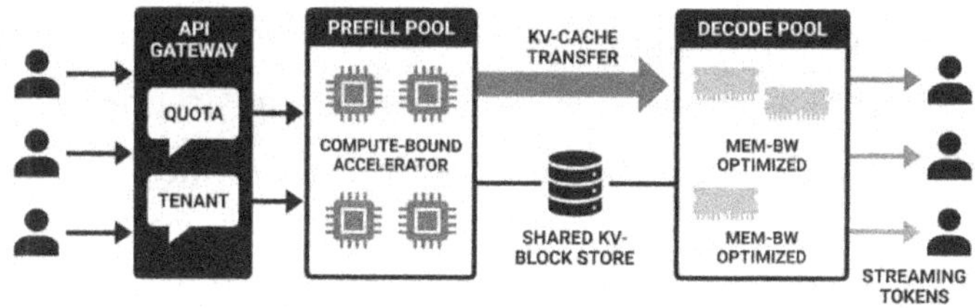

Retrieval-Augmented Generation as a Distributed Pattern

Retrieval-augmented generation (RAG) extends an LLM with external knowledge retrieved at inference time. From a distributed systems perspective, RAG transforms a single-model serving problem into a multi-component distributed serving problem with its own latency budget, failure modes, and scaling dimensions. The vector index is the core infrastructure component: documents are encoded into dense vectors, indexed in a distributed vector store, and retrieved via approximate nearest neighbor (ANN) search at query time. ANN search over one billion vectors requires a distributed index: the index is sharded across nodes, queries are broadcast to all shards, and the top-K results from each shard are merged and reranked (scatter-gather).

Hybrid search combines dense vector search with sparse keyword search (e.g., BM25). Dense retrieval excels at semantic similarity; sparse retrieval excels at exact-match recall for technical terms, proper nouns, and product identifiers. Running both searches in parallel, merging results with reciprocal rank fusion, and passing the merged top-K to a cross-encoder reranker almost always outperforms either modality alone. The reranker's latency must fit within the LLM generation latency budget; a typical production target is total retrieval and reranking under 200 milliseconds. Prompt routing sits upstream of retrieval in a multi-domain deployment, directing different query intents to different knowledge bases or bypassing retrieval entirely for general questions. The router's latency must be budgeted as an additional serving component and tested independently under load.

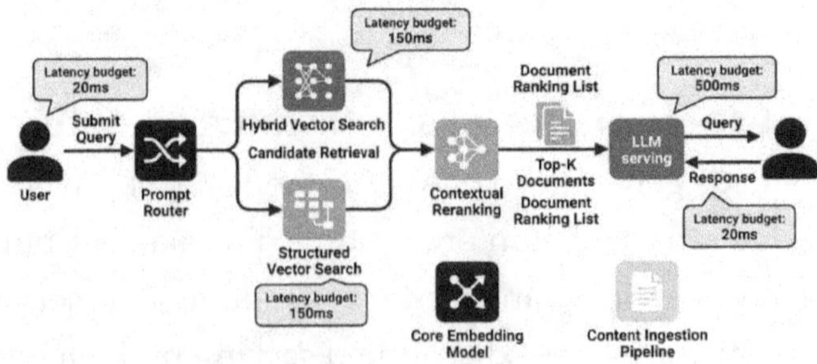

Agentic Workloads and the Multi-Step Inference Pattern

Agentic workloads extend LLM inference from a single forward pass to a multi-step plan in which the model decides at each step whether to produce a final response or to call an external tool to reason over the result. The execution graph is a tree or DAG of inference steps interleaved with tool calls and intermediate state accumulation. This is not a classical model serving; it is a distributed workflow with the LLM as orchestrator.

The operational profile differs from single-step inference in three important ways. Latency is unbounded by construction: a poorly prompted agent may take 20 steps to complete a task that should take 1. Cost is runtime-determined: a single task can consume hundreds of thousands of tokens across multiple inference calls if unconstrained. Failure modes are more complex: a tool returning an error may trigger indefinite retries or cascading calls. Cost ceilings (total token budget per task) and step budgets (maximum tool calls) are the primary controls; both must be enforced in the orchestration layer rather than trusted to the model. Retry budgets (per tool, with exponential backoff) prevent transiently unavailable tools from exhausting the step budget without progress.

The intermediate state must be durably stored so that if the orchestrator crashes mid-task, the task can resume from the last completed step rather than restart from scratch. A distributed task queue with checkpoint-and-resume semantics is the canonical pattern. Safety guardrails for agentic workloads must be applied at every tool-call boundary, not only at the final output. A model with access to a code interpreter can take harmful intermediate actions invisible in the final response. An output safety filter on every model output that may result in a tool call, a content policy check on every tool call argument, and a tool call audit log recording every action are prerequisites for production deployment, not optional features to add later.

Agentic Workload Orchestration with Cost and Safety Controls

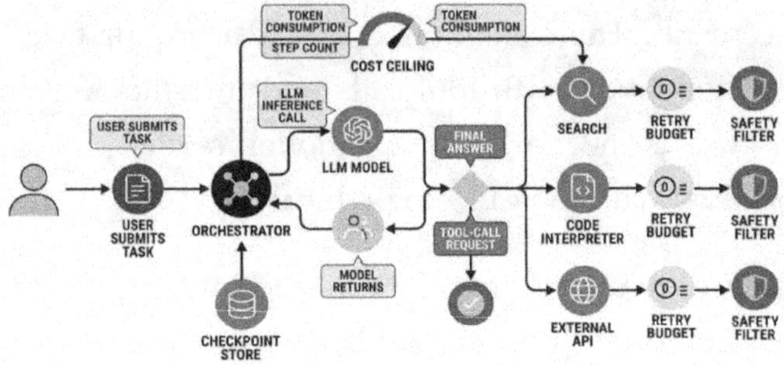

Evaluation Harness and Safety Filters at Scale

Running a standard benchmark suite for a large language model can take days on a single model replica. A distributed evaluation harness shards the benchmark across evaluation workers, each running a model replica and evaluating its assigned subset in parallel, with results aggregated by a coordinator. The harness must set random seeds per example to ensure reproducibility, record the exact checkpoint hash for every run, and flag benchmark contamination for any item that overlaps the training set. LLM-as-judge evaluation uses a stronger or differently aligned judge model to assess quality on subjective dimensions (helpfulness, harmlessness, honesty) that are expensive to annotate at scale. Judge models exhibit systematic biases (e.g., a preference for longer or their own outputs) that must be calibrated before scores are used for model selection.

Safety filters are applied at multiple points in the request lifecycle. Input filters check the user's prompt for jailbreak attempts, personal information, and restricted content before passing it to the model. Output filters check the model's response before returning it. Cascade filtering, using a fast, inexpensive filter first and a more accurate, slower filter only when the fast filter flags the input, keeps total safety filter latency within the serving budget. Abuse prevention at the API level adds rate limiting at the user, tenant, and IP levels. Prompt injection detection addresses the attack pattern in which adversarial instructions are embedded in the content the model processes. Defense-in-depth through input sanitization, strict output parsing schemas, and tool call argument validation reduces the attack surface. A production LLM platform without these controls is not responsible by design.

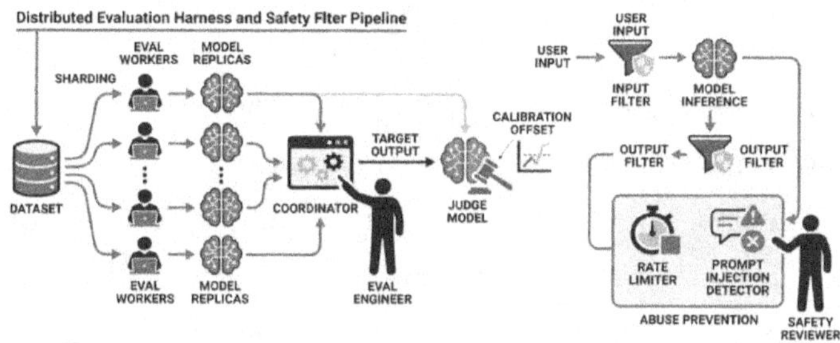

Continued Pretraining and Domain Adaptation

Continued pretraining extends the foundation model's training on a domain-specific corpus after the initial pretraining phase, bridging a general-purpose base model and a domain-specialized model without the cost of training from scratch. The infrastructure is nearly identical to that used in the initial pretraining. The differences are the data mix (heavily weighted toward the target domain) and the learning rate schedule (starting from a lower rate to avoid catastrophic forgetting). Catastrophic forgetting is the central failure mode: training exclusively on domain data gradually overwrites general capabilities. The canonical mitigation is to mix five to twenty percent general-domain data into the ongoing pretraining corpus.

LoRA (Low-Rank Adaptation) makes domain adaptation feasible for teams without pretraining-scale compute. LoRA inserts trainable rank-decomposed matrices into selected layers and trains only those adapters, keeping the base model frozen. For a one-hundred-billion-parameter model with LoRA rank sixteen applied to all attention projection matrices, the trainable parameter count is roughly one hundred million, a thousand times smaller than the full model. Adapter checkpoints are small enough to version for many domain variants of the same base model. Multi-adapter serving loads and unloads adapters per request based on tenant configuration, enabling a shared base model replica to serve multiple domain specializations. Prefix tuning and prompt tuning are lighter-weight alternatives with smaller parameter counts and faster training times, but at the cost of lower representational capacity. The pilot-to-production pathway for PEFT-based adaptation is to validate the adapter on held-out domain benchmarks, confirm that general capability regression is within acceptable bounds, and then deploy it via a multi-adapter infrastructure.

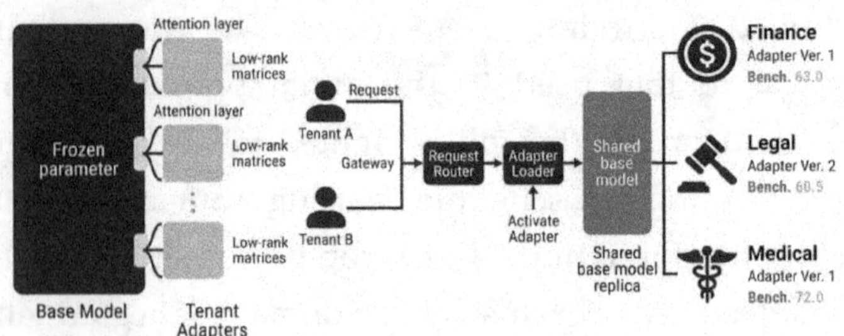

Technical Checklist

Use this checklist as a release gate before declaring an LLM platform component production-ready.

- The team has chosen a sharded-data-parallel strategy (ZeRO-1, ZeRO-2, ZeRO-3, or FSDP) deliberately, documented the decision rationale, and recorded the alternative that was declined and why.
- The 3D or 4D parallelism configuration (TP degree, PP degree, DP degree, optional SP degree) is sized to the cluster topology and verified by a profiling run at one-tenth scale before the full production run.
- Activation checkpointing policy is documented: the memory budget with checkpointing enabled fits within per-accelerator limits, and recompute overhead has been measured and accepted.
- Post-training pipelines (SFT, RLHF, or DPO) have an explicit orchestration plan for concurrent model copies, including which models are pinned in GPU memory and which can be offloaded to CPU.
- Preference data and demonstration data are provenance-tracked with annotator metadata, version identifiers, and an audit log producible on request.
- KV cache is treated as a first-class infrastructure resource: capacity has been planned using the per-token size formula, eviction policy is documented, and paged attention or an equivalent is in place.
- Prefix caching is deployed for any system prompt appearing in more than ten percent of requests, and the cache hit rate is tracked as a serving metric.

- Continuous batching or equivalent throughput patterns are in place for decode-phase serving, and GPU utilization during decode is tracked above a defined floor.
- Long-context training configuration is planned around explicit memory and cost budgets; sequence parallelism degree is validated; ring attention or equivalent is in place for context lengths above thirty-two thousand tokens.
- Pretraining data has documented deduplication, decontamination, and provenance pipelines. Decontamination results are recorded for each benchmark and each training run.
- MoE training configuration (capacity factor, load-balancing loss weight, expert parallelism degree) is documented and validated against expert utilization metrics from a profiling run.
- Agentic workloads have explicit cost ceilings, step budgets, per-tool retry budgets, and safety guardrails at every tool-call boundary before reaching production.
- Safety filters are deployed at input and output stages, latency is within the serving budget, and cascade filtering is in place when any filter exceeds twenty milliseconds.
- The evaluation harness is distributed, checkpoint-versioned, and decontamination-aware. Each evaluation run records the model checkpoint hash and benchmark contamination status.
- Domain adaptation adapters (LoRA or equivalent) are version-controlled per domain, validated against domain and general benchmarks, and deployed via multi-adapter serving infrastructure.

Team Conversation

Use these questions in a team meeting to pressure-test alignment on the patterns in this chapter.

1. Which of the LLM-specific patterns in this chapter are we already using by accident, and which ones are we missing? Where would the gap hurt us first?
2. Where in our serving stack are we treating the KV cache as a side effect rather than as infrastructure? What would need to change to give it formal capacity planning and eviction policies?
3. Have we ever shipped a post-training pipeline (SFT, RLHF, or DPO) to production? What did we learn about its operational shape, and what would we do differently with the patterns in this chapter?
4. Is our long-context strategy a research artifact or a production capability? Have we measured activation memory and communication overhead at our target context length on production hardware?
5. How would our pretraining data pipelines stand up to a regulator's audit? Can we produce a provenance trace for any document in the training corpus within a business day?
6. What is the cost ceiling for our most ambitious agentic workload, and who owns enforcing it? Does that person have the authority to terminate a runaway task mid-execution?
7. Which of the distributed patterns from earlier chapters compose naturally with our LLM workloads, and which ones break? What is the breakage mechanism?
8. How are we validating model quality after each post-training stage? Is our evaluation harness fast enough to catch regressions before a new model version reaches production?
9. Have we applied safety filters at every tool-call boundary in our agentic systems, or only at the final output? What is the worst-case intermediate action an agent could take without triggering a filter?

10. If we had to cut our LLM serving GPU cost by fifty percent next quarter without reducing user-facing quality, which patterns in this chapter would we apply first, and in what order?

Key Takeaway

Large language models and foundation models are the new mainstream of distributed ML. Every pattern this book covers, from compute parallelism to fault-tolerant checkpointing to autoscaling to observability, appears in the LLM stack. What changes is the composition: sharded data parallelism at ZeRO-3 depth, 3D or 4D configurations that map model shape to cluster topology, and post-training pipelines that require four concurrent model copies and careful orchestration. The engineer who understands the primitives from earlier chapters and the LLM-specific compositions from this chapter can recognize and name any pattern they encounter on any framework and any cloud.

The KV cache is infrastructure. It is not a model implementation detail or a serving engine internal. It is a stateful resource that requires formal capacity planning, eviction policies, tenant isolation, and operational clarity on par with a production database. Teams that implement paged attention, prefix caching, and admission control before traffic arrives experience qualitatively lower incident volume than teams that add them reactively. The same is true of inference throughput patterns: continuous batching and speculative decoding bend the GPU cost curve more than any hardware upgrade, and both are architectural choices that must be made before the serving cluster is sized.

Agentic workloads are the frontier where these patterns compose in the most complex ways. An agent is a distributed workflow with the LLM as orchestrator, each tool call as a remote procedure call, and intermediate state as a distributed checkpoint. Cost ceilings, step budgets, retry budgets, durable intermediate state, and safety guardrails at every tool-call boundary are not features to layer on after the agent is built. They are the architecture. A team that reaches production with those controls in place has built a responsible-by-design system. A team that defers them has scheduled an incident for a future date.

12 Case Studies and Reference Architectures

Opening Scenario

A new engineering director arrives at a company with five well-funded ML teams, each of which has built and operated distributed training and serving pipelines for two or three years. Her first question is simple: What is our reference architecture for a training job? The answers come back as five different documents: three out of date, two in slide format, and none written in the same vocabulary. One team calls its coordination layer an orchestrator. Another calls the same concept a launcher. A third does not name it at all, because the concept lives entirely in a shell script that one engineer wrote and nobody else can read.

She spends her first month mapping the real architectures and finds that every team has independently solved the same five problems: how to shard data across workers, how to checkpoint a long-running job, how to autoscale serving replicas under load, how to route traffic between model versions, and how to detect that something is wrong before a user files a ticket. The solutions differ not because the problems differ, but because no one compared notes. By her estimate, the total duplicated engineering effort amounts to roughly two team-years of work that could have been spent on shared infrastructure or documentation.

Her first deliverable is a reference-architecture document that the rest of the company can use as the default starting point for a new production ML workload. Writing it forces her to make the implicit decisions explicit: which parallelism pattern fits which model size, where consistency must be traded for latency, and what the cost envelope looks like at each scale tier. The document does not prescribe a single stack, but it names the patterns from which any stack can be assembled.

This chapter is the long form of what that document should contain. It presents four composite reference architectures, each shaped by a real production workload class. No real company names appear because the goal is pattern literacy, not vendor endorsement. Each architecture is presented with its business context, binding constraints, the patterns it composes from earlier chapters, its key trade-off decisions, and the lessons that survive the incident retrospectives. The chapter closes with a template the reader can adapt without copying the case studies wholesale.

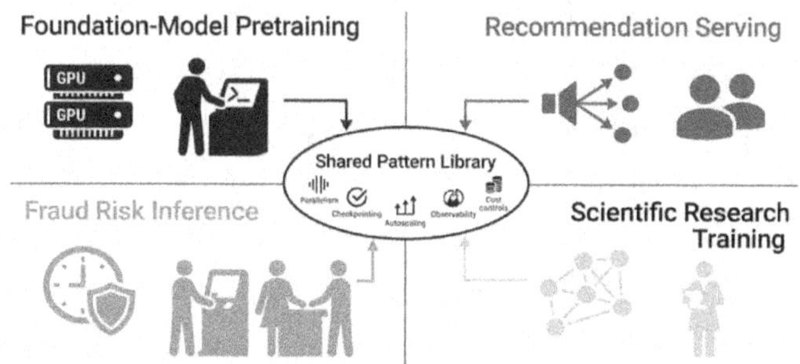

Why It Matters

A vocabulary of isolated patterns is not the same as an architecture. An architecture is a composition, and the skill of composing patterns correctly is what the field most consistently underestimates. Teams that choose the wrong parallelism strategy for their model size find that training throughput is capped by a bottleneck that no hardware scaling can remove. Teams that route serving traffic without a clear latency budget find that the first time a model call is on the critical path of a user-facing request, every other team is blocked waiting for the ML team to debug a P99 regression.

The mission-aligned goal of this chapter is to short-circuit that cycle. Reference architectures are how an organization scales its pattern knowledge beyond the founding team. They are the highest-leverage artifacts a technical team can produce because they encode decisions that would otherwise be rediscovered at a cost during every new project. The trade-off analysis embedded in each architecture shows exactly which decisions deserve explicit documentation versus which can be left to local judgment. A team that maintains written reference architectures with named owners, explicit trade-off sections, and a regular refresh cadence operates with operational clarity that a team without them cannot match.

Case Study 1: Foundation-Model Pretraining at Scale

The first composite organization is a research group pretraining a dense transformer language model with nearly 70 billion parameters on a multi-region GPU cluster spanning two cloud regions and one collocated on-premises site. The business context: a foundation model the organization plans to release as a hosted API and as a fine-tuned backbone for downstream products. The training budget is fixed, the launch timeline is public, and the cost of a restart from scratch is measured in weeks of calendar time and millions of dollars of compute.

The binding constraints are: no single region has enough high-bandwidth-memory accelerators to train the full model, the data corpus is geographically distributed and cannot all be moved before training begins, the organization must demonstrate RLHF readiness at the end of pretraining, and the cluster includes both recent-generation and prior-generation GPU types with different memory capacities and peak FLOP rates.

The parallelism strategy, developed using the framework from Chapter 2, combines three-dimensional parallelism: tensor parallelism within a single accelerator node, pipeline parallelism across nodes within a region, and data parallelism across regions. The team sized the tensor parallelism degree to match the intra-node interconnect bandwidth, sized the pipeline parallelism degree to fill the remaining model memory requirement, and treated the data parallelism degree as the residual covering the full batch-size target. Getting those three numbers wrong in either direction costs model FLOP utilization that cannot be recovered by adding hardware.

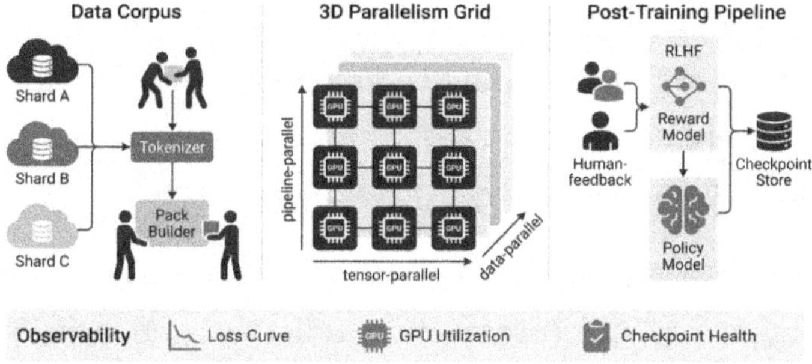

The data pipeline applied the interleaved data-parallelism pattern from Chapter 3, pre-tokenizing each shard into fixed-length sequence packs and distributing the packs across workers using a deterministic shuffle with a documented seed. The shuffle seed and pack-to-worker mapping were both persisted in the checkpoint, so a restart resumed exactly the data order the interrupted run would have seen.

Checkpointing followed the fault-tolerant training patterns from Chapter 5 in two streams: a hot checkpoint to a fast NVMe-backed object store every 500 steps, and a cold checkpoint to a cross-region durable store every 5,000 steps. The hot checkpoint absorbed individual node failures, which occurred roughly once every 8 hours on a cluster of this size. The cold checkpoint protected against full-region disruptions and silent corruption in the checkpoint store.

The RLHF readiness requirement drove two architectural decisions. First, the team reserved dedicated inference replicas alongside the training cluster for reward model evaluation during pretraining, allowing the reward model to be exercised and debugged before pretraining was completed. Second, the team used the multi-phase training pipeline pattern from Chapter 10, treating pretraining, supervised fine-tuning, and RLHF as three sequential phases of a single DAG-scheduled job. That design cut the wall-clock time from pretraining completion to the first RLHF checkpoint by approximately 6 days compared to treating each phase as an independent launch.

The cost architecture from Chapter 7 relied on spot instances for the outer data-parallel replicas (preemption rolls back at most 500 steps) and on accelerator sharing for the reward-model inference replicas, which ran at low utilization during pretraining and could colocate with non-training inference workloads. The combination reduced the effective compute cost by roughly 30 percent compared to reserved on-demand instances. Key observability lesson: pipeline bubble ratio is a leading indicator of throughput degradation and should be alerted at the same priority as a loss spike, because a bubble-ratio regression caused by a slow pipeline-stage node collapses training throughput without producing any visible loss anomaly.

Lessons learned: 3D parallelism sizing decisions made before the first training run are very expensive to change mid-run, so the team should invest in a small-scale 1,000-step profiling run before committing to the full-scale configuration. Cross-region communication for the data-parallel gradient all-reduce is the single largest cost line in the network budget and should be sized with a 2x headroom buffer. The RLHF readiness requirement should be part of the architecture from day one.

Case Study 2: High-Throughput Recommendation Serving

The second composite organization operates a consumer-scale recommendation platform processing hundreds of thousands of inference calls per second at peak. The business context is a multi-stage ranking pipeline in which multiple models contribute scores to a final ranked list. The serving platform must hit a 50-millisecond end-to-end latency budget at the 99th percentile. Missing that budget is directly correlated with measurable drops in real-time user engagement metrics.

The architecture is a multi-model serving stack of the type described in Chapter 6, with four stages: retrieval (fetching a candidate set from an approximate nearest-neighbor index), lightweight scoring (a small model filtering the candidate set), heavy ranking (a large model scoring the filtered set), and business-logic post-processing (a deterministic rule layer applying editorial constraints). Each stage has its own serving tier and autoscaling policy because the stages' compute and latency profiles differ by an order of magnitude.

High-Throughput Recommendation Serving Pipeline

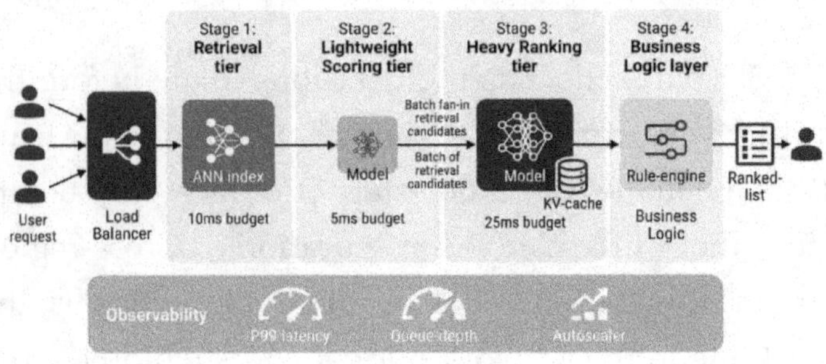

The KV-style cache layer operates on embedding lookups for user and item features in a two-tier structure: an in-process memory cache for the hottest 10 percent of user-feature vectors, backed by a distributed key-value cache for the next 40 percent. The bottom 50 percent go directly to the feature store. The in-process cache hit rate was 92 percent at steady state, meaning the heavy-ranking model saw pre-fetched features for nearly all high-traffic requests and did not wait on a feature-store round-trip.

Autoscaling policy differed across stages. The retrieval tier scaled on QPS because its bottleneck is I/O throughput. The lightweight-scoring tier is scaled based on CPU utilization because its model is small enough to run on CPU, and its per-request compute is bounded. The heavy-ranking tier was scaled based on GPU memory pressure rather than GPU utilization because the heavy-ranking model saturates GPU memory before compute, and scaling on utilization would under-provision the tier at peak. The business logic layer is scaled based on the depth of the request queue because the request queue is CPU-bound and bursty.

The key trade-off is cache freshness versus latency. User features change continuously, but refreshing the in-process cache too aggressively can drive feature-store read traffic beyond the feature store's capacity. The team settled on a time-to-live of 30 seconds, the longest TTL at which feature staleness had no detectable impact on the engagement metric. That decision is explicitly documented in the reference architecture, because it may look arbitrary in code but represents weeks of experimental work. Lessons learned: latency budgets cascade and must be allocated to individual stages before the architecture is built, not discovered through load testing after deployment.

Case Study 3: Real-Time Fraud and Risk Inference

The third composite organization is a financial services platform running ML-based fraud and risk scoring on every payment transaction in real time. The latency requirement is 10 milliseconds at the 99.9th percentile, with a hard deadline that cannot be negotiated. If the model does not return a score before the payment authorization times out, the transaction is either declined by default or approved without a score. From a distributed systems perspective, the architecture is the most constrained of the four case studies.

Chapter 6 names this pattern class 'hard-deadline inference' and distinguishes it from the soft-deadline inference of the recommendation case. In hard-deadline inference, the system must shed load, degrade gracefully, or route to a fallback model before it misses the deadline. It cannot queue the request and serve it late. That constraint propagates through every layer of the architecture.

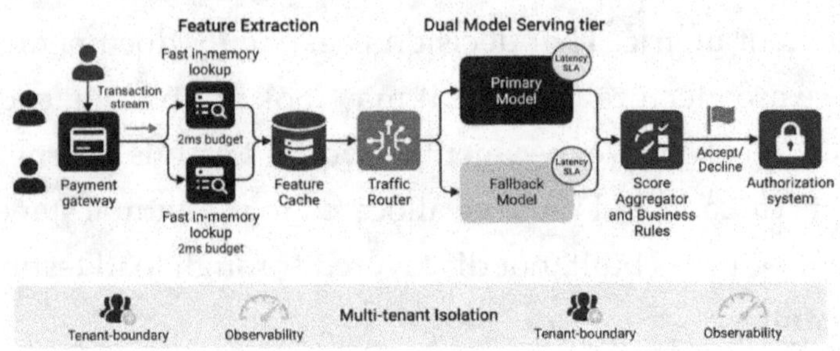

The serving pattern is a primary-plus-fallback dual-model design. The primary model is a gradient-boosted ensemble that can score a transaction in under 5 milliseconds when features are pre-fetched. The fallback model is a shallow logistic regression that scores within 2 milliseconds even when features must be fetched from scratch. The circuit-breaker pattern from Chapter 6 monitors the primary model's P99.9 latency in a 10-second rolling window and automatically switches traffic to the fallback when the primary exceeds 7 milliseconds. The fallback has lower recall for complex fraud patterns, but the business explicitly accepted that degraded recall is preferable to missing a hard deadline.

The multi-tenant dimension from Chapter 9 arises because the platform serves multiple merchant clients, each with a client-specific adaptation layer on top of the shared base model. Each incoming request carries a tenant identifier; the routing layer maps it to the correct model variant, and the serving tier enforces compute quotas per tenant, ensuring that no single high-volume merchant starves the capacity available to smaller merchants. The consistency trade-off is explicit: model updates are canary-gated over 24 hours to allow two full diurnal traffic cycles to validate latency before full promotion. Faster updates risk latency spikes during model warm-up that could trigger hard-deadline misses.

Lessons learned: hard-deadline inference architectures must be designed with the fallback path in mind from the start, not as an afterthought. Teams that add a fallback model after a production incident typically find that the fallback's feature dependencies do not match those of the primary model, and that aligning them requires rebuilding the fallback from scratch. Multi-tenancy at the model level requires tenant quotas at the serving layer, not just at the API gateway, because enforcing quotas at the gateway does not prevent a single tenant from monopolizing GPU memory within the serving pod.

Case Study 4: Scientific Research Training

The fourth composite organization is a research computing group at a large academic institution or a national laboratory. The workload class is long-running distributed training for scientific ML models, including physics simulation surrogates and climate downscaling models. Defining characteristics: training runs lasting days or weeks, a heterogeneous cluster spanning multiple GPU generations alongside FPGAs and custom inference accelerators, and an allocation model in which compute time is budgeted by a committee rather than purchased on demand.

The cost of failure is measured not in revenue but in allocation time. A training run that crashes after ten days and cannot resume from a checkpoint does not lose money directly, but it loses ten days of scarce allocated compute that cannot be recovered within the current cycle. The fault-tolerance requirement is therefore more stringent than in a commercial setting where spot instances can be repurchased.

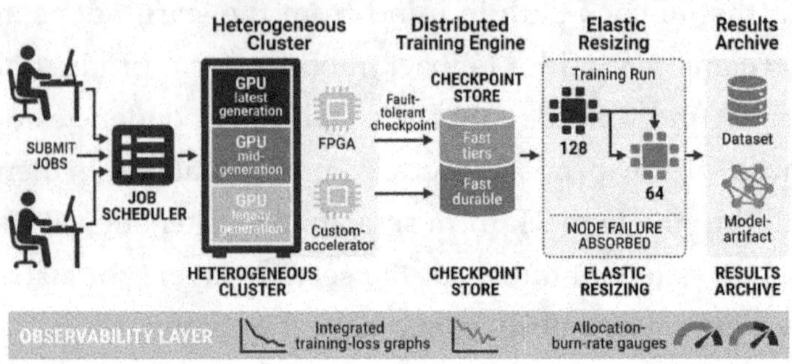

The parallelism strategy cannot be fixed at job submission time because the effective cluster size changes as nodes fail. The team applied the elastic training pattern from Chapter 5, allowing a job to shrink from its initial node count to a minimum viable count without restarting. When a node fails, the job repartitions its data-parallel replica set across surviving nodes, rebalances the gradient all-reduce topology, and continues from the most recent checkpoint. The checkpoint interval is dynamically adjusted: shortened as the cluster approaches the minimum viable count (because the probability of a second failure increases) and lengthened when the cluster is healthy (because checkpointing competes with training for I/O bandwidth).

The heterogeneous accelerator problem, named in Chapter 2 as straggler amplification, required accelerator-aware job placement: the scheduler placed compute-intensive pipeline stages on the fastest accelerator tier and embedding-lookup stages (I/O-bound) on slower tiers. This reduced the pipeline bubble fraction from 18 percent under naive round-robin placement to 7 percent under the optimized placement. The team also switched to asynchronous gradient aggregation with a 2-step staleness bound for the data-parallel dimension, reducing the effective wait time in the slow tier while maintaining convergence quality within acceptable bounds.

Checkpoint design used the layered checkpoint pattern from Chapter 5: a local in-memory checkpoint every 200 steps (recoverable in under 30 seconds), a local NVMe checkpoint every 2,000 steps, and a durable object-store checkpoint every 20,000 steps. A node failure rolls back at most 200 steps, a rack failure at most 2,000 steps, and only a catastrophic storage failure requires rolling back more than 20,000 steps. Experiment provenance, a requirement born of an incident in which a result could not be reproduced because the training configuration had been modified in place, is now a first-class output: every run records its configuration hash, dataset version, software stack version, and node-topology map in the same object store bucket as the durable checkpoint.

Lessons learned: Elastic training is not optional for long-running scientific workloads on shared clusters. A fixed-topology job that cannot shrink will be canceled by the first node failure on a multi-week run, making it essentially unschedulable on any cluster with a realistic mean time between failures. Heterogeneity is a permanent feature of scientific clusters, not a temporary condition, because hardware refresh cycles are longer than model development cycles.

Case Study 5: A 50-Person ML Team Scaling from 1 GPU to 100 GPUs

The fifth composite organization is a Series B fintech with a fifty-person machine learning team. Eighteen months ago, the entire practice ran in a single research notebook attached to one A100, and the early models — fraud scoring, transaction categorization, a customer-support classifier — were trained one at a time by whoever owned them. The business has since outgrown that arrangement on two axes at once: it now needs to train several models in parallel rather than serially, and it must serve those models in production at sub-100-millisecond latency on live transaction traffic. The reserved budget is roughly one hundred GPUs across a deliberate mix of on-demand, reserved, and spot capacity. The defining tension of this case is not scale in the absolute sense — one hundred GPUs is a modest fleet — but the mismatch between an ambitious workload and a small team with no platform specialists to operate it. This is the most common situation in which the patterns of this book are actually applied, and it is the one that the foundation-model and hyperscale literature serves least well, because the constraint that dominates every decision is human rather than computational.

The organization's trajectory is worth stating precisely, because it determines which mistakes are tempting. The team grew from five engineers to fifty over eighteen months, and the instinct that served it at five — build whatever is needed, because nothing exists yet — is exactly the instinct that will sink it at fifty, because every component built is a component someone must operate forever. The workload grew in the same period from one model trained occasionally to roughly a dozen models trained on overlapping schedules. It served continuously, and the serving side now carries real money: the fraud model scores live payment authorizations, so a serving outage is a revenue and trust event, not a degraded research experience. The architecture described below is the team's response to that growth, and its governing principle throughout is that the team's time, not its hardware budget, is the scarcest resource.

The binding constraints follow directly from that mismatch. Headcount is finite, and every engineer is a model developer first; there is no dedicated ML platform team, only one site-reliability engineer split part-time across this and other obligations. The company operates in a regulated industry and must maintain SOC 2 and PCI compliance postures, which require data handling, network segmentation, and audit logging from the outset rather than as a later retrofit. The hardware is heterogeneous, spanning two accelerator generations acquired at different times, complicating any topology that assumes uniform device capabilities. And there is a hard cost ceiling: the fleet is sized to a budget the finance organization will not raise, so every efficiency decision is also a survival decision.

The orchestration choice was to run on managed Kubernetes augmented with a lightweight machine learning scheduler layered on top, explicitly rejecting the temptation to build a custom platform. The framework from Chapter 4 makes the argument plainly: a custom scheduler is justified only when the workload's scheduling requirements exceed what an off-the-shelf layer can express, and at fifty people and one hundred GPUs, they do not. The half-time SRE can operate a managed control plane and a thin scheduling layer; that same person cannot maintain a bespoke platform without becoming a single point of failure for the entire company. The decision not to build is the most consequential architectural decision in the case.

The thin scheduling layer is worth describing because the temptation to over-engineer it is strong. Its only jobs are gang-scheduling the workers of a distributed job, so they start together or not at all, enforcing per-team resource quotas so one team's sweep cannot starve another's training run, and bin-packing jobs onto the heterogeneous fleet so the older-generation devices are not left idle while the newer ones queue. Each of these is a capability the off-the-shelf scheduler ecosystem already provides; the team's discipline was to configure those capabilities rather than reimplement them, and to resist the recurring proposal to build a richer internal scheduler that would, in the words of one engineer who argued against it, become the most important and least loved system in the company. The orchestration layer is deliberately boring, and that is the point: a boring orchestration layer is one that the part-time SRE can reason about during an incident at two in the morning.

The training topology is data-parallel only for the largest model the team trains, with no tensor and no pipeline parallelism. This is a deliberate selection of the simplest viable point in the Chapter 2 framework rather than the most capable one. Tensor and pipeline parallelism would enable training a larger model. Still, each adds an operational surface — topology-aware placement, pipeline-bubble tuning, cross-stage failure handling — that a team this size cannot operate safely while also shipping product. The failure modes of pipeline parallelism, in particular, are subtle: a single slow stage collapses throughput without producing any visible loss anomaly, and diagnosing it requires watching the pipeline-bubble ratio, a metric the team would have to learn to instrument, alert on, and interpret. That is a competence the team could acquire, but doing so has an opportunity cost measured in shipped models, and the analysis in Chapter 2 makes it clear that the cost is not worth paying until the model genuinely cannot be trained any other way.

To fit the largest model into device memory under pure data parallelism, the team applied ZeRO-2 optimizer and gradient sharding, which recovers most of the memory headroom that tensor parallelism would have provided without introducing cross-node tensor communication on the critical path. ZeRO-2 was deliberately chosen over the more aggressive ZeRO-3: the additional parameter sharding of ZeRO-3 buys more memory at the cost of an extra all-gather of parameters on every forward pass, and the team's largest model fits comfortably under ZeRO-2, so the extra communication would be paid for with no benefit. The model is sized to what data parallelism with ZeRO-2 can hold, and the product roadmap is written to that constraint rather than against it. When a product proposal arrives that would require a model too large for this regime, the architectural conversation is not about how to add tensor parallelism, but about whether the larger model is worth the operational step change it would require. So far, the answer has been to find a smaller model that fits, because the framework makes the alternative's cost explicit, and the business has consistently judged it too high.

The cost architecture follows Chapter 7 directly and is the reason the fleet fits the ceiling. Sixty percent of capacity is reserved to cover the steady-state serving floor and the always-running training jobs, whose interruption would be expensive. 30% is on-demand, absorbing bursts when several teams train at once. The final ten percent is spot capacity, used exclusively for non-critical hyperparameter sweeps and experimental runs whose loss on preemption is bounded. Checkpoint amortization is what keeps the spot tier economically positive: every spot job checkpoints frequently enough that a preemption rolls back only a small, known amount of work. Hence, the discount on spot capacity exceeds the expected cost of re-running the lost steps. Without that amortization discipline, the spot tier would be a false economy; with it, it is the company's cheapest computer.

The three-way split is not a fixed ratio handed down once but a managed position that the team revisits quarterly against actual usage. The reserved fraction is sized to the load the fleet carries, even at its quietest — the serving floor never drops below it, so reserving that capacity at a discount is unambiguously correct. The on-demand fraction is the buffer for predictable bursts, and the team tolerates its higher unit price because the alternative, reserving for the peak, would leave expensive capacity idle most of the time. The spot fraction is the experimental margin, and its size is bounded not by what spot is available but by how much preemptible work the team genuinely has; padding the spot tier with work that cannot tolerate preemption would convert a discount into a reliability problem. The single most important discipline underneath the whole structure is that the team measures model FLOP utilization per dollar, not raw GPU-hours consumed, because a reserved GPU running at low MFU is more expensive per unit of useful work than a spot GPU running at high MFU. A cost architecture that optimizes GPU-hours rather than useful work optimizes the wrong quantity.

The heterogeneous fleet introduces a complication that the team handles by policy rather than by clever scheduling. Because the cluster mixes two accelerator generations with different memory capacities and peak FLOP rates, a naive placement that spreads a single data-parallel job across both generations turns the slower devices into stragglers that hold up the gradient all-reduce on every step, dragging the whole job to the speed of its slowest member. Rather than build the topology-aware placement that a larger team might attempt, this team adopts a simpler rule: a given training job runs entirely on a single generation, never mixed, so that within a job, every replica advances at the same rate. The older generation is reserved for the smaller models and the experimental sweeps, while the newer generation is reserved for the largest model and the latency-sensitive serving tier. This wastes a little capacity — a job confined to one generation cannot borrow idle devices from the other — but it eliminates straggler amplification without any scheduling sophistication, which is exactly the trade a small team should make.

Observability is deliberately consolidated rather than comprehensive. There is one shared dashboard for the whole practice, not a dashboard per team, because a fifty-person organization cannot afford the cognitive cost of fragmented observability, and a single SRE cannot maintain a proliferation of them. Model FLOP utilization is the primary efficiency metric across both training and serving, chosen over raw GPU utilization for the reasons outlined in the observability chapter — it is the metric that actually tracks whether the expensive hardware is performing useful arithmetic. A single on-call rotation covers both training and serving; the team is too small to staff separate rotations, and the unified rotation has the side benefit of keeping training and serving concerns visible to the same people. The consolidation is itself a defense against an anti-pattern the observability chapter names: dashboards built for a steady state that fail during incidents. The shared dashboard's primary view is organized around the questions an on-call engineer asks when something is wrong — which job died, which rank stalled, whether a deployment moved the cost-per-request curve — and the steady-state efficiency panels are relegated to a secondary view consulted when nothing is broken. The instrumentation deliberately includes the leading indicators the observability chapter argues for — gradient-norm trajectory and NaN counts — so that a divergent training run is caught before it consumes a night of compute, which on a fixed budget is a cost the team feels directly.

Security is implemented through namespace-per-team isolation, centralized secrets management, and the absence of a custom container network interface. Each team operates in its own namespace with RBAC scoped to that namespace, which provides the tenant isolation the SOC 2 and PCI posture requires without the operational burden of a fully segmented multi-cluster design. Secrets are centralized in a managed secret store and injected rather than baked into images or mounted indiscriminately. The team explicitly declined to deploy a custom CNI, opting for the default networking with identity-aware policies layered on top, because a bespoke network plane is exactly the kind of high-operational-cost component a team without a platform group should avoid. The namespace boundary is the unit of blast-radius containment: a misconfiguration or compromise in one team's workload is bounded to that team's namespace rather than exposing the whole fleet, which is precisely the cluster-level-RBAC-without-isolation anti-pattern the security chapter warns against. Service-account tokens are not auto-mounted into pods that do not call the control plane, closing the lateral-movement path that default token mounting leaves open. The PCI requirement could, in principle, have justified a heavily segmented custom network. Still, the team judged — correctly, given its headcount — that it could satisfy the requirement with managed networking and workload-identity policies that authorize specific service-to-service connections rather than trusting everything within a namespace. The auditors accepted the design because it demonstrably enforced the required isolation; the team accepted it because one part-time SRE could operate it.

The serving side, which carries the sub-100-millisecond latency requirement, follows the same restraint as the training side. The fraud model is served from a pool of replicas behind a load balancer, with per-model queue-depth signals driving autoscaling rather than aggregate request volume, so that a surge to the fraud model scales the fraud model itself rather than its neighbors — the head-of-line-blocking anti-pattern. The serving chapter names are avoided by construction. Batching is continuous rather than static, admitting requests into a running batch so that no single payment authorization waits for a batch to fill, which would blow the latency budget. The team deliberately does not shard any served model across nodes, keeping each model resident on a single device or node, because the models are small enough to fit, and cross-node inference would put network latency on the critical path of every score. The cost-per-request curve is a first-class canary metric: a new model version is promoted to full traffic only after it has demonstrated, on a fraction of live traffic, that it has not regressed latency, error rate, or cost per request, the last of which a quality-only canary would miss. The serving architecture is, like everything else in the case, the simplest point in the catalog that meets the requirement.

The compromise is best understood by naming the negative space — what the team chose not to do, and why. There is no tensor parallelism, because the team cannot operate it safely. There is no multi-region deployment, because the latency and consistency benefits do not justify the operational complexity at this scale, and the regulatory posture is simpler to maintain in a single region. There is no custom scheduler, because the managed layer suffices. There is no self-built feature store because a managed store meets the requirement, and a bespoke one would consume the engineers the company needs for product work. Each of these is a capability the team could have built but deliberately did not, and the discipline of declining them kept the practice operable. There is no automated multi-model canary infrastructure of the kind a hyperscaler would build, because the team's deployment cadence is low enough that a deliberate, partly manual promotion gated on cost and latency is adequate and far cheaper to maintain. Each of these is a deliberate refusal rather than a backlog item, and each is justified by the same calculus: the capability's benefit at the team's current scale is smaller than the perpetual operational cost of owning it. Naming the negative space is the lesson of this case, because the failure mode for a team like this is not choosing the wrong pattern but choosing too many patterns, accumulating a platform's worth of bespoke infrastructure that would require a platform team's worth of people to run. The architecture is defined as much by its omissions as by its inclusions, and the omissions are the harder and more valuable engineering.

The lessons learned generalize beyond this particular fintech. A small team scales further with patterns than with platforms — it advances by selecting the right pattern from a catalog and operating it well, not by building infrastructure it then has to maintain. The framework from Chapter 2 still applies in full; the team chose the simplest viable point within it at every decision, accepting a lower ceiling on individual workload scale in exchange for an operational surface that fifty people and one part-time SRE can actually run. The organization that tries to build a platform before it has the headcount to operate it does not scale faster; it scales until the platform's maintenance costs consume the team that built it. This team avoided that fate by treating restraint as an architectural principle. The framework's value to a small team is not that it points to the most capable architecture, but that it makes the cost of capability explicit, so the team can see precisely what it is declining and why. The case stands as the counterexample to the assumption, common in the field, that scaling a machine learning practice means building a machine learning platform. It can instead mean choosing patterns well enough that no platform is needed yet.

Case Study 6: Regulated Healthcare Distributed Inference On-Premises

The sixth composite organization is a U.S. regional hospital system that runs distributed inference for clinical decision support across 40 hospital sites. The models assist clinicians with tasks such as triage scoring, imaging triage, and deterioration prediction, and they run almost entirely on-premises, with a strictly limited cloud burst reserved for non-PHI inference workloads that do not carry protected health information. The fleet is roughly sixty inference GPUs distributed across two on-premises data centers. The business context inverts the assumptions of most distributed machine learning literature: this is not a training-dominated organization scaling up a model, but an inference-dominated one whose central engineering problem is operating a clinical service safely, continuously, and auditably under a regulatory regime that treats every model change as a governed event. The device count is comparable to that of the previous fintech case, but almost every other parameter is reversed. Where the fintech optimized for the velocity of a small team, this organization optimizes for the defensibility of every decision to a regulator, and where the fintech treated cost as the binding ceiling, this organization treats compliance as the binding ceiling and accepts a cost premium as the price of meeting it. The lesson the two cases share, read together, is that the device count says almost nothing about the architecture; the operating constraints say almost everything.

The binding constraints are unusually hard and almost entirely non-negotiable. No protected health information may leave the data centers, which rules out any architecture that depends on cloud training over patient data. The hospital network is segmented, so the inference platform must operate within network boundaries that it does not control or flatten. Audit records must be retained for ten years, which shapes logging and storage from the first design decision. The model lifecycle is FDA-regulated: every retrain is a regulatory event with documentation, validation, and approval obligations, so models change rarely and deliberately rather than continuously. And the service must be available twenty-four hours a day with no maintenance windows, because the hospitals it serves do not close, and a clinical decision-support outage during a night shift is a patient-safety event, not an inconvenience. These constraints do not trade off against one another as cost and performance do; they are conjunctive requirements, each of which must hold simultaneously, and an architecture that satisfies four of them and violates the fifth satisfies none of them in the eyes of the regulator. That conjunctive character is what makes the design problem hard. There is no slider to tune between compliance and efficiency along which an optimum can be found; the constraints set a hard feasibility boundary, and the engineering happens entirely inside it.

The prohibition on moving patient data shapes the training architecture. There is no multi-region training and no corpus centralization. Instead, the system performs periodic federated fine-tuning across the two data centers, aggregating gradients rather than moving raw data, so that each site's protected health information never leaves its own boundary. At the same time, the model still benefits from the combined signal. This is a direct application of the principle that the architecture must conform to the data's legal location rather than the other way around. Gradient aggregation is the only cross-site data movement the design permits, and even that is scrutinized for leakage risk before it is allowed, because gradients computed over a small cohort can themselves leak information about that cohort. The privacy review treats the aggregation channel as a potential disclosure path rather than as a safe internal pipe. The federated arrangement is not adopted for the usual federated-learning motivation of distributed ownership; it is adopted because the law forbids the simpler centralized alternative. The aggregation itself runs over a dedicated, audited channel between the two data centers rather than over a general-purpose link, and the design treats that channel as part of the regulated boundary, logging every aggregation round with the model versions and site identifiers involved so that the provenance of any deployed model can be reconstructed years later. The team also deliberately keeps the federated topology to exactly two participants — the two data centers — rather than pushing federation down to the level of individual hospitals, because forty federated participants would multiply both the coordination cost and the privacy-review surface without a commensurate gain in model quality. The choice of two participants is the simplest topology that satisfies the data-residency constraint, and simplicity here is not a convenience but a requirement, because every additional moving part is something the regulator will ask the team to justify. That distinction matters for the architecture, because the team is not trying to match the convergence quality of centralized training and accepting federation as a compromise — it is treating federation as the only legal topology and then doing the engineering necessary to make its convergence acceptable, which includes longer aggregation intervals and careful handling of the heterogeneity between sites' data distributions.

Because inference is the dominant workload and training is occasional, the platform is sized for steady-state inference rather than for training peaks. This is the opposite of the foundation-model case, where the cluster is provisioned for the training run, and inference is the afterthought. Here, the sixty GPUs are sized to handle the forty sites' continuous inference load, with headroom for failover, and the rare training jobs are scheduled to fit around that floor rather than drive the capacity plan. Sizing for the steady state rather than the peak is the correct decision precisely because the peak is rare and the steady state is the patient-facing service that cannot be allowed to degrade. The contrast with the foundation-model case study earlier in this chapter is instructive: there, the cluster existed to complete a training run, and inference was provisioned around it, while here the cluster exists to serve clinicians, and training is provisioned around that. The same patterns from the catalog appear in both, but they are composed in opposite priority orders, and the priority order is dictated by which workload carries the consequence. In a hospital, the consequence lives in inference, so inference owns the capacity plan, and training takes what is left.

The availability requirement, with no maintenance windows, drives a specific inference topology. The forty sites are served from the two data centers in an active-active arrangement, so that the loss of either data center degrades capacity but does not interrupt service. Model updates are rolled out one replica at a time behind a load balancer that drains traffic from a replica before it is taken down and returns traffic only after the replacement reports healthy. There is never a moment when the service as a whole is offline, because there is never a moment when both data centers are simultaneously unavailable to a given site. This is the same rolling-update discipline a cloud-native serving stack would use, applied on-premises and constrained by the requirement that each data center must be able to carry the clinical load alone during the window when the other is updated. Sizing each data center to carry the full load rather than half is part of why per-GPU efficiency is lower than it would be in a cloud deployment, and the team accepts that overhead as the direct cost of the no-maintenance-window guarantee.

Orchestration is hybrid. Kubernetes runs the inference services, where its scheduling, health-checking, and rolling-update mechanics map well onto a long-lived serving fleet. A separate batch scheduler handles the training jobs, which do not need the inference cluster's resources and would only contend with patient-facing traffic if scheduled there. Separating the two control planes keeps the training workload from ever competing with inference for the GPUs that serve clinicians, and it lets each scheduler be tuned for its own workload class rather than being compromised to serve both. The inference scheduler optimizes for tight latency and high availability, with health checks and rolling updates that never take the whole service down; the batch scheduler optimizes for throughput and fair allocation of the occasional training job, with no requirement to be latency-sensitive because no clinician is waiting on it. Attempting to serve both disciplines from a single scheduler would force a compromise that degrades the property each workload most needs, and the team judged the modest cost of operating two control planes well worth the clean separation of concerns it buys.

The deployment path is built to be air-gap-compatible. Model artifacts are signed in the build environment, outside the data center, and those signatures are verified inside the data center before any model is deployed, with no external registry pulls at runtime. The principle from the security chapter — that the model artifact is an attack surface equal to the container — is enforced literally here, because a clinical model loaded from an unverified source is both a safety risk and a compliance failure. Nothing is pulled from the internet at runtime; everything that runs inside the data center was verified before it entered. This is the model-artifact-registry anti-pattern from the security chapter taken to its strictest conclusion: the team treats the model file as an attack surface fully equal to the container, applies the same signing and verification discipline to both, and assumes that anything entering the air-gapped environment is hostile until its signature proves otherwise. In a clinical setting, the stakes of a poisoned model are not merely a security breach but a patient-safety failure, because a backdoored or corrupted clinical model could produce dangerous, seemingly plausible recommendations, so the verification step is as much a safety control as a security one. The build environment that signs the artifacts is itself audited, closing the obvious gap that an attacker who compromised the signer could sign malicious models.

Tenant isolation is implemented per hospital. Each hospital operates in its own namespace with separate encryption keys, and network policies are enforced by workload identity rather than by namespace, so that trust is granted to specific services rather than to anything that shares a namespace. This is the identity-based network policy that the security chapter prescribes, applied because the regulated environment cannot tolerate the lateral movement risk that namespace-wide trust would introduce. Separate per-hospital keys also mean that a key compromise is contained to a single site rather than exposing the whole system, which is the same blast-radius reasoning the previous case study applied to namespaces, here extended to encryption boundaries because the regulatory environment demands it. The workload-identity network policies are stricter than the namespace-based policies that a less regulated organization might accept, because the team explicitly rejected the network-policy anti-pattern in which everything sharing a namespace inherits a common trust. In this design, a connection between two services is authorized only if both services' identities are on an explicit allow-list for that connection, so a compromised pod inherits no network reach beyond what its specific identity was granted. The lateral movement that namespace-wide trust would permit is structurally impossible.

Observability is built around the FDA post-market surveillance obligation as much as around operations. Every prediction is logged with its model version, a hash of the input, and the clinician's override outcome — whether the human accepted, modified, or rejected the model's recommendation. That record feeds the post-market surveillance pipeline required by the regulatory lifecycle, turning routine inference logging into the evidence base for demonstrating that the deployed model performs as approved. The instrumentation, therefore, serves two masters at once: the on-call engineer who needs to know the service is healthy, and the regulatory function that must prove, years later and across a ten-year retention window, exactly which model produced which recommendation and how the clinician responded. The input hash, rather than the raw input, is deliberately logged because logging the raw clinical input would replicate protected health information in the observability store and multiply the surface that must be protected; the hash preserves the ability to detect input distribution shift and to correlate predictions without retaining the PHI itself. The clinician-override signal is the most valuable field in the record, because it is the ground-truth feedback that post-market surveillance depends on: a model whose recommendations are increasingly overridden is a model whose real-world performance is drifting from its approved behavior, and that drift is precisely what the FDA lifecycle exists to catch before it harms a patient.

The compromise is an explicit acceptance of lower per-GPU efficiency in exchange for compliance and audit posture. A cloud-native deployment would achieve higher utilization through elastic scaling, larger shared pools, and aggressive spot usage, and the team knows it. They accept the lower efficiency because the regulated environment requires data residency, air-gapped deployment, ten-year audit retention, and per-site isolation that a maximally efficient cloud design cannot provide. They also defer scale-out training entirely; the most aggressive training the organization runs is fine-tuning a seven-billion-parameter model on a single eight-GPU node, which fits comfortably without any of the multi-node parallelism required by the earlier case studies. The decision to cap training ambition at single-node fine-tuning is not a limitation the team resents; it is a scope choice that keeps the regulated lifecycle tractable. Every retrain is a regulatory event with documentation and validation obligations. Hence, the marginal cost of a more ambitious training regime is not merely computational but regulatory overhead, and a multi-node training run that produced a marginally better model would still have to clear the same approval bar while introducing more operational surface to audit. The team reasoned that the clinical benefit of a larger or more frequently retrained model did not justify that overhead. So it deliberately accepts a less aggressive training posture than its hardware could support. The accepted lower per-GPU efficiency, similarly, is not a failure to optimize but a consequence of constraints the team would violate the law to remove: elastic cloud scaling, shared pools across tenants, and aggressive spot usage are all efficiency levers that the data-residency and isolation requirements forbid.

The lessons learned restate a theme the series returns to repeatedly. Distributed machine learning patterns work as well at sixty GPUs as at six thousand; the patterns a team selects reflect its operating constraints, not the device count, and a small regulated fleet exercises the catalog as fully as a large research cluster does. The FDA lifecycle is the single most constraining requirement in this case, and the decisive architectural insight is that it is never negotiated around — it is designed for from day one, treated as a fixed input to which every other decision conforms. An organization that bolts regulatory compliance onto a system designed for efficiency will rebuild that system; one that designs for the lifecycle first, as this team did, pays a steady efficiency tax and keeps its approval. The deeper point, which connects this case to the maturity model of Chapter 12, is that maturity in a regulated environment is measured not by the sophistication of the parallelism strategy but by the defensibility of the whole system to an auditor who arrives years after a decision was made and asks why it was made. A team that can answer that question for every component — why this isolation boundary, why this checkpoint interval, why this signing step — is mature regardless of whether it runs sixty GPUs or six thousand. This organization's architecture is unglamorous by the standards of the foundation-model literature, and it is exactly right for what it must do, which is the only standard that matters.

A final lesson concerns the relationship between the federated training topology and the regulatory lifecycle, because the two interact in a way that is easy to miss. Each federated fine-tuning round produces a new model, and each new model is a regulatory event that must be documented, validated, and approved before it can serve a clinician, which means the cadence of training is bounded not by how fast the gradients can be aggregated but by how fast the regulatory process can clear a new version. The team therefore trains far less frequently than its hardware would allow, batching improvements into infrequent, well-documented releases rather than retraining continuously, because a continuous-retraining posture would generate regulatory events faster than the organization could process them. This is the inverse of the fintech in the previous case, which trains often and cheaply; here, the binding constraint on training frequency is the regulator's calendar, not the cluster's throughput, and the architecture is shaped to that reality. An engineer arriving from a commercial setting, where faster retraining is almost always better, must unlearn that instinct in a regulated one, where the cost of a model change is dominated by the governance it triggers rather than the compute it consumes.

Multi-Tenant ML Platform Reference Architecture

The four case studies share an implicit fifth architecture: the multi-tenant ML platform that hosts them all. In organizations with more than one ML team, some version of this platform emerges, either by deliberate design or by accretion. Deliberate design is cheaper. This section synthesizes the cross-cutting platform components from all four case studies into a single reference architecture that a platform team can use as its default starting point.

The platform has five horizontal layers. The compute layer manages the accelerator fleet through a quota-based scheduler, following the patterns from Chapters 4 and 7. The storage layer provides fast, ephemeral storage for checkpoints and data staging, as well as durable archival storage for model artifacts and experiment provenance, following the patterns in Chapter 3. The serving layer provides shared inference infrastructure, including the autoscaling, batching, and canary-release patterns from Chapter 6. The observability layer provides centralized metrics, logs, and traces for all tenants, following the patterns from Chapter

8. The security layer enforces tenant isolation, identity federation, and secrets management, following the patterns from Chapter 9.

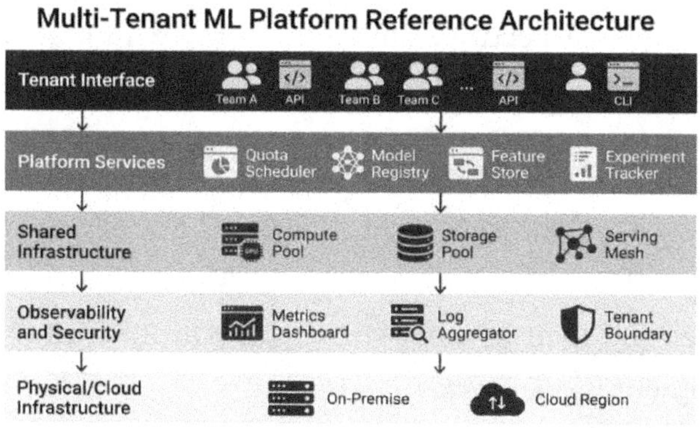

Multi-Tenant ML Platform Reference Architecture

Tenant Interface: Team A, API, Team B, Team C, ..., API, CLI
Platform Services: Quota Scheduler, Model Registry, Feature Store, Experiment Tracker
Shared Infrastructure: Compute Pool, Storage Pool, Serving Mesh
Observability and Security: Metrics Dashboard, Log Aggregator, Tenant Boundary
Physical/Cloud Infrastructure: On-Premise, Cloud Region

The tenancy model defines the isolation boundary across three levels. Namespace isolation limits resource consumption through quota enforcement while allowing tenants to share physical nodes, which is appropriate for training workloads within the same security domain. Node-level isolation dedicates a set of nodes to a single tenant during the allocation period, which is appropriate for tenants with sensitive model weights or data privacy requirements. Cluster isolation provides tenants with a logically separate cluster within the platform's management plane, suitable for those with regulatory or contractual requirements that prevent hardware sharing.

The scheduler balances three competing objectives: individual-tenant fairness, cluster-level utilization, and latency predictability for small jobs behind large ones. Chapter 4 covers gang scheduling, priority classes, and preemption patterns. The reference architecture recommends a three-class priority scheduler: best-effort (preemptable, no latency guarantee), standard (non-preemptable, FIFO within class), and high-priority (non-preemptable, jump-queues ahead of standard), with hard quota caps per tenant per priority class. The model registry provides the platform with architectural coherence: it links every model artifact to its provenance record, evaluation metrics, and deployment status, and enables the security layer to enforce model access policies, ensuring that a tenant's model is private by default.

Cross-Case Patterns and Lessons Learned

Across the four case studies, seven patterns appear in every architecture regardless of workload class. Their recurrence is evidence that they are load-bearing structural patterns rather than incidental implementation choices.

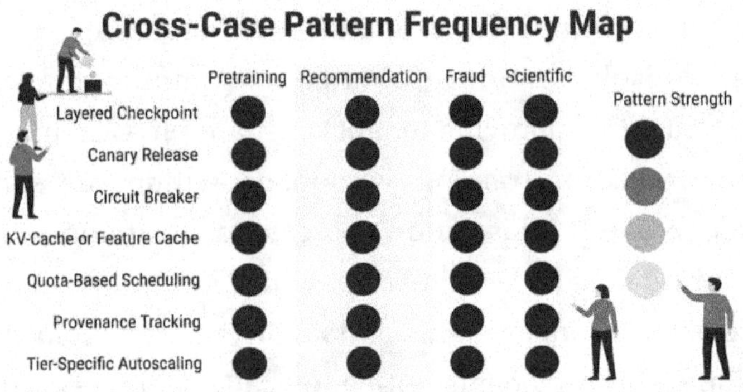

The layered checkpoint pattern appears in all four architectures. Every production-grade distributed training system needs at least two checkpoint tiers: a fast local tier that absorbs individual-node failures with minimal rollback, and a durable remote tier that absorbs catastrophic failures. The specific media and retention policies differ, but the two-tier structure is invariant. Chapter 5 names this pattern; it confirms that it is not optional in any workload class.

The canary release pattern appears in all four architectures, though gate criteria differ. In recommendation serving, the gate is an engagement proxy metric. In fraud inference, it is a P99.9 latency. In pretraining, it is the training loss trajectory. In scientific training, it is allocation efficiency. The shared structure: a new artifact is exposed to a small fraction of traffic or compute before full promotion, gated on measurable success criteria. Chapter 6 covers this in the serving context; it applies equally to changes to training configuration and to platform software upgrades.

The circuit-breaker pattern appears in all four architectures, though instantiated each time differently. In recommendation serving, it routes away from a slow model tier. In fraud inference, it routes from primary to fallback. During pretraining, it triggers a checkpoint rollback when loss spikes. In scientific training, it triggers elastic resizing on node failure. The invariant: detect a failure signal, automatically transition to a degraded-but-functional state, and alert a human to investigate.

The KV-cache or feature-cache pattern appears in all four architectures. In pretraining, the pre-tokenized pack cache prevents re-tokenization on restart. In scientific training, the in-memory checkpoint prevents full-disk I/O at every step. The shared insight is that caching intermediate representations close to the compute that consumes them is a structural performance lever that does not require changing the model or training objective. Provenance tracking, quota-based scheduling, and tier-specific autoscaling complete the seven, each appearing universally for the same reasons: provenance makes post-incident investigation tractable; quota scheduling prevents cross-tenant starvation; tier-specific autoscaling matches the policy to the bottleneck resource of each specific workload tier.

Reading the Trade-Off Triangle in Real Architectures

Chapter 1 introduced the consistency-performance-cost trade-off triangle as a conceptual frame for distributed ML decisions. The four case studies provide four concrete instances of how that triangle is navigated in practice. Revisiting it here shows that the triangle is not a theoretical construct but a diagnostic tool with direct operational utility.

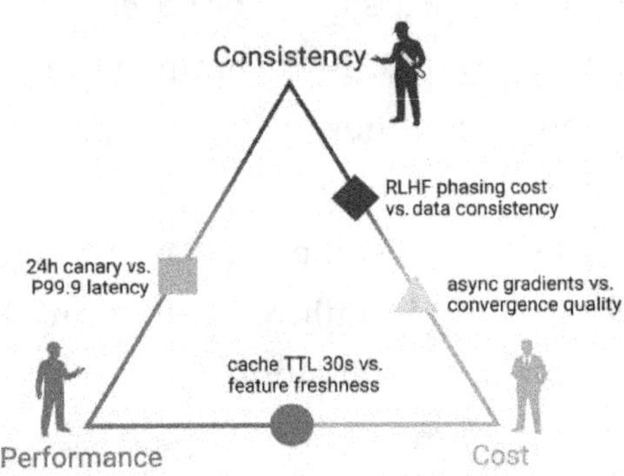

The pretraining architecture sat closest to the consistency-cost edge. The primary trade-off was between the cost of running a separate RLHF-ready reward model evaluation alongside pretraining and the risk of discovering, after pretraining was completed, that the reward model was not ready. The team chose to pay the cost premium to preserve consistency between phases, a textbook instance of paying more to avoid a failure whose recovery cost would have been far higher.

The recommendation architecture sat closest to the performance-cost edge. The primary trade-off was cache TTL: a shorter TTL improves feature freshness at the cost of increased feature-store read volume and higher latency on cache misses. The team empirically tuned the TTL to find the maximum value at which freshness degradation was imperceptible in the engagement metric.

That is the correct way to resolve a three-way trade-off: identify the metric that converts all three dimensions into a single observable outcome, then tune to the optimum of that metric.

The fraud architecture sat closest to the edge of performance consistency. The 24-hour canary gate protects the latency SLA at the cost of slower model updates. That decision is non-negotiable as long as the hard deadline is harder than the model update cadence. The scientific architecture revisited the consistency-cost edge from a different angle: synchronous gradient aggregation maximizes convergence consistency but stalls on the slowest worker. In contrast, asynchronous aggregation maximizes cluster utilization at the cost of bounded gradient staleness. The team chose asynchronous after empirically verifying that the difference in convergence quality was within the tolerance of their scientific domain. The triangle is most useful as a shared vocabulary for architecture reviews: naming the trade-off axis is the first step toward resolving it.

A Reference-Architecture Template

The four case studies share a common structure that can be extracted into a reusable template. This template is the deliverable that the engineering director from the Opening Scenario was looking for. It has eight sections, each forcing the team to make an explicit decision that would otherwise remain implicit. A reference architecture that omits any of the eight sections is incomplete, because it has left at least one category of consequential decision undocumented.

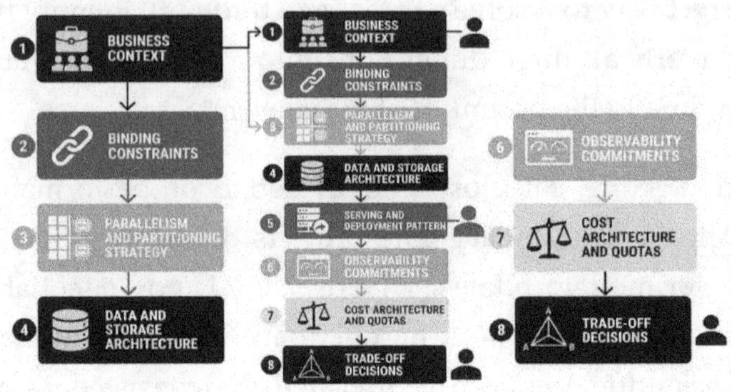

Section one, Business Context, documents why the system exists, what business outcome it produces, and what the cost of failure is in business terms. It answers the question: if this system fails, what is actually broken? Making the answer explicit forces the team to size the investment in reliability correctly. Section two, Binding Constraints, documents what the architecture cannot negotiate away: hard latency deadlines, regulatory data-residency requirements, budget caps, and organizational ownership boundaries. Binding constraints are inputs to the trade-off triangle, not its outputs.

Section three, Parallelism and Partitioning Strategy, names the parallelism patterns from Chapter 2 that the system uses, to what degree, and why, and documents the data sharding strategy from Chapter 3, including how assignment changes on restart. Section four, Data and Storage Architecture, documents the full data path from raw input to model artifact, every storage tier, read and write patterns, retention and deletion policy, and the checkpoint architecture with its tested recovery procedure. Section five, Serving and Deployment Pattern, names the canary release policy, traffic-split schedule, gate criteria, rollback trigger, and rollback procedure. For multi-model pipelines, it names the latency budget and circuit-breaker policy for each stage.

Section six, Observability Commitments, specifies the metrics, logs, and traces the system produces; the alerting policy for each; the golden signals for this workload class; the SLOs governing the system; and the runbooks for the most likely incident types. Section seven, Cost Architecture and Quotas, documents the cost per training step or inference request, expected monthly cost at median and peak load, available cost levers, and the quota policy. Section eight, Trade-Off Decisions, is the section that distinguishes a reference architecture from a system diagram: it documents every decision that could have gone a different way, along with the reason the team chose the option it did and the signal that would trigger revisiting it. The 24-hour canary gate, the 30-second cache TTL, the 2-step staleness bound, and the RLHF co-training decision are all examples of what belongs here.

Observability Across All Four Architectures

Every case study identified observability as a first-class architectural requirement. The patterns from Chapter 8 appear in each architecture, but are instantiated differently across workload classes. Comparing the four instantiations surfaces principles that generalize across all distributed ML workloads.

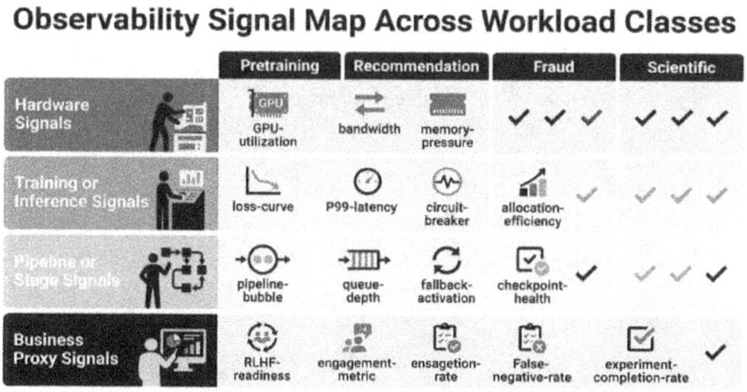

All four architectures instrument the hardware layer with the same signals: GPU utilization, memory pressure, interconnect bandwidth, and temperature. Hardware signals are the foundation because they are causal. All four define workload-specific golden signals beyond the generic four (latency, traffic, errors, saturation): for training workloads, throughput in tokens or samples per second; for serving workloads, P99 or P99.9 latency at the pipeline level; for scientific workloads, allocation efficiency.

All four require leading indicators alongside lagging metrics. The pipeline bubble ratio in pretraining, the circuit-breaker activation rate in fraud, the cache hit rate in recommendation, and the in-memory checkpoint health metric in scientific training are all leading indicators whose specific relevance was learned by operating through at least one incident. All four also require a business proxy metric connecting infrastructure health to a business outcome. A platform team that cannot connect its infrastructure metrics to a business proxy will eventually have its budget cut because it cannot demonstrate business value.

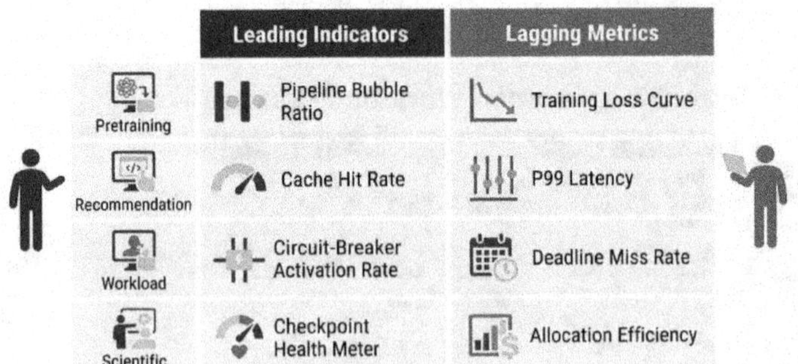

Cost Architecture Patterns Across Case Studies

Chapter 7 covers cost optimization patterns in depth. The four case studies apply those patterns in different combinations, and the cross-case comparison reveals which cost patterns have the highest leverage in which workload class.

Spot or preemptible instances are highest-leverage for pretraining and scientific training, because those workloads can tolerate a checkpoint rollback in exchange for a 30 to 70 percent reduction in compute cost. They are not available for fraud inference because a preemption during an active transaction window causes a hard-deadline miss. They are available in limited form for recommendation serving at the retrieval and lightweight-scoring tiers, which restart quickly, but not at the heavy-ranking tier, which has a long model-weight loading warm-up.

Cost Pattern Leverage **by Workload Class**

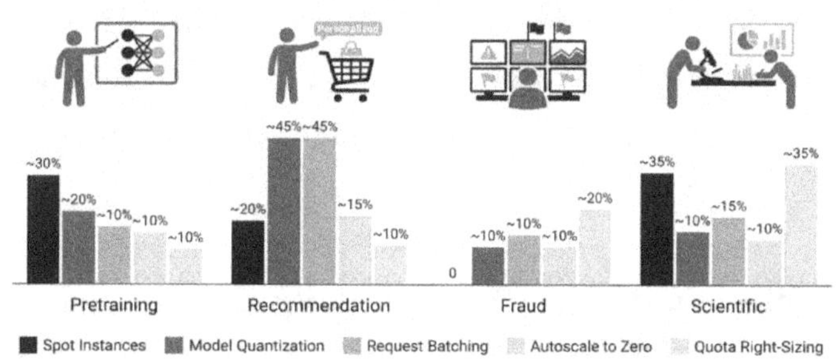

Model quantization offers the highest leverage for recommendation and fraud inference because continuous high-throughput serving means that the memory footprint reduction from INT8 quantization directly translates into more model instances per GPU. For pretraining, quantization does not apply to the main training loop because training requires FP32 or BF16 precision for numerical stability. Request batching is the highest-leverage recommendation approach because request volume is high, and sequence-length uniformity makes micro-batching efficient, raising GPU utilization from 40 to 78 percent at median load in the case study. For fraud inference, batching is constrained by the hard deadline: a micro-batch cannot wait for additional requests to fill if an in-flight request is within 2 milliseconds of its deadline.

Quota right-sizing, reducing allocated quota to match actual usage, is the highest-leverage in multi-tenant platforms where over-provisioning is common. The scientific case showed the largest opportunity: researchers systematically requested 20 to 30 percent more allocation than they actually used, driven by uncertainty about job duration. Making allocation efficiency visible at the job level reduced over-provisioning more than committee-level right-sizing alone did, because it provided researchers with a direct feedback loop. Platform cost governance, enforced through a monthly showback mechanism that assigns actual accelerator hours consumed to each team, reduced platform-wide idle time by making the cost of idle reservations visible to the teams responsible for them.

Multi-Tenant Platform FinOps Feedback Loop

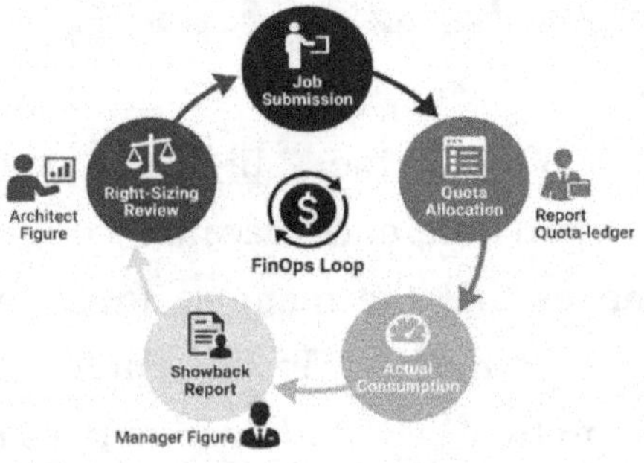

Technical Checklist

Use this checklist as a release gate before declaring a distributed ML system production-ready against the patterns of this chapter.

- The team has a written reference architecture for each of its production workload classes, in a format a new engineer can read and understand without additional explanation.
- Each reference architecture names the parallelism pattern from Chapter 2 that it uses (tensor, pipeline, data, or a combination), the degree of each dimension, and the rationale for those numbers.
- Each reference architecture documents its data sharding strategy from Chapter 3, including how shards are assigned to workers and how that assignment is preserved across checkpoint restarts.
- Each reference architecture documents its checkpoint architecture from Chapter 5: the number of tiers, the checkpoint interval for each, the retention policy, and the tested recovery procedure.
- Each serving reference architecture names the latency budget for each stage in the pipeline and the circuit-breaker policy for each inter-stage call, following the patterns of Chapter 6.
- Each architecture has explicit consistency, performance, and cost trade-off decisions written down, with the signal that would trigger revisiting each decision.
- Each reference architecture documents the multi-tenancy model and quota enforcement mechanism from Chapter 9, including the isolation boundary for each tenant class.
- The observability section names at least one leading indicator per workload class, not just lagging metrics, following the instrumentation patterns of Chapter 8.
- Each architecture documents a business proxy metric connecting infrastructure health to a business outcome visible to non-technical stakeholders.

- The cost architecture section documents the cost per training step or inference request, expected monthly cost at median and peak load, and at least two levers available to reduce that cost.
- Lessons learned from previous incidents have been explicitly folded back into the reference architecture, with the incident date and the section that was updated.
- The reference architecture template is the mandatory starting point for any new production workload, enforced through the team's architecture review process.
- Cross-team architecture reviews are a standing forum with a regular cadence, not triggered only by incidents.
- Reference architectures are versioned and dated, have a named owner, and have a documented refresh cadence no longer than 12 months.
- All seven cross-case patterns (layered checkpoint, canary release, circuit breaker, KV-cache, quota scheduling, provenance tracking, tier-specific autoscaling) are explicitly named and present in every production architecture where they apply.

Team Conversation

Use these questions in architecture reviews, sprint retrospectives, or team meetings to pressure-test alignment on the patterns of this chapter.

- Do we have a written reference architecture for our most important production workload, and would two engineers describe it the same way without looking at the document?
- Which of the four case studies in this chapter is closest to our most demanding production workload, and what does it reveal about a pattern we are not currently applying?

- Where have we made trade-off decisions implicitly, and what would it take to write those decisions into section eight of the reference architecture template?
- Which of the seven cross-case patterns (layered checkpoint, canary release, circuit breaker, KV-cache, quota scheduling, provenance tracking, tier-specific autoscaling) are present in our architectures, and which are conspicuously absent?
- When was the last time we updated a reference architecture in response to an incident, and can we name the section that was changed?
- Do our reference architectures include business proxy metrics that translate infrastructure health into outcomes a non-technical stakeholder can understand?
- If we had to onboard a new team next quarter, which reference architecture would we hand them first, and how long would it take that team to become operational?
- What is our current checkpoint recovery time for our most critical training workload, and have we tested that recovery time in the past six months?
- Which cost patterns from Chapter 7 are we applying, and which are we leaving on the table because we have not profiled the workload carefully enough to know whether they are safe to apply?
- What is our standing forum for cross-team architecture review, and when did it last meet?

Key Takeaway

Patterns are most valuable when they are named, composed deliberately, and documented in a form that survives the engineers who built the system. The four case studies in this chapter are not four different problems; they are four different compositions of the same pattern library, assembled against different binding constraints and different trade-off priorities. The foundation-model pretraining case assembled 3D parallelism, multi-tier checkpointing, RLHF co-training, and cost-aware spot scheduling. The recommendation serves a case assembled multi-model pipeline design, KV-cache feature caching, tier-specific autoscaling, and canary-gated deployment. The fraud inference case assembled hard-deadline circuit breaking, primary-plus-fallback serving, multi-tenant quota enforcement, and latency-gated canary release. The scientific training case assembled elastic resizing, asynchronous gradient aggregation, heterogeneous accelerator placement, and experiment provenance tracking. In each case, the architecture is less about the specific tools chosen and more about the conscious assembly of patterns into a composition that satisfies binding constraints and makes trade-offs explicit.

The reference architecture template is a high-leverage artifact because it forces conscious assembly. A team that fills in all eight sections has, by the act of filling them in, made every consequential decision explicit, identified the failure modes it accepts in exchange for the ones it does not, connected its infrastructure to a business outcome, and produced a document a new team member can read and become operational against. A team that skips the template and builds the system from implementation up will probably produce a working system, but without the documentation that makes it maintainable, debuggable, and transferable. The pilot-to-production pathway for any new distributed ML workload runs directly through the template: start with business context and binding constraints, then select patterns from the preceding 10 chapters to satisfy them.

Reference architectures are living documents and need an owner and a refresh cadence. The best time to assign that owner and schedule that cadence is before the system is built, not after the first incident, which makes the absence of documentation painful. The trade-off triangle from Chapter 1, the parallelism patterns from Chapter 2, the data patterns from Chapter 3, the scheduling patterns from Chapter 4, the fault-tolerance patterns from Chapter 5, the serving patterns from Chapter 6, the cost patterns from Chapter 7, the observability patterns from Chapter 8, the security patterns from Chapter 9, and the foundation-model patterns from Chapter 10 all belong in the template. Using it does not constrain the team to a single stack; it constrains the team to making its choices explicit, which is the only constraint a patterns-first approach needs to impose. That constraint, consistently applied, is what turns a collection of capable engineers into an organization with operational clarity.

13 The Modern Distributed ML Platform: Putting It All Together

Opening Scenario

Priya inherited the ML platform on a Tuesday. The previous platform lead had left six weeks earlier, the documentation was scattered across three wikis and a shared drive that nobody had organized in two years, and the VP of Engineering had a board presentation in three weeks that required an honest answer to a single question: where does this platform stand relative to what a modern distributed ML shop should look like?

Priya spent the first two days reading incident retrospectives and cost reports. The picture that emerged was uncomfortable. Training jobs were running on a shared cluster with no admission controls, which meant a single runaway experiment could starve a production fine-tuning job. The serving fleet had three inference runtimes deployed by three teams, each with its own monitoring dashboard and on-call rotation. A data quality incident six months earlier had cost the team four days of debugging because there was no lineage tracking between the feature store and the training pipeline. The security team had flagged two open issues regarding model artifact access controls that were unresolved in the backlog. None of this was catastrophic in isolation, but taken together, it described a platform that had grown by accretion rather than by design.

What Priya lacked was a shared vocabulary to describe the gap between what existed and what was needed. Without that vocabulary, every conversation with leadership became a negotiation over priorities rather than a structured assessment of risk. She needed a way to say: here is the stack, here is where we are on each layer, and here is the ordered list of investments that will move us forward. She needed, in short, a maturity model that was specific enough to be diagnostic and honest enough to be trusted.

By the end of the third week, she had one. The maturity model she built from her assessment became the artifact that the VP used in the board presentation. More importantly, it became the shared vocabulary the platform team used over the next two quarters to prioritize work, resolve disagreements about scope, and explain to domain ML teams why certain requests were deferred. This chapter is the model Priya should have had on day one.

The Full Distributed ML Platform Stack: From Infrastructure to Product

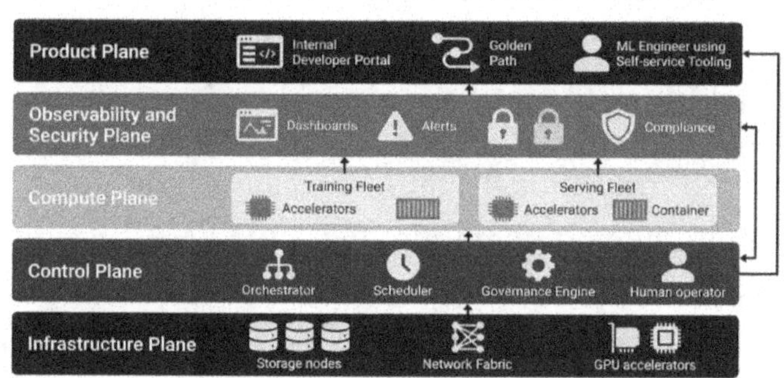

Why It Matters

The previous eleven chapters taught the individual patterns: how compute is parallelized, how data is sharded, how orchestrators schedule jobs, how training is made fault-tolerant, how inference is optimized, how cost is managed, how observability is instrumented, how security is enforced, how large language models are handled at scale, and how case studies reveal the compositions that appear repeatedly in production. Each of those patterns is valuable in isolation. The problem is that a platform is not a collection of individual patterns. It is a system, and systems have properties that their parts do not. Coherence, reliability, and operational clarity emerge from the interactions between layers, not from any single layer in isolation.

The cost of incoherence is concrete. When the training and serving layers use different feature representations, bugs can arise that unit tests on either layer alone cannot catch. When the observability layer covers training jobs but not inference endpoints, production incidents are diagnosed by guessing rather than by reading signals. When the security posture differs between the data plane and the model artifact store, compliance reviews surface gaps that require emergency remediation. Each of these failures has a well-known pattern solution, covered in the chapters that precede this one. The synthesis question is: how does a team know which solutions to apply first, and when the platform has reached a level of maturity that allows it to address the next class of problems?

The answer is a maturity model that is mission-aligned with the actual work of building and operating distributed ML systems. A good maturity model does two things. First, it gives a team a shared vocabulary for describing its location. Second, it gives the team a concrete picture of what the next stage looks like, so that investment decisions are grounded in a destination rather than a direction. This chapter provides both and connects each stage of the model back to the specific patterns from earlier chapters, so the reader knows exactly where to look for implementation guidance.

The Modern Distributed ML Platform Stack

A modern distributed ML platform is not a single system. It is a composition of six planes, each with its own responsibilities, failure modes, and patterns that govern how it interacts with the planes above and below. Understanding the stack as planes rather than as tools is the first step toward architectural coherence. Tools change. The responsibilities of each plane do not.

The data plane covers everything that happens to data before it reaches a compute process: storage, sharding, ingest pipelines, and the lineage tracking that connects raw data sources to the tensors a training job consumes. A working data plane is invisible. Training jobs start on time, features are consistent between training and serving, and data quality incidents are caught before they corrupt a checkpoint.

The control plane covers orchestration, scheduling, and governance. This is the layer that decides which workloads run where and when, enforces resource quotas, manages access policies, and maintains the registry of models, datasets, and experiments. The patterns from Chapter 4 live here: gang scheduling, bin packing, preemption policies, and fair-share queuing. A shared cluster with no admission controls is a control plane that has not yet been built.

The compute plane covers the actual execution of workloads: the training fleet with its accelerators and checkpoint storage, and the serving fleet with its inference runtime, batching logic, and autoscaler. This plane is the focus of Chapters 2, 5, and 6. It is also the plane most teams build first, creating the common failure mode of a capable compute plane sitting atop an underdeveloped control plane. Compute without governance is expensive and fragile.

The observability plane makes the entire stack legible. It covers distributed tracing, metrics collection, log aggregation, alerting, and the dashboards that let an on-call engineer understand what is happening across all layers simultaneously. An observability plane that stops at the infrastructure boundary and does not extend into model behavior is incomplete in the ways that matter most during incidents.

The security plane enforces the access, isolation, and compliance guarantees that the organization requires. It covers identity and authentication for workloads and users, network isolation between tenants, encryption at rest and in transit, and the audit logging that compliance reviews depend on. The security plane is frequently treated as an afterthought and retrofitted under deadline pressure.

The product plane is the layer that most distinguishes a mature platform from a capable cluster. It covers the internal developer portal, the golden paths that let ML engineers start a new project without making a dozen infrastructure decisions, the documentation that makes the platform self-service, and the feedback mechanisms that tell the platform team where to extend the paths. A platform that requires every ML engineer to understand cluster networking is not a product. A platform that exposes a clean project template, a one-command training submission, and a model deployment workflow with sensible defaults is.

The Six Planes and Their Canonical Patterns

Product Plane	Developer Portal	Golden Path	
Security Plane	Lock	Identity Badgee	Audit Log
Observability Plane	Metrics	Trace	Alert
Compute Plane	GPU Accelerator	Training Job	Serving Endpoint
Control Plane	Orchestrator Node	Scheduler Queue	Policy
Data Plane	Storage	Shard	Lineage

The six planes are not independent. Changes to the data plane affect the correctness guarantees that the control plane can offer. Changes to the compute plane affect the observability instrumentation needed to cover new workload types. Changes to the security plane affect the developer experience delivered by the product plane. Managing these dependencies is the central architectural challenge of a mature platform.

Three coherence patterns appear most often in healthy platforms. Shared primitives mean that the same data format, authentication token, and job identifier flow through every layer without translation. Translation layers between planes are where bugs accumulate and where latency is introduced. Shared observability contracts mean that every component emits the same set of golden signals in the same format, so the observability plane can be built once and maintained as a unit rather than as a collection of per-component dashboards. Shared vocabulary means the words the platform team uses to describe its components match those ML engineers use to describe their workloads. A platform that uses "job" to mean one thing in the control plane and another in the developer portal creates confusion that compounds over time.

The Maturity Model: Five Stages from Ad Hoc to Autonomous

The five-stage maturity model presented in this chapter is grounded in the patterns covered by the preceding eleven chapters. Each stage is defined by concrete criteria rather than aspirational descriptions so that a team can place itself honestly. The stages are named to reflect the primary characteristic of each level: ad hoc, repeatable, standardized, self-service, and autonomous. The model is a shared vocabulary, not a leaderboard. A team in stage two that knows it is in stage two and has a plan to reach stage three is in a better position than a team that claims stage four without the criteria to support it.

Stage one is ad hoc. The platform consists of scripts and shared credentials familiar to one or two engineers. There is no formal cluster, no admission controls, and no documented procedures. Models are deployed manually. There is no lineage tracking, no guarantee of reproducibility, and no cost accounting.

Stage two is repeatable. The team has a shared cluster, a shared storage layer, and a minimal orchestration setup. Jobs can be submitted without manual intervention and produce comparable results when re-run. Cost is tracked at the cluster level but not allocated to teams or projects. The failure mode at this stage is the shared cluster becoming a contention zone as the team grows, due to the lack of implemented admission controls and quotas.

Stage three is standardized. The platform has a formal control plane with admission controls, resource quotas, and a model registry. Observability covers training jobs and inference endpoints with a consistent set of golden signals. Security policies are defined and enforced. Data lineage is tracked for production pipelines. The failure mode is that the platform is correct, but not ergonomic. ML engineers can use it, but doing so requires understanding too many infrastructure details.

Stage four is self-service. ML engineers can start a new project, submit a training job, register a model, and deploy it to production without writing infrastructure code. The platform team operates on a product cadence with a roadmap and feedback mechanisms. Cost is allocated at the project level. The failure mode is that the golden paths cover common cases well, but edge cases poorly.

Stage five is autonomous. The platform detects optimization opportunities and acts on them without human intervention, within defined policy bounds. This includes automatic spot instance reclamation, dynamic resource reallocation, and proactive scaling of the serving fleet. The platform enforces compliance policies at admission time rather than through post-hoc auditing. This stage is achievable but rare. Most teams should aim for stage four before considering the investments required for stage five.

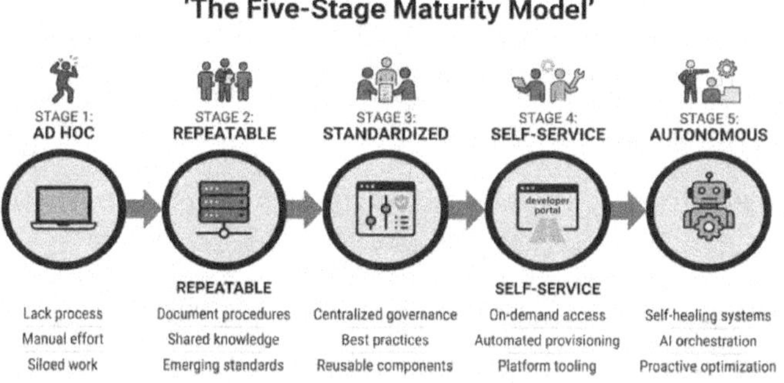

The model is deliberately simple. Complexity in a maturity model is the enemy of honest assessment. Teams that debate which stage they are in are not advancing. The right question is not "which stage are we at?" but "what are the two or three specific criteria from the next stage that we do not yet meet, and what would it cost to meet them?"

Stage Diagnostics: Placing Your Team Honestly

The most common failure mode in applying a maturity model is overrating. Teams with strong engineers and a history of shipping tend to rate their platforms higher than the criteria support. The reason is usually that the team is judged by its best-case behavior rather than by its typical behavior. A platform that runs smoothly when the principal engineers are available but degrades significantly when they are on leave is not a stage-three platform, even if it has the technical components of one. Operational maturity is measured at the median, not the maximum.

Four diagnostic questions reliably distinguish stage two from stage three. First: Can a new engineer submit a training job without being walked through the process by a senior team member? If the answer is no, the platform is at most stage two, regardless of what the documentation says. Second: can the on-call engineer determine the root cause of a training job failure within thirty minutes using only the observability tooling, without escalating to the engineer who wrote the job? If the answer is no, the observability plane is incomplete. Third: can the team produce a cost report broken down by team and project within one business day, without manual data wrangling? If the answer is no, the FinOps posture is at most stage two. Fourth: Are security policies enforced automatically at admission, or are they enforced only through periodic manual review? Manual review is a stage-two control.

The diagnostic questions for stage three to stage four shift from correctness to ergonomics. What is the median time from new project creation to first training run? Stage three platforms typically require two to five days; stage four platforms typically require two to four hours. What fraction of production deployments follow the standard golden path without modification? If the answer is less than 80%, the golden paths need to be extended. When an ML engineer needs to do something the platform does not support, is there a documented off-ramp that keeps them within the security and observability perimeter? If the answer is no, the product plane is incomplete.

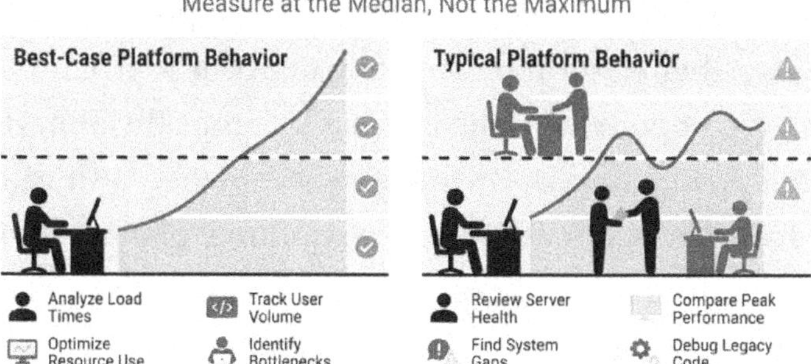

The temptation to rate one stage too high is amplified when the rating has organizational consequences. A VP who asks "show me the onboarding time for the last five new ML engineers" gets a more honest answer than one who asks "are we at stage three?" The maturity model is most useful when it is grounded in measurable outcomes rather than architectural descriptions.

The failure modes characteristic of each stage deserve explicit treatment, because recognizing them is often faster than running through the full diagnostic. At stage one, the failure mode is key-person dependency: the platform breaks when the one engineer who understands it is unavailable. At stage two, the failure mode is resource contention: a single large job can starve all other workloads due to the absence of quota enforcement. At stage three, the failure mode is ergonomic friction: engineers work around the platform rather than through it because the documented path is too slow. At stage four, the failure mode is off-path brittleness: the golden paths work well, but anything outside the standard use case causes engineers to go dark and build ad hoc solutions. At stage five, the failure mode is automation opacity: the platform takes automated actions that engineers cannot explain, leading to distrust and manual overrides that undermine the automation.

Architectural Coherence Across Planes

A platform with capable components in each plane but no coherence between them is harder to operate than one with simpler, tightly integrated components. This is not a theoretical claim. It is the pattern that appears in post-incident reviews when root cause analysis traces a production failure across three planes before landing on the actual cause. Coherence is the property that makes a multi-plane platform legible as a single system.

Three architectural patterns produce coherence across planes in distributed ML platforms. The first is the universal job identifier. Every workload that enters the platform is assigned a unique identifier at admission time, and that identifier propagates through every plane. The training job's identifier appears in the scheduler logs, training metrics, checkpoint paths, model registry entry, serving deployment, and billing report. When an incident occurs, the on-call engineer starts with the job identifier and can traverse the entire history of that workload without switching tools or schemas. Platforms that do not enforce this pattern produce incident timelines that require manual correlation of timestamps and hostnames, a process that is slow and error-prone.

The second coherence pattern is the shared contract for golden signals. Every component in every plane emits four categories of signal in a consistent format: throughput (how much work is being done), latency (how long it takes), error rate (how often it fails), and saturation (how close to capacity the component is). These are the four golden signals adapted from site reliability engineering for ML workloads. When every component uses the same signal categories and emission format, the observability plane can be implemented as a uniform layer rather than as a collection of per-component adapters. Components that deviate from the contract require custom instrumentation, which is expensive to maintain and fragile during version changes.

The third coherence pattern is the platform vocabulary contract. The platform team publishes a controlled vocabulary of terms that defines exactly what is meant by "job," "model," "artifact," "deployment," "tenant," and the other nouns that appear throughout the system. This vocabulary is enforced across the developer portal, API schemas, and documentation. When vocabulary drifts, two kinds of costs arise. The first is communication cost: conversations between the platform team and ML teams require continuous disambiguation. The second is integration cost: systems that use different definitions of the same term require translation layers at their boundaries, where data loss and bugs accumulate.

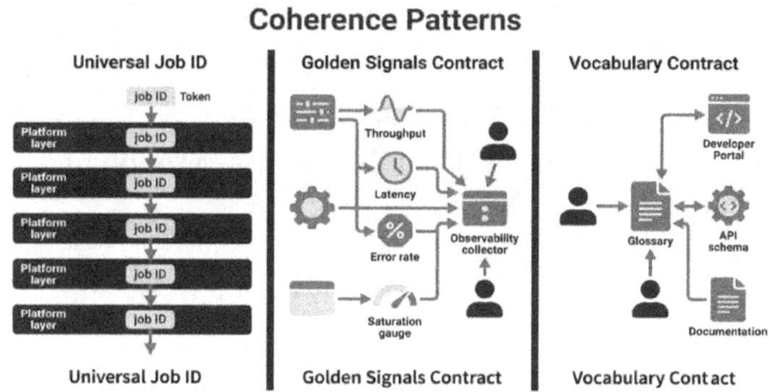

Architectural coherence is not free. Enforcing a universal job identifier requires that every component be modified to accept and propagate the identifier; components managed by external vendors may not support this without wrapper logic. Enforcing the golden signals contract requires documentation and review. Enforcing the vocabulary contract requires the platform team to decline proposals that introduce competing terminology, which is as much a social challenge as a technical one.

The cost of coherence is a one-time investment that produces compounding returns. The cost of incoherence is a recurring tax paid at every incident, every integration, and every onboarding. Teams that have operated on incoherent platforms for several years estimate that twenty to thirty percent of engineering time is spent on translation and correlation tasks that would be unnecessary on a coherent platform.

Build, Buy, and Open-Source: A Decision Framework

Every distributed ML platform team faces recurring decisions about whether to build a component, buy a commercial product, or adopt an open-source solution. These decisions are often made once and then implicitly remade every time a component is updated, extended, or replaced. The cost of implicit remakes is high: they consume engineering time without producing a documented rationale, and the next team facing the same decision must start from scratch rather than build on the previous team's analysis.

The build-buy-open-source decision has four primary inputs. The first is differentiation: does the component represent a capability that distinguishes the platform? If yes, the building is worth considering. If not, building is usually the most expensive path. A custom in-house log aggregation pipeline is rarely better than a well-maintained open-source alternative.

The second input is the time horizon. A component needed in six weeks should rarely be built from scratch. Build decisions made under time pressure tend to produce technical debt that accumulates faster than the team can service it, while build decisions made with a clear two-to-three-year roadmap tend to produce components the team genuinely understands and can extend.

The third input is the total cost of ownership. A commercial product whose annual license is lower than the fully loaded engineering cost of building and maintaining the equivalent component is usually the right choice, even if the build option would produce a technically superior result. The trap is underestimating the build option: it includes not just initial engineering time but documentation, testing, monitoring, version upgrades, and knowledge transfer.

The fourth input is vendor lock-in risk. A commercial component that requires proprietary data formats or specific cloud provider APIs constrains future architectural choices. The pattern for managing lock-in is to place adapter layers between the platform's internal contracts and the vendor's external interfaces, so the vendor can be replaced without changing internal APIs.

Build-Buy-Open-Source Decision Framework

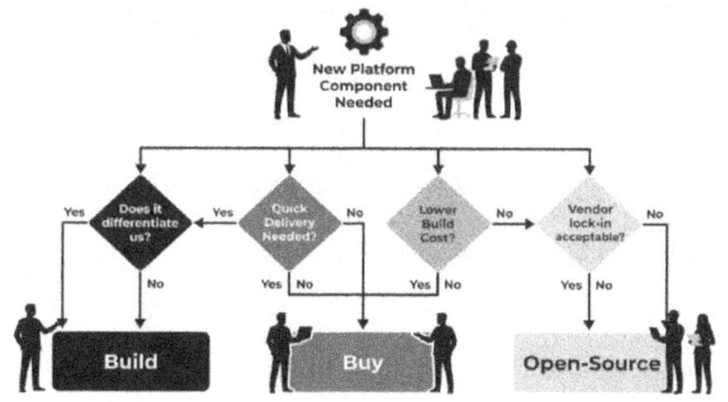

Open-source adoption carries its own risk profile. An open-source component maintained by a large community with commercial backing is a low-risk choice. A component maintained by a single organization for its own internal use and released as a secondary activity is high-risk. The evaluation pattern assesses release cadence, maintainer responsiveness, contributor diversity, and the presence of commercial entities that have staked products on the component's reliability.

The build-buy-open-source decision should be reviewed on a known cadence: annually for stable components and at major version boundaries for actively evolving ones. Reviews triggered only by failures or vendor price increases are reactive and often lead to rushed decisions.

Org Design for Platform Maturity

Technical patterns do not implement themselves. The organizational structure around a distributed ML platform determines whether the technical patterns described in this book are applied consistently across teams that touch the platform. Org design is a force multiplier: the right structure amplifies the value of the technical investments, and the wrong structure neutralizes them.

The org design that matches a stage-three or stage-four platform has three distinct team types with well-defined interfaces. The first is the platform infrastructure team, which owns the compute, control, and foundational data planes. Its success metrics are cluster utilization, job queue latency, and storage reliability. If this team does not exist as a distinct unit, platform reliability is owned by whoever is debugging the next failure, a stage-one characteristic.

The second is the ML platform team, which owns the observability plane, the security plane, the product plane, and the higher-level data plane capabilities such as the feature store and the model registry. Its success metrics are time-to-first-run for new projects, off-path incident rate, and net promoter score among ML engineers. This is the team most often described as a platform-as-a-product organization.

The third is the domain ML team, which owns specific ML applications or problem domains. These teams consume the platform rather than build it. Domain ML teams that have good platforms build features. Domain ML teams that have poor platforms spend engineering time on infrastructure.

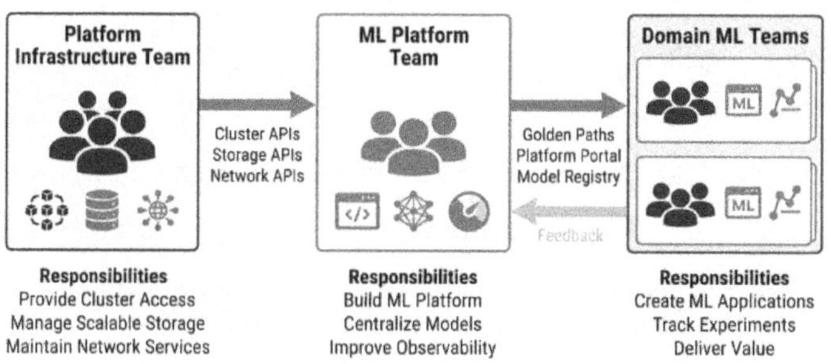

The seams between these three team types are where most platform quality problems originate. Three failure patterns at the seams are worth naming. The first is the "ticket over the wall" pattern: the infrastructure team implements a capability to its own specifications without engaging the downstream use case. The fix is a lightweight RFC process requiring the infrastructure team to describe the interface before building. The second is the "shadow platform": a domain ML team builds its own tooling because the platform does not meet its needs. Shadow platforms grow when the feedback loop between domain teams and the ML platform team is broken; the fix is a regular feedback forum. The third is "platform as gatekeeper": the ML platform team becomes a bottleneck because compliance policies are enforced through human review rather than automated controls. The fix is policy-as-code, a stage-three-to-stage-four transition.

Patterns from organizations that have operated mature platforms suggest a rough ratio: for every ten to fifteen engineers on domain ML teams, one full-time equivalent on the ML platform team and half a full-time equivalent on the infrastructure team are needed to maintain a stage-four platform.

Platform as Product: Operating the Developer Experience

A platform that is not operated as a product will eventually become a liability. This is the pattern that appears when the ML platform team has strong technical instincts but no product discipline. The result is a technically capable platform that ML engineers avoid because using it requires more expertise than solving the problem independently. Platform-as-a-product is not a management philosophy. It is an operating model with specific practices.

The first practice is to treat internal ML engineers as customers and measure their satisfaction. This means running periodic surveys, tracking leading indicators such as time-to-first-run and the off-path incident rate, and using those metrics to drive the platform roadmap. A platform team that cannot name its top three customer complaints is not operating as a product team. A platform team that can name them and has a plan to address them is.

The second practice is maintaining a public roadmap. Internal ML teams need to know what is coming so they can decide when to adopt a new capability rather than build their own workaround. A quarterly-updated roadmap, communicated to domain ML teams, prevents shadow platforms by giving engineers a reason to wait for the official capability rather than building their own.

The third practice is to define and enforce a deprecation policy. Platforms that add capabilities indefinitely without removing outdated ones become increasingly complex to operate and to explain. A deprecation policy that gives engineers six months of notice before removing a capability, provides a migration path, and is enforced consistently signals that the platform team takes backward compatibility seriously without treating every old interface as permanent.

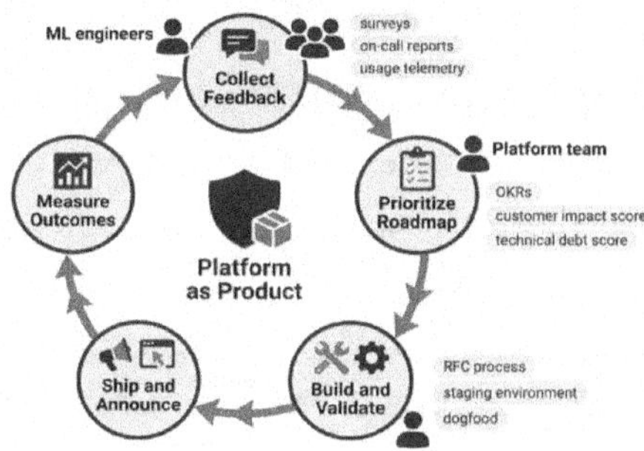

The fourth practice is dogfooding. The ML platform team should run its own training jobs and serve its own models using the platform it maintains. This is not a symbolic act. It is an early-warning system for ergonomic problems that do not appear in bug reports because ML engineers work around them rather than reporting them. Platform teams that dogfood their own platforms consistently produce better developer experiences than those that do not, because the feedback loop is immediate rather than mediated by a survey or a support ticket.

The fifth practice is managing the golden paths as a product asset. Golden paths are not static templates. They are evolving artifacts that need to be extended as new workload types emerge, updated as platform capabilities improve, and deprecated when the underlying technology they depend on is replaced. A golden path more than 18 months old that hasn't been updated recently is probably no longer golden. The test is simple: does the path describe the fastest, safest way to accomplish the task, or does it describe how the task was accomplished eighteen months ago?

A Ninety-Day Plan to Advance One Maturity Stage

The maturity model is most useful when it is connected to a concrete plan. The ninety-day plan described in this section is structured for a team that has completed an honest assessment and identified itself as being at stage two or stage three. The plan advances the team one stage in ninety days, not by attempting to implement every criterion of the next stage simultaneously, but by focusing on the two or three specific gaps that have the highest impact on the team's current failure modes.

The plan has four phases. Phase one, the first two weeks, is the assessment phase. The team runs the diagnostic questions to produce a specific list of criteria gaps. Each gap is written as a named problem with a measurable current state and a measurable target state: "Onboarding time averages four days; target is one day." Gaps that cannot be stated in measurable terms are refined until they can be expressed in measurable terms. This step distinguishes a plan from a wish list.

Phase two, weeks three through six, is the foundation phase. The team addresses the infrastructure prerequisites for the identified gaps. If the primary gap is onboarding time, the foundation phase installs a project template system and documents the end-to-end workflow. If the primary gap is on-call diagnosis time, the foundation phase audits observability coverage and adds the missing signal sources. Foundation work is unglamorous but necessary. Teams that skip building features on an unstable foundation produce platforms that impress in demos but fail in production.

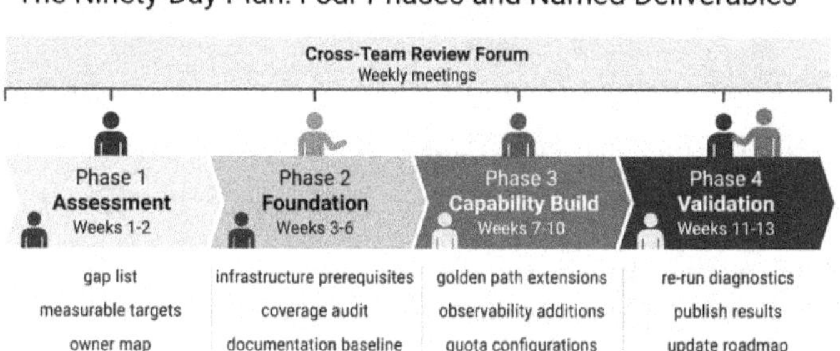

Phase three, weeks seven through ten, is the capability build phase. The team implements capabilities that address the identified gaps: new golden-path templates, quota configurations, extended observability coverage, and manual security controls converted to policy-as-code. The key discipline is scope control. Every capability added beyond the specific gaps from phase one risks the ninety-day deadline.

Phase four, weeks eleven through thirteen, is the validation phase. The team re-runs the diagnostic questions with new measurements. Gaps that still do not meet their target states are documented as carry-forward items for the next ninety-day plan. The validation results are published to the ML platform team, domain ML teams, and leadership, creating both accountability and a historical record for future investment requests.

The single cross-team review forum holds the plan together. This is a weekly meeting lasting 30 to 45 minutes that includes the ML platform team lead, one representative from the platform infrastructure team, and one representative from each domain ML team. It is a decision-making meeting: it surfaces blockers, resolves priority conflicts, and approves scope changes. A plan without this forum tends to lose momentum in weeks four through six as competing priorities emerge and blockers go unresolved.

FinOps for Distributed ML: Making Cost a First-Class Signal

Cost management in distributed ML is not a finance function that operates in parallel with engineering. It is an engineering discipline with operational properties similar to those of reliability and security. Platforms that treat cost as a first-class signal build it into the control plane and the product plane rather than reporting it as a periodic summary. The patterns for FinOps in distributed ML are covered in Chapter 7; this section connects them to the maturity model.

At stage two, cost is visible at the cluster level. The team knows how much compute it is consuming in aggregate, but cannot attribute that cost to specific projects, teams, or model versions. The transition to stage three requires project-level cost attribution, which in turn requires implementing and connecting the universal job identifier pattern to the billing layer.

At stage three, cost is attributed and visible to team leads. The transition to stage four requires surfacing cost information to individual ML engineers in the developer portal alongside their training job submissions. An engineer who sees the estimated cost of a job before submitting it makes different decisions than one who receives a monthly report. The feedback loop between action and cost must be deliberately constructed in shared cluster environments.

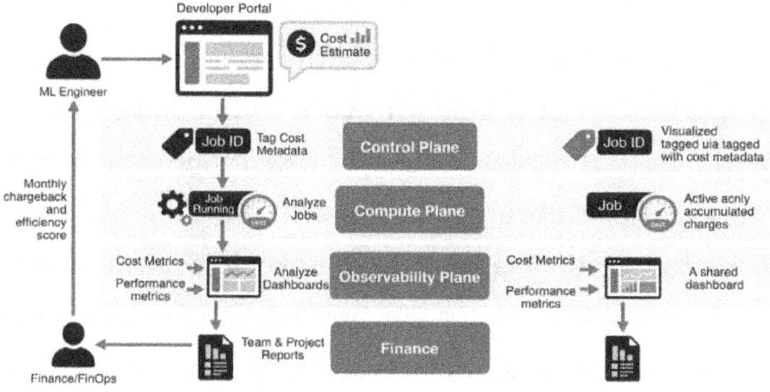

At stage four, cost efficiency is a feature of the developer experience. Engineers receive budget allocations per project, cost anomaly alerts when a job runs significantly over its estimate, and efficiency scores that compare their workloads to platform-wide benchmarks. This is the same category of feedback experienced engineers use in other infrastructure contexts: a developer who sees that their database query is scanning the full table changes the query; an ML engineer who sees zero GPU utilization after epoch one changes the data pipeline.

Spot instance management is the highest-leverage FinOps capability for most distributed ML platforms. Training jobs that can tolerate interruption and resume from checkpoints can run on spot or preemptible capacity at sixty to eighty percent cost reduction without any reduction in training quality. The enabling pattern is transparent checkpointing combined with a spot reclamation handler that triggers the checkpoint before the instance is reclaimed. Platforms that have not implemented this pattern are paying on-demand prices for interruptible workloads, the most common single source of ML cost inefficiency at stage two and stage three.

Observability Plane: The Nervous System of the Platform

Observability in a distributed ML platform is the mechanism by which every operator in every plane knows whether the system is behaving as intended. A complete observability plane enables on-call engineers to diagnose training failures in under 30 minutes, capacity planners to predict saturation before it occurs, model owners to detect performance degradation within hours of a bad deployment, and cost analysts to identify budget anomalies within 1 business day.

The observability plane has four components that must be implemented together. The first is metrics collection: the platform aggregates time-series metrics from every plane with sufficient resolution to distinguish transient spikes from sustained trends. The second is distributed tracing: a trace context is propagated through every operation, so the full execution path of a training job can be reconstructed after the fact. The third is structured logging: every component emits logs in a consistent structured format that can be queried programmatically. The fourth is alerting with defined ownership: every alert has a named owner, a defined severity, and a runbook. Alerts without runbooks degrade on-call quality over time.

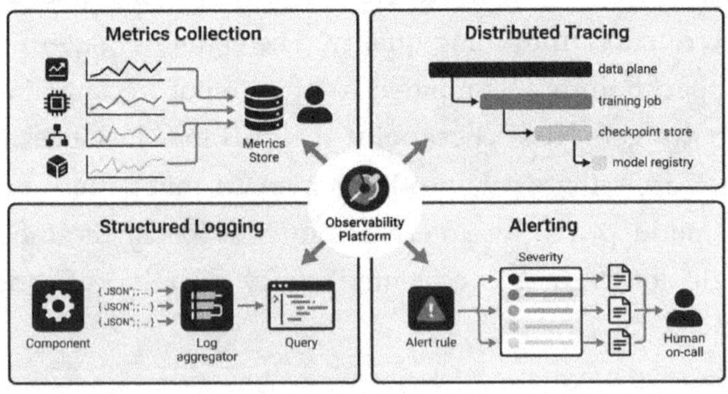

The ML-specific observability signals that distinguish a distributed ML observability plane from general-purpose infrastructure monitoring are covered in Chapter 8. They include training loss curves with anomaly detection, gradient norm tracking, inference latency at p50/p95/p99, and token throughput for LLM serving endpoints. These signals must be first-class citizens in the observability plane, not afterthoughts.

Model performance observability deserves special attention. A model serving predictions with high infrastructure reliability but degraded prediction quality is failing silently in a way that infrastructure metrics cannot detect. The platform-level pattern is to integrate model-level evaluation into the deployment pipeline, gating every deployment on an evaluation result against a held-out validation set. This shifts model quality from periodic human review to automated control.

Security Plane: Responsible by Design

A distributed ML platform concentrates three categories of high-value assets: sensitive training data, proprietary model weights, and expensive compute infrastructure. These three assets require three distinct security control patterns, and a platform that protects one without protecting the others has a security posture that is, at best, incomplete and, at worst, actively misleading. Responsible by design means that security controls are built into the platform's normal operating paths rather than applied as an overlay that can be bypassed under time pressure.

Training data security requires access controls at the dataset level, not the cluster level. The pattern is a data access policy enforced at the storage layer and connected to the identity of the requesting workload. Workloads run with service account identities, and those identities are granted specific dataset permissions at job submission time. This is a stage-three requirement; without it, data access controls are perimeter-based and cannot support compliance audits.

Model artifact security requires that trained model weights be treated with the same access controls as the training data used to train them. Organizations with strong data access controls and weak model artifact controls expose model weights to principals who would not be permitted to access the training data directly. The pattern is a model registry with fine-grained access controls and cryptographic signing, so every model in production can be traced to a specific training run and verified through a chain of custody.

Compute infrastructure security requires network isolation between tenants, quota enforcement to prevent denial-of-service by runaway workloads, and logging of all privileged operations. The admission controller is the enforcement point for compute limits and must be configured with conservative defaults.

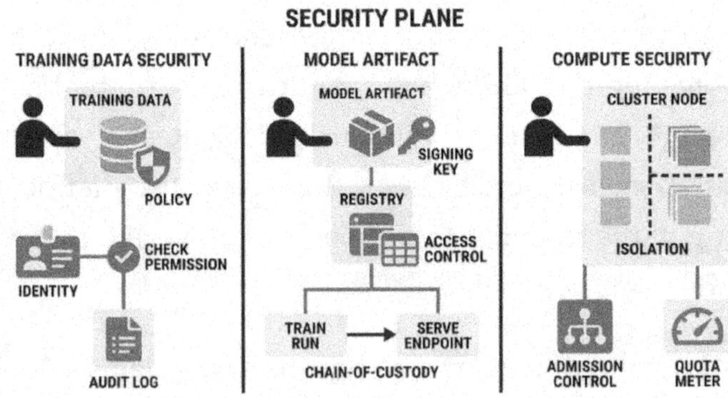

The security plane's compliance posture must be linked to the audit logging produced by the observability plane. A compliance query like "which workloads accessed dataset X between dates A and B, and which produced models currently in production?" should be answerable in minutes by querying the audit log and model registry. Platforms that cannot answer this query in under an hour are not ready for production compliance audits.

Looking Ahead: Patterns That Outlast the Next Platform Cycle

The patterns in this book were selected for durability. The test applied to each one was: Will this pattern still be relevant in five years, regardless of which training framework, inference runtime, or cloud provider is dominant? The answer is yes, because they describe structural problems of distributed systems rather than artifacts of any specific implementation.

Five patterns are the durable investments any team building a distributed ML platform should prioritize. The first is transparent checkpointing. Any distributed system that runs long-duration workloads needs the ability to interrupt and resume without data loss. The implementation details change; the requirement does not.

The second is the universal job identifier with full lineage tracking. As ML systems become more regulated, the ability to trace a prediction back to the model version, the training run, and the training data that produced it becomes a regulatory requirement rather than a debugging convenience. Teams that build lineage tracking early create an asset whose value compounds over time.

The third is the golden signals contract. ML workloads have specific observability needs that generic infrastructure monitoring does not address. Teams that define and enforce a golden signals contract create a shared observability layer that is robust to team turnover and tool upgrades. The contract is the durable artifact; the monitoring tool that implements it is replaceable.

The fourth is the admission controller pattern. As ML workloads grow in scale and number, the shared cluster becomes an increasingly contested resource. Admission controls are the mechanism by which the platform expresses its priorities under contention. Platforms without admission controls become progressively less reliable as they scale.

The fifth is platform as product. Treating internal ML engineers as customers and operating the platform on a product cadence is the pattern most likely to determine whether the technical investments in this book are actually put to use. A technically excellent platform that engineers avoid produces less value than a technically simpler platform that engineers adopt enthusiastically.

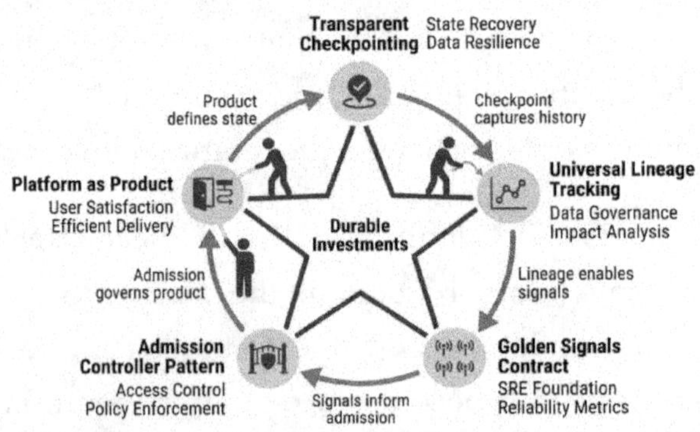

The next three to five years will bring changes that make some implementation details obsolete. Hardware architectures will change. Framework APIs will be revised. What will not change is the need to parallelize compute across multiple accelerators, govern shared resources under contention, make distributed systems legible through observability, and deliver platform capabilities to engineers without requiring them to become infrastructure experts. These are the problems any team building a distributed ML platform in 2030 will still be solving, using different tools but with the same principles.

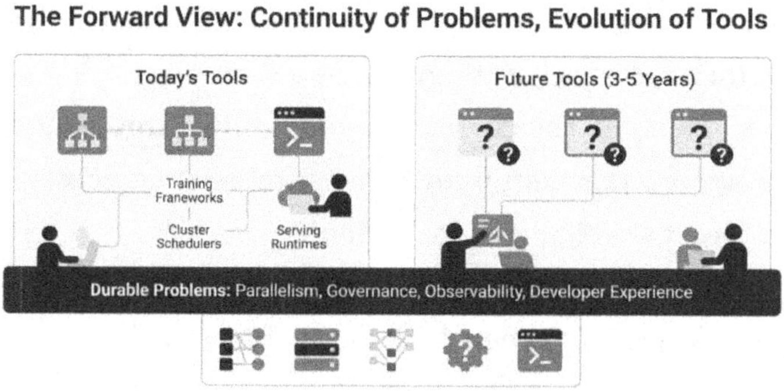

The pilot-to-production pathway for a distributed ML platform is not a straight line. Organizations that have scaled from stage two to stage four consistently describe the journey as a series of crises resolved by investing in the pattern that would have prevented each one. The value of the maturity model is to compress that timeline: to give teams the vocabulary and the framework to make those investments proactively rather than reactively, and to make the pilot-to-production pathway a designed path rather than an improvised one.

Technical Checklist

- The team has a written description of its distributed ML platform stack, organized by the six planes: data, control, compute, observability, security, and product.
- Each plane in the stack names at least two canonical patterns it implements and identifies which chapter of the platform's internal documentation (or a reference like this book) describes the pattern in detail.

- The team has placed itself on the five-stage maturity model using the diagnostic questions from this chapter, not by self-assessment of architectural sophistication.
- The current maturity stage is supported by at least three measurable criteria: onboarding time for new ML engineers, on-call root-cause-determination time, and cost attribution resolution (cluster, project, or engineer level).
- The biggest gap between the current stage and the target stage is written as a named problem with a measurable current state, a measurable target state, and a single named owner.
- Build-buy-open-source decisions for all components in the control and compute planes have been reviewed within the last twelve months and are documented with the four-factor analysis: differentiation, time horizon, total cost of ownership, and vendor lock-in risk.
- Org design matches the maturity stage of the platform: stage-three and stage-four platforms have distinct platform infrastructure and ML platform teams with defined interfaces.
- The ML platform team has a public internal roadmap updated at least quarterly and shared with domain ML teams.
- A ninety-day plan exists with named deliverables, a phase structure, a single cross-team review forum, and a validation step that re-runs the diagnostic questions.
- The observability plane covers all four components: metrics collection, distributed tracing, structured logging, and alerting with runbooks and named owners.
- The security plane enforces data access controls at the dataset level (not just the cluster level), model artifact access controls in the model registry, and compute isolation through admission controls.

- Cost attribution is implemented at least at the project level, with cost estimates visible to engineers at job submission time and anomaly alerts configured for jobs running more than fifty percent over their estimated cost.
- The universal job identifier pattern is implemented, and the job identifier propagates through the scheduler, training logs, checkpoint paths, model registry, serving deployment, and billing layer.
- The platform has a defined deprecation policy with a minimum notice period and a documented migration path requirement.
- The team has identified the top three customer complaints from domain ML engineers within the last quarter, and each has an owner and a target resolution date.

Team Conversation

1. Which stage of the maturity model are we honestly at, and which stage would we claim if asked by a peer organization? What explains the gap between those two answers?
2. If we ran the four diagnostic questions for stage-two-to-stage-three today, which one would we fail, and why?
3. Where in our stack is incoherence costing us the most? Do we have the observability to measure that cost, or are we estimating it from incident memory?
4. Which build-buy-open-source decisions have we silently remade in the last year, and did those remakes produce documented rationale or ad hoc choices?
5. Does our org design match our platform's maturity? Which team type is most understaffed relative to the platform's current stage?
6. What is the single deliverable that, if completed in the next ninety days, would have the largest impact on our team's productivity? Who owns it today?

7. Which patterns from this book do we already implement well, and which ones have we avoided? What drove the avoidance?
8. If we needed to pass a compliance audit that required tracing a production model back to its training data within four hours, could we do it today? What would break?
9. What would it take to get the median onboarding time for new ML engineers below one day? Who would need to be involved, and what would need to be built?
10. Which of the five durable patterns do we think will matter most to our organization in three years, and are we investing in it now?

Key Takeaway

A modern distributed ML platform is a layered system, and the layers must be coherent. The six planes described in this chapter (data, control, compute, observability, security, and product) each have distinct responsibilities, distinct failure modes, and distinct canonical patterns. No single plane can substitute for a missing one. A compute plane that is powerful but sits atop an underdeveloped control plane is expensive and fragile. An observability plane that covers infrastructure but not model behavior is blind to the failures that matter most. The coherence patterns that connect the planes (the universal job identifier, the golden signals contract, and the shared vocabulary) are investments that compound in value as the platform matures.

The five-stage maturity model is a shared vocabulary for honest assessment, not a leaderboard for competitive positioning. The diagnostic questions in this chapter are more useful than self-rating because they measure behaviors rather than architectural intentions. A team that can answer diagnostic questions honestly, identify the two or three gaps that matter most, and design a 90-

day plan to close them is practicing operational clarity. The maturity model does not tell a team what to build. It tells the team where it is and what the next step looks like, so that investment decisions are grounded in a destination rather than in a direction.

The patterns in this book are durable because the problems they solve are structural. Parallelism, governance, observability, and developer experience are the enduring concerns of any team building distributed ML systems at scale. The tools that implement these patterns will change. The frameworks, cloud providers, and hardware architectures in 2030 will differ from those of today. But the engineer who understands transparent checkpointing, lineage tracking, the golden signals contract, the admission controller pattern, and platform-as-product will be effective on any platform, with any stack, at any scale. That is the pilot-to-production pathway this book was written to provide.

Closing

Twelve chapters ago, this book opened with a simple proposition: distributed machine learning is no longer the exotic edge of the field. Training a large language model across hundreds of accelerators or serving a recommendation model to tens of millions of users are problems every serious ML organization now faces. The patterns that solve those problems are not locked inside any single framework or vendor. They are compositions of decisions about parallelism, data movement, scheduling, failure recovery, cost, observability, and security. This book names those decisions, explains the trade-offs behind each one, and gives architects and engineers a vocabulary they can carry into any cluster, on any platform, in any team. The platform cycle rotates every three to five years; the patterns persist.

What You Have Learned

Chapters 1 and 2 established the foundational vocabulary. Chapter 1 introduced the three trade-offs that govern every distributed ML decision: consistency, performance, and cost. Every pattern in the book is a specific resolution of that tension. Chapter 2 names the compute parallelism primitives: data, model, pipeline, and tensor parallelism. These are the primary design vocabulary for any system that trains or serves at scale, and every later chapter uses them as a shared language.

Chapters 3 and 4 moved into data and orchestration. Chapter 3 established that distributed training is often bottlenecked by the data pipeline rather than the network fabric. Sharding strategies, I/O optimization, and separating storage from compute are the practices that prevent the pipeline from becoming the training bottleneck. Chapter 4 showed that gang scheduling, queue-based resource management, and ML-aware schedulers, when combined with generic Kubernetes primitives, make clusters usable for production workloads.

Chapters 5 and 6 form the core production-engineering spine: reliable training across thousands of accelerators amid constant hardware failures, followed by continuous batching and KV-cache management for serving at scale. Chapter 7 showed that most GPU spend is avoidable with deliberate resource management, and Chapter 8 built the observability layer that makes every other chapter debuggable.

Chapter 9 established security and multi-tenancy as first-class platform concerns: workload isolation and supply-chain integrity are load-bearing walls. Chapter 10 showed that LLM workloads are characterized by the familiar parallelism and serving patterns, applied to a new model class with a different cost and latency profile. Chapter 11 grounded everything in four end-to-end case studies with adaptable reference architectures. Chapter 12 synthesized the whole into a layered platform view and a five-stage maturity model that the reader can use to assess their organization and act on the result.

The Patterns That Matter Most

Some patterns have survived at least two framework cycles and will survive the next one as well. These are the patterns worth internalizing deeply, encoding in platform primitives, and defending when a fashionable alternative arrives.

Data parallelism with gradient synchronization is the first. Every technique for training at scale, from ring-allreduce to ZeRO-3 to pipeline-parallel variants, is a refinement of the core insight that independent replicas can cooperate through a structured synchronization protocol. The synchronization protocol changes; the underlying pattern does not. An engineer who understands why gradient synchronization works can evaluate any new framework's communication primitive on first principles rather than on benchmarks alone.

Checkpoint-driven fault tolerance is the second. At scale, hardware failure is not an edge case; it is a scheduled event. Writing consistent checkpoints, resuming from the most recent consistent state, and designing training loops that are idempotent across resume boundaries is the foundation of reliable large-scale training. No amount of hardware improvement eliminates this requirement; the failure rate per job shifts from hardware faults to software faults as the cluster matures, but it does not disappear.

Continuous batching for inference is the third. A serving system that allows requests to enter and leave an in-flight batch dynamically maximizes accelerator occupancy without sacrificing tail-latency guarantees. The specific implementation varies with each serving runtime release; the scheduling principle remains stable. Shared observability primitives are the fourth: consistent, structured traces and metrics across every platform layer are what separate a system that can be governed from one that can only be heroically debugged. Teams that instrument observability as a design concern resolve incidents in minutes; teams that bolt it on afterward resolve the same incidents in hours.

Platform-level paved roads are the fifth. The most effective way to propagate good patterns across a large organization is not by policy mandates but by making the correct pattern the easiest path. A job template that enforces checkpoint-writing and produces structured logs by default is a paved road. When the platform team makes the correct path, the fast path, standards propagate organically, and violations become visible immediately as deviations from the path.

The consistency-performance-cost triangle from Chapter 1 is the sixth: the pattern beneath every other pattern. Every distributed ML decision resolves that tension, and the engineer who has internalized it makes better decisions faster because they know which question to ask first. Security and multi-tenancy as platform-layer primitives are the seventh. Treating workload isolation and supply-chain integrity as platform defaults rather than team-level responsibilities is the only model that scales as a platform grows from one team to dozens.

What Comes Next

The distributed ML frontier is moving on several dimensions simultaneously, and understanding where it is moving helps an architect decide which platform investments to make today so the platform is not caught flat-footed by the next wave.

Heterogeneous accelerator clusters are arriving. GPU-only clusters are giving way to environments that integrate GPU architectures, specialized matrix-multiplication accelerators, and custom silicon on a single fabric. Scheduling and placement patterns that work in homogeneous clusters break down when device types have different memory hierarchies and communication primitives. Teams that invest now in accelerator-agnostic abstraction layers and topology-aware placement policies will adapt more readily when the mix changes, as it will change on a two- to three-year cycle.

Decentralized training routes around the bandwidth constraints of centralized all-reduce topologies by allowing partial gradient exchange across geographically distributed nodes. The pattern is already being applied in federated learning contexts, where data sovereignty prevents the centralization of training data. As foundation models grow and inter-datacenter bandwidth costs fall, decentralized training will move from research prototype to production pattern. Multi-modal foundation models are extending distributed serving patterns into new territory: a model that jointly processes text, images, and audio has a different memory footprint and batching strategy than a text-only transformer, and its failure modes are less well documented in the current literature.

Agentic workloads at scale present a new infrastructure problem. An LLM agent that plans, calls tools, observes results, and replans over multiple steps generates a workload profile that is fundamentally different from that of a single-shot inference request. The serving system must manage long-horizon context windows, variable compute demand, and state persistence across steps. The underlying disciplines are familiar: resource isolation, fault-tolerant state management, and structured observability are as important for an agentic workflow as for a training job. Sustainable ML is also becoming a first-order concern: carbon-aware scheduling and workload placement that prefer lower-carbon-intensity regions are early forms of what will become a standard platform capability.

A Final Word to Architects

Platform thinking is a discipline, not a role. It is the practice of designing systems that serve not just the current workload but the next 10, making good patterns easy and bad patterns visible, and that accumulate organizational capability rather than individual heroics. Every engineer who has owned a production system understands the difference between a system that was designed and one that grew. The designed system addresses a failure mode that has not happened yet. The system that grew has a hero who knows where all the undocumented decisions are buried, and a crisis when that hero is unavailable.

The courage this work requires is the courage to choose a simple pattern over a fashionable one when the simple pattern is sufficient. A two-node data-parallel job does not need tensor parallelism. A team of eight does not need a multi-tenant GPU cluster with a full MLOps platform. The right answer to a complex architecture proposal is often a durable question: what problem does this actually solve that the simpler approach does not? The most consistent failure mode across large distributed ML organizations is not under-investment in technology. It is over-investment in complexity at the wrong stage of maturity. Advance one stage at a time, with named deliverables and a single owner, and the platform compounds. Skip stages, and the platform collapses under its own weight.

The architect's primary obligation is operational clarity: the state in which every person on the team knows what the system is supposed to do, how to tell whether it is doing it, and what to do when it is not. Operational clarity is not achieved by writing a document. It is achieved by building a system that makes its own state legible through structured observability, fails loudly and specifically when something is wrong, and recovers gracefully because recovery was designed in rather than retrofitted. Design the boring things carefully: a consistent naming convention for distributed jobs, a mandatory health-check endpoint on every training pod, and a single agreed-upon definition of training throughput are not glamorous decisions. They are load-bearing walls, and they compound.

Where to Go From Here

The most immediate next step is not to read another book. It is to audit the system you own against one chapter of this book and produce a specific, actionable gap list. Pick the chapter that names the problem you are currently experiencing. Work through the technical checklist at the end of that chapter against your actual system. The gap between the checklist and reality is the most valuable backlog item you can bring to your next planning session because it is grounded in the specific constraints of your cluster, team, and workload.

The companion repository contains the practical artifacts referenced throughout the book: Kubernetes manifests for distributed training and serving workloads, framework configuration templates annotated with the decisions they embody, profiling and benchmarking scripts, cost calculator worksheets for unit-economics conversations, and the maturity self-assessment worksheet from Chapter 12. These artifacts are tool-neutral and structured for adaptation rather than verbatim copying. The README in each subdirectory explains which patterns the artifact illustrates and what modifications to expect when adapting it to a specific cluster.

The highest-signal references are organized in Appendix D by chapter. Foundational distributed systems papers on consensus, consistent hashing, and distributed storage remain directly applicable to scheduling and fault-tolerance problems in training, even though they predate modern ML infrastructure. The OSDI, SOSP, and MLSys conference proceedings publish the papers that become the next cycle's production patterns. The MLOps Community, the Practitioners track at major ML conferences, and SRE-for-ML working groups produce practitioner-grade knowledge that moves faster than any publication cycle.

The highest-leverage internal investment for technical leads is the design review process. A review that asks the same structured questions for every new distributed ML workload, explicitly names trade-offs, and treats observability and fault tolerance as non-optional will raise the team's average output quality faster than any individual training program. The questions this book poses at the end of each chapter are a starting point for that review template. Hardware will change, frameworks will rotate, and others will succeed the specific platforms named in this book's illustrative examples. The patterns will not change at the same rate. The patterns are a durable investment. The frameworks are the vehicle you use to deliver them.

The engineers, architects, and technical leads who build reliable distributed ML systems are not supporting their organization's mission. They are the mission. The model that flags a fraudulent transaction, routes an emergency vehicle, or accelerates a scientific discovery does none of those things reliably without the distributed infrastructure discipline behind it. That discipline is neither glamorous nor accidental. It is designed, practiced, and defended, sprint by sprint, code review by code review, pattern by pattern. Build systems that earn trust. Build platforms that outlast the engineers who built them. Build practices that a new team member can inherit and immediately use to do mission-aligned work. That is the standard the field deserves.

14 Appendix E: Distributed Systems Foundations Refresher

Why a distributed-systems refresher belongs in an ML book

Machine learning practitioners frequently arrive at distributed training through mathematics rather than systems. They have built and trained models on a single machine; they understand gradients, optimizers, and learning-rate schedules, and then one day the model no longer fits on a single device. They are abruptly operating a distributed system — usually without having read Lamport, absorbed the CAP framing, or thought carefully about what a network actually guarantees. The result is a class of failures that look like machine learning bugs but are distributed-systems bugs in disguise: a training run that hangs on a collective, a checkpoint that cannot be restored consistently, a gradient that is silently corrupted by a failing link. This appendix supplies the minimum distributed-systems vocabulary the rest of the book assumes, defining each concept before it is used and building intuition before reaching for formalism. It is not a substitute for a distributed-systems course; it is the bridge a machine learning practitioner needs to cross to read the rest of this book without gaps. The intended reader has trained models and is now responsible for training them across many machines. The appendix is organized so that each concept is defined in plain terms, illustrated with an intuition a practitioner already holds, and only then connected to the formal idea that names it. A reader who already commands this material should skip the appendix; a reader who has felt the unease of operating something larger than a single process will find here the vocabulary that turns that unease into precise questions.

The network as a first-class concern

The first thing a single-node practitioner must internalize is that, in a distributed system, the network is not a detail; it is often the dominant cost, and reasoning about it requires distinguishing between two properties that beginners conflate. Bandwidth is how much data can move per unit time, measured in gigabytes per second; latency is how long it takes a single message to travel from one endpoint to another, measured in microseconds. These are independent. A link can have enormous bandwidth and poor latency, like a freight train that carries vast cargo but takes a long time to arrive, or low bandwidth and excellent latency, like a motorcycle courier who arrives quickly but carries little. Distributed training is sensitive to both, and which one dominates depends on the operation: a large-gradient all-reduce is bandwidth-bound, while frequent small synchronizations are latency-bound.

The second distinction is between intra-node and inter-node communication, and the gap between them is enormous. Within a single accelerator node, devices are connected by a dedicated high-bandwidth interconnect — on the order of hundreds of gigabytes per second per link — designed precisely for the tight coupling that tensor parallelism demands. Between nodes, communication crosses a data-center network that is typically an order of magnitude slower and several times higher in latency. This single fact explains most of the parallelism guidance in this book: tensor parallelism belongs inside a node because it puts communication on the critical path of every layer. In contrast, data parallelism tolerates crossing nodes because its communication is batched into a single gradient exchange per step.

Network topology is the structure that connects the nodes, and three families recur. A fat-tree provides full bisection bandwidth between any pair of nodes by widening the links toward the root of the tree, which makes communication patterns predictable but expensive to build at scale. A dragonfly groups nodes into densely connected clusters with sparser long-haul links between groups, trading uniform bandwidth for lower cost and good locality. A torus connects nodes in a regular grid that wraps around at the edges, which favors communication patterns where each node talks mostly to its neighbors — exactly the pattern a ring all-reduce produces. The topology is not an abstraction the architect can ignore: it determines which parallelism strategy is efficient, because a strategy whose communication pattern matches the topology runs near peak. In contrast, one that fights the topology bottlenecks on the links overuses them.

The practical intuition to carry away is that distance is expensive and that the unit of distance is the boundary crossed, not the meter traveled. A message that stays within a node is cheap; a message that crosses to another node is an order of magnitude more expensive; a message that crosses to another rack or region is even more expensive. Every parallelism decision in this book can be read as an attempt to keep the most frequent communication on the cheapest links and to push the rare communication onto the more expensive ones. Tensor parallelism, which communicates on every layer, is confined to the cheapest tier; data parallelism, which communicates once per step, is permitted to cross the expensive tiers; pipeline parallelism sits between them. A practitioner who internalizes only the principle that communication cost is dominated by the boundaries it crosses will make most topology decisions correctly without needing the formal models at all.

Consistency models

A consistency model is a contract about what a reader of shared state is guaranteed to see relative to what writers have written. The strongest is strict consistency, in which every read returns the most recent write instantaneously across the entire system; it is a useful ideal and physically unachievable across a network, because instantaneous agreement contradicts the finite speed of communication. Sequential consistency relaxes this: all participants agree on some single total order of operations. That order is consistent with each participant's own program order, but the order need not match real time. Causal consistency relaxes it further, guaranteeing only that operations related by cause and effect are seen in the right order while permitting concurrent, unrelated operations to be observed in different orders by different participants. Eventual consistency is the weakest of the commonly named models. In the absence of new writes, all replicas eventually converge to the same value, with no promise about what is observed in the interim.

These models matter for machine learning because different components assume different ones. A classic parameter server providing synchronous updates approximates sequential consistency: every worker reads a parameter version consistent with a single global update order. Fully Sharded Data Parallel assumes a stronger coordination. Each step gathers the full parameter set, computes, and re-shards. Hence, every rank operates on the same parameter version within a step, which is closer to the strict ideal enforced by explicit synchronization. The reason training can tolerate weaker consistency than a database is that the optimization process is robust to bounded staleness in a way that a financial transaction is not. A stochastic gradient computed against parameters that are a step or two out of date still points roughly downhill, so asynchronous training with a bounded staleness window converges; a bank balance read a step out of date is simply wrong. The architect's job is to know which guarantee each component provides and to confirm that the training algorithm tolerates the staleness that the chosen guarantee permits.

The intuition that unlocks this section is that consistency is a cost, not a free good, and that systems buy only as much of it as they need. Strict consistency is the most expensive because it requires coordination on every operation; eventual consistency is the cheapest because it requires almost none. A database serving financial transactions pays for strong consistency because a wrong answer is catastrophic; a training system pays for only as much as convergence requires because a slightly stale gradient is merely a slightly noisier one, and the optimizer was already designed to tolerate noise. This is why asynchronous training with bounded staleness is a legitimate engineering choice rather than a corner cut: the algorithm's tolerance for noise is a budget, and staleness spends from that budget rather than from the correctness budget. The mistake to avoid is assuming a component provides a stronger guarantee than it does and building an algorithm that silently depends on the guarantee that was never there.

Collective communication primitives

Distributed training is built from a small set of collective operations, each of which coordinates data movement across all participating ranks at once. An all-reduce combines a value from every rank — typically by summing gradients — and delivers the combined result back to every rank, which is the operation at the heart of data-parallel training. An all-gather collects a distinct piece of data from each rank and delivers the full concatenation to all ranks, which is how a sharded parameter set is reassembled for a forward pass. A reduce-scatter is the complement: it combines values across ranks and leaves each rank with a distinct slice of the result, which is how gradients are reduced and resharded in one step. A broadcast sends one rank's data to all others and is used to distribute an initial state. An all-to-all has every rank send a distinct message to every other rank, which is the communication pattern that mixture-of-experts routing and certain forms of parallelism require.

The cost of a collective is a function of three quantities: the message size, the world size — how many ranks participate — and the topology that connects them. The intuition that matters most concerns the contrast between a ring and a tree algorithm for all-reduce. A tree all-reduce combines values up a hierarchy in several communication steps proportional to the logarithm of the world size, which is excellent for small messages where latency dominates, because fewer hops means less accumulated latency. A ring all-reduce arranges the ranks in a circle and passes data around it so that each rank sends and receives the same amount regardless of world size, making its bandwidth cost independent of the number of ranks. For the large messages typical of gradient exchange, the ring wins, because the operation is bandwidth-bound and the ring keeps every link saturated with useful traffic rather than concentrating it near a tree's root. This is why production training stacks default to ring-based all-reduce for large gradients and reserve tree algorithms for small, latency-sensitive collectives.

The intuition worth holding is that there is no single best collective algorithm, only a best algorithm for a given message size, world size, and topology, and that a communication library worth using will switch algorithms automatically as those parameters change. A practitioner does not usually choose the algorithm by hand. Still, a practitioner who understands the trade-off can diagnose the symptom when the library chooses badly — a collective that should be bandwidth-bound spending its time on latency, or vice versa — and can size messages and group memberships. Hence, the library has a good algorithm available to pick. The deeper lesson is that the cost of a collective is not incidental overhead to be ignored; in a large data-parallel run the gradient all-reduce can be the single largest line in the time budget, and treating it as a first-class quantity to be measured and optimized, exactly as the case studies in this book do, is what separates a run at high model FLOP utilization from one that wastes half its hardware waiting on the network.

Failure modes in distributed systems

A failure model is a precise statement of how components are allowed to fail, and choosing the right one determines what fault tolerance even means. A fail-stop failure is the friendliest: a component stops entirely, and other components can detect that it has stopped, which is the model most cluster fault-tolerance machinery assumes. A fail-silent failure is one where a component stops producing output without announcing it, so detection requires a timeout or a liveness check rather than an error signal. An omission failure is the loss of some messages while the component otherwise functions, as a flaky network link can produce. A timing failure is a response that is correct in content but arrives too late to be useful, which matters acutely in hard-deadline inference. A Byzantine failure is the most general and most malicious: a component behaves arbitrarily, including sending different or incorrect information to different peers, whether due to a bug, corruption, or an attacker.

In an ordinary machine learning cluster, fail-stop and fail-silent failures are the realistic everyday cases — a node crashes, a process hangs, a GPU falls off the bus — and the fault-tolerance patterns in Chapter 5 are built to detect and recover from them. Full Byzantine tolerance is rarely engineered into training systems because the overhead is high and the adversarial threat model usually does not justify it. However, federated and multi-party settings are exceptions. The failure that deserves the most respect is silent corruption, a fail-silent variant in which a component continues operating but produces subtly wrong results — a bit flipped in memory, a gradient corrupted by a failing link, a checkpoint written incompletely. It is the most dangerous failure precisely because it is the hardest to detect: nothing crashes, no alarm fires, and the corruption propagates into the model where it may not surface until the loss diverges thousands of steps later or, worse, until a degraded model reaches production. This is why the observability guidance in this book treats gradient-norm trajectories and NaN counts as leading indicators and why durable checkpoints are validated rather than merely written.

The intuition to carry forward is that the cost of a failure is set as much by how it is detected as by how often it occurs. A fail-stop failure is loud and frequent but cheap, because detection is immediate and recovery is mechanical. A silent corruption is rare, quiet, and expensive because the gap between when it happens and when it is noticed can span thousands of steps, and everything computed in that gap may have to be discarded. The engineering response, visible throughout this book, is to convert silent failures into loud ones by instrumenting leading indicators that deviate before the loss occurs, and to defend against undetectable corruption by validating durable state rather than trusting that a write that returned success actually wrote the correct bytes. A practitioner who assumes the friendly fail-stop model and is surprised by a silent corruption has made the most common and most expensive distributed-systems mistake in machine learning.

The CAP framing, gently

The CAP theorem is among the most-cited and most misunderstood results in distributed systems, so it is worth carefully stating what it does and does not claim. CAP says that when a network partition occurs — when some nodes cannot communicate with others — a system must choose between consistency, meaning every read sees the latest write, and availability, meaning every request receives a response. It does not say a system must sacrifice consistency or availability in the absence of a partition; in normal operation, a well-built system provides both. The theorem is a statement about behavior during partitions, specifically, not a general license to abandon one property, and the common misreading that a system can only have two of the three obscures that the choice between C and A only arises once P has occurred.

A training cluster is best understood as a CP system — one that preserves consistency and sacrifices availability during a partition — that tolerates partitions through retries rather than by continuing in a divided state. When a worker becomes unreachable, the correct behavior is not to proceed with a partial set of gradients, which would corrupt the optimization, but to halt the affected step, detect the failure, and recover from the most recent consistent checkpoint. The cluster chooses consistency: it would rather pause and resume correctly than continue available but wrong. AP framing, in which the system remains available by accepting divergent states, is appropriate for a globally distributed datastore serving reads but not for a synchronous training step, because a training step has no meaningful notion of a correct result computed from an inconsistent parameter set. Understanding the cluster as CP-with-retries clarifies why the fault-tolerance machinery is built around detection and rollback rather than graceful degradation.

The gentle version of the lesson, for the practitioner who has heard CAP invoked as a slogan, is this: CAP is not a menu from which an architect picks two items once and for all, but a description of an unavoidable choice that arises only at the moment a partition occurs, and a well-designed system spends almost all of its life in a regime where the choice does not arise at all. The useful question is therefore not whether a training cluster is CP or AP in the abstract, but what it should do in the specific instant when a worker becomes unreachable. The answer for synchronous training is unambiguous: stop, because a training step computed without all of its gradients is not a degraded result but a wrong one. Recognizing that the right behavior is to halt and recover, rather than to limp along, is what makes the fault-tolerance patterns of this book coherent rather than arbitrary.

Time, ordering, and checkpoints

Distributed systems force a distinction between two notions of time that a single machine lets practitioners conflate. Physical time is the wall-clock reading of a node's local clock, which drifts and cannot be perfectly synchronized across a network, so two events stamped by two different clocks cannot be reliably ordered by their timestamps alone. Logical time, introduced by Lamport, abandons wall-clock readings in favor of counters that capture the only ordering that matters — the causal order in which events could have influenced one another. The practical lesson is that "which event happened first" is not always a well-defined question across nodes, and a system that requires a consistent global order must construct one explicitly rather than relying on timestamps.

This is why a checkpoint is fundamentally a global snapshot problem rather than a simple save. A correct checkpoint must capture a state that the system could actually have been in at a single instant — every rank's parameters, optimizer state, and data position taken from a consistent cut across the whole job — so that a restore resumes from a coherent point rather than from a mixture of states that never coexisted. Coordinated checkpointing achieves this by pausing computation, quiescing communication, and writing a globally consistent snapshot, which is correct but expensive at scale because every rank waits for the slowest writer, stalling the whole job. Asynchronous and staggered checkpointing approaches reduce stalls by allowing ranks to write without a global pause. Still, they concede something in exchange: either a more complex protocol to reconstruct a consistent cut after the fact, or a weaker guarantee that the restored state is only approximately consistent and the run must tolerate the small inconsistency. The architect's decision is which concession is acceptable for a given workload, and the fault-tolerance chapter's layered checkpoint designs are precisely an engineering response to this tension.

The intuition that makes checkpointing comprehensible is that a checkpoint is a photograph of a moving system, and that the photograph is only useful if every part of the system was captured at the same instant. A photograph assembled from pieces taken at different moments — one rank's parameters from now, another's from a step ago — depicts a state the system was never actually in, and restoring from it can produce subtle, hard-to-trace divergence. This is why the apparently simple act of saving state is, in a distributed system, the global snapshot problem in disguise, and why the engineering effort that the fault-tolerance chapter spends on checkpoint design is not over-engineering but a direct response to a genuine distributed-systems difficulty. The layered designs in this book — frequent cheap local checkpoints backed by occasional expensive durable ones — are the practical resolution: pay for a globally consistent snapshot rarely, and absorb most failures with cheaper, more frequent saves whose consistency requirements are easier to meet.

Where to read more

A practitioner who wants to build genuine fluency in these foundations should start with Leslie Lamport's "Time, Clocks, and the Ordering of Events in a Distributed System," the 1978 paper that introduced logical clocks and remains the clearest entry point to reasoning about ordering without synchronized time. For breadth, Tanenbaum and van Steen's *Distributed Systems* is the standard comprehensive text, covering consistency models, fault tolerance, and coordination in depth and at a level a machine learning engineer can absorb without a prior systems background. For communication primitives specifically, the original ring-all-reduce treatment — popularized for deep learning through Baidu's work and grounded in the high-performance computing literature on bandwidth-optimal collectives — explains why the ring algorithm scales as it does and is worth reading before tuning a training stack's communication layer. For the CAP framing and its frequent misreadings, Eric Brewer's later reflections on the theorem and Martin Kleppmann's *Designing Data-Intensive Applications* together correct the common misunderstandings and connect the theory to the systems an engineer actually operates.

These concepts are not academic ornaments; each appears in operational form somewhere in this book. The network distinctions of this appendix underwrite the parallelism guidance of Chapter 2 and the topology-aware placement of the case studies in Chapter 11. The consistency models govern the parameter-staleness decisions in Chapters 2 and 3. The collective primitives are the substrate of the data-parallel and sharded-training patterns throughout Chapters 2, 3, and 10. The failure models drive the fault-tolerant training patterns of Chapter 5 and the observability leading indicators of Chapter 8. The CAP framing explains why Chapter 5 builds recovery around detection and rollback. And the snapshot view of checkpoints is the theoretical core of every checkpoint design in Chapter 5. A reader who finishes this appendix and then meets these ideas again in their chapters should recognize them not as new material but as the operational expression of the foundations laid out here.

www.ingramcontent.com/pod-product-compliance
Lightning Source LLC
LaVergne TN
LVHW010555100826
845148LV00014B/2718

* 9 7 9 8 9 0 4 9 8 0 0 3 0 *